Jan Harmsen, René Bos
Multiphase Reactors

Also of Interest

Product and Process Design.
Driving Innovation
Harmsen, de Haan, Swinkels, 2018
ISBN 978-3-11-046772-7, e-ISBN (PDF) 978-3-11-046774-1

Process Intensification.
Breakthrough in Design, Industrial Innovation Practices, and Education
Harmsen, Verkerk, 2020
ISBN 978-3-11-065734-0, e-ISBN (PDF) 978-3-11-065735-7

Process Technology.
An Introduction
De Haan, Padding, 2022
ISBN 978-3-11-071243-8, e-ISBN 978-3-11-071244-5

Advanced Reactor Modeling with MATLAB.
Case Studies with Solved Examples
Tesser, Russo, 2020
ISBN 978-3-11-063219-4, e-ISBN (PDF) 978-3-11-063292-7

Technology Development.
Lessons from Industrial Chemistry and Process Science
Stites, 2022
ISBN 978-3-11-045171-9, e-ISBN (PDF) 978-3-11-045163-4

Jan Harmsen, René Bos

Multiphase Reactors

Reaction Engineering Concepts, Selection, and Industrial Applications

DE GRUYTER

Authors
Jan Harmsen
Harmsen Consultancy B.V
HoofdwegZuid 18
2912 ED Nieuwerkerk a/d Ijssel
The Netherlands
info@harmsenconsultancy.nl

René Bos
Shell Global Solutions International B.V.
Grasweg 31
1031 HW Amsterdam
The Netherlands
rene.bos@shell.com

Ghent University
Laboratory for Chemical Technology
Technologiepark-Zwijnaarde 125
B-9052 Gent
Belgium

ISBN 978-3-11-071376-3
e-ISBN (PDF) 978-3-11-071377-0
e-ISBN (EPUB) 978-3-11-071384-8

Library of Congress Control Number: 2022951751

Bibliographic information published by the Deutsche Nationalbibliothek
The Deutsche Nationalbibliothek lists this publication in the Deutsche Nationalbibliografie; detailed bibliographic data are available on the Internet at http://dnb.dnb.de.

Cover image: Idea by Jan Harmsen and René Bos, Design by Ion Jonas
Typesetting: Integra Software Services Pvt. Ltd.
Printing and binding: CPI books GmbH, Leck

www.degruyter.com

Contents

Part D: **Education**

Preface

We, the authors, enjoy a combined 80+ years of experience in reaction engineering both in work and in education. Our work in industry comprises design, development, and commercial scale implementation of reactors and processes. We also organize and conduct courses. The courses are given inside Shell and in universities. For these courses we not only use textbooks but also make our own education means, such as exercise problems to "solve."

We noticed that course partitioners have different barriers to acquiring reaction engineering concepts. These differences are found in university students as well as in industry. They are likely connected to intrinsically different ways of learning. Some find the mathematical approaches a barrier, while others find barriers in analogy examples or constructed academic problems with one and only one correct answer (being not the real stuff and not reflecting industrial practice).

There are a lot of excellent textbooks both on reaction engineering in general and multiphase reaction engineering. It was in no way our intention or completely unrealistic ambition to write "a better" textbook. However, we arrived at the conclusion that there is a need for a textbook that largely avoids these barriers, to teach people reaction engineering as well as to provide a concise and easily accessible reference work describing a variety of aspects of the field. This is then a main motivation for our book.

The secondary motivation is that the subject of multiphase reactors design theory is considered to be hard to master and is, in general, only given as advanced reaction engineering courses. We think that multiphase reactors are indeed more complex, but by approaching them with a focus on conceptual understanding, rather than by detailed mathematical modeling, such teaching can be done even in introductory courses. We also think that this is needed because in industrial practice the majority of reactors is not in single phase but multiphase. So single phase has little practical use, although of course many of the principles are still important and must be well understood also for multiphase reactors. When students later enter into industry and then find out that their knowledge is of little use for industry, they could become disappointed about their education.

We also provide information to teachers of chemical reaction engineering (CRE) on hypotheses why many students find the subject hard to master and provide additional education methods to test these hypotheses in practice. We also include a short history of chemical reaction engineering with its founding fathers in Chapter 1. Historic descriptions give additional insight and provide lessons about hurdles to overcome. We provide below our own history with chemical reaction engineering.

https://doi.org/10.1515/9783110713770-203

Some personal reflections by Jan

I entered the University of Twente in 1969. The first course in chemical reaction engineering was in my third year. The course was given by Ton Beenackers and Wim van Swaaij. Wim van Swaaij had just become a professor and gave one or two lectures, while the majority of lectures were given by Ton Beenackers. I was immediately fascinated by the subject and spent hours at home working out the exercises provided in the course book. The course book appeared to be sections of the manuscript of the book published in 1984. I passed the exam along with some 30% of the students.

One year later, I did my BSc research subject by Beenackers on his PhD subject mass transfer and chemical reaction in a cyclone reactor. Both the theory of mass transfer with fast reaction and experimental work on the cyclone reactor in which gaseous SO_3 reacted with benzene in a solvent. The reaction took place in the liquid film surrounding the gas bubbles. A consecutive reaction of the product with the SO_3 also took place in the film. The mass transfer coefficient k_l in the cyclone reactor is much higher than in mechanically stirred vessels. This resulted in far higher selectivities in the cyclone reactor. I did my MSc research project with Professor Lyklema on oxidation inside mud layers of lakes. I did experiments with a mud layer in a specially designed piece of equipment inside a thermostat room which was continuously kept at 18 °C. The experimental results were compared to a model of mass transfer with reaction in which the mass transfer model was obtained from the penetration theory of Danckwerts. The derived Hatta number was so a function of penetration time.

In 1977 I joined the Shell KSLA research in Amsterdam and came into the group reaction engineering. I worked on modeling of the hydro-demetallization bunker reactor using large cold models to study the gas, liquid, and solids flow and validate these models. For the gas and liquid flow, I made a 3D model based on potential flow and gravity. The reaction engineering model was a segregated model based on the residence time of each streamline.

In 1981 I entered Shell Biosciences in Sittingbourne, UK, and worked on several bioreactors. I made a simple growth model for the occurrence of an unwanted mutant for polysaccharide production and was able to show that below an averaged residence time in the backmixed reactor the mutant could not overtake the main strain microorganism. I could explain it to the microbiologists by using their term: dilution rate, which is the inverse of the averaged residence time, and indicating that the dilution rate was higher than the specific growth rate, so that the mutant was washed out of the reactor.

In 1984 I rejoined KSLA and became project leader on an oil from shale project. The process concept based on fluidization was made by Heinz Voetter senior, one of the most experienced and best chemical engineers inside Shell. At KSLA we experimented with several cold models, among others with a 20 ton/h cold model to study all solids transport. In 1987 I went to Shell Chemicals in Pernis as an advising process technologist for chemical plants. I worked among others on a micromixing model for a very fast reaction of olefins with gases. The olefins had a fast dimerization. By making a Zwietering micromixing model I could explain how the product distribution could be tuned.

In 1992 I moved to Shell central office to become a senior reactor engineer providing advice to process engineers designing processes. Very soon I also became an internal process design consultant. In 1996 I moved back to Shell research Amsterdam as principal reaction engineer and a leader of a reaction engineering network, providing courses and inviting external experts for lectures and courses. In 1997 I was appointed part-time professor of sustainable chemical technology at TU Delft. In 2010 I retired from Shell and became an independent consultant in Sustainable Process Innovation. I consulted many companies and gave courses in industry and academia.

I also started to write textbooks. In 2019 Karin Sora, Director of Science, Technology, Engineering, and Materials of De Gruyter publishers asked me: “Do you know someone who could write a book on multiphase reactors?” I said: “Yes, René Bos. But it will be hard to convince him to find the time to write it.” Then I said: “But I can convince him. I even may ask him to co-write it with me.” So I did. With this book as a result.

Some personal reflections by René

Well, when Jan asked me the first time about co-writing this book, I indeed said something like “maybe, but not now.” I was just transitioning from a management role to a Principal Scientist role while also still Lead Generation program manager, and I had just started as a part-time professor of Industrial Reaction Engineering at Ghent University. “Ask me again in a year or so.” And so Jan did that in 2020. I then said “Yes, under two main conditions.” These conditions were:

1) We really write it together; and it also reflects our difference in approaches.
2) It should not be just another textbook on Chemical Reaction Engineering; inevitably this would not even be getting close to those from our own teachers or other great ones.

In retrospect, I think that may have been Jan’s starting point already and the reason why he believed he could convince me (or trick me into it).

Jan and I first met in the early nineties, when I was having a great time at the “Koninklijke Shell Laboratorium, Amsterdam”: KSLA. I had joined Shell in 1991 while finishing my PhD from Twente University under the stimulating guidance of Prof. dr. Roel Westerterp. This was on “Reactor and Catalyst Dynamics and Stability: The Selective Hydrogenation of Ethyne in Ethene” (“hydrogenation of acetylene in ethylene” I would write nowadays). In retrospect, it feels/almost scary that my PhD committee or "committee of highly learned opponents" comprised Professors Gerhard Eigenberger, John Geus, Henk van den Berg, Julian Ross, and Wim van Swaaij.

When I took the master's course Reactor Engineering in 1984, the Westerterp, van Swaaij, and Beenackers textbook that Jan mentioned was supposed to come out just before the classes started. It did not arrive in time. So, I probably learned it from the same "manuscript" as Jan did a few years earlier. Three years later, one of my tasks was – together with some of the other PhD's and academic staff – to go through the first edition from late 1984. We should find all the printing errors in there, so these could be corrected for the re-prints. Reaction Engineers from Twente still refer to this book as the "green bible of CRE." Of course, from an educational point of view, most will rightly so point to the world-famous book by Octave Levenspiel.

Remarkably, I started at Shell in the same department that Jan had worked in a decade earlier: Equipment Engineering, section Reactor Engineering EE/3. This department has been the inspiring workplace of illustrious scientists in the field of Chemical (Reaction) Engineering well before my time, most of them becoming professors at renowned universities next to or after their Shell work: Cor Ouwerkerk, Dick Darton, Hans Wesselingh, Jan van Deemter, Harry van den Akker, René Oliemans, Kees Rietema, and last but not least, R. Krishna and Wim van Swaaij, with whom I had the privilege to have actually worked with.

Jan became my "counterpart" from the central office in The Hague. Clearly, we shared not only our Alma Mater and our teachers but also our passion for Chemical Reaction Engineering, and particularly applying this scientifically exciting, relevant, and fun field in industrial practice. Combining business and pleasure so to speak. It was also a pleasure to see how we applied sometimes opposite (reaction engineering) methods and still landed on the same – or almost the same – conclusions. We later also adopted this "several ways leading to Rome" approach in our Reaction Engineering and Conceptual Process Design course at Shell (see Chapters 1 and 12).

After my first five years as Reactor Engineer at KSLA, I moved to Shell Chemicals in Pernis where I had a fantastic time as an advising technologist for the so-called "Versatic Acids" plant. I still remember Wim van Swaaij getting really excited when I told him that I would be leaving R&D to work for Versatic Acids. He said something like: "You will get not only a multiphase reactor with complex chemistry and the whole range of micro- and macro mixing, but in fact the whole of Chemical Engineering in just one plant." So indeed, I learned a lot – not just Chemical Engineering! Also how fabulous ideas "on paper" or "in the lab" may not work so well in the "real world."

Taking all these learnings on board, several increasingly more senior technical roles followed, including an expat assignment in Houston before returning to my base in Amsterdam. Mostly, these were roles within process development, both in the early and late stages of the respective programs and plant support. Notably, roughly five years each within Ethylene Oxide/Ethylene Glycols, Styrene Monomer/Propylene Oxide, and Fischer-Tropsch/GTL. After that I moved into a team lead role within what we called Emerging Technologies, that is, more in the highly experimental and exploratory early-phase innovation space. There I learned that even in that early innovation space CRE can make a poor idea into a great idea – and vice versa!

Almost in a full circle fashion, from 2018 onward CRE returned more prominently in my working life: I was appointed Guest Professor of Industrial Reaction Engineering at Ghent University, among other things taking over the course "Chemical Reactor Fundamentals and Applications" from Professor Guy Marin. This one-day-a-week secondment as well as my new role as Senior Principal Science Expert enabled me to say "Yes" to Jan regarding the co-writing of this book.

So here it is!

We spent two years writing this book. In some periods we were completely absorbed by it. We thank our wives, Mineke Spruijt and Anette Bos, for their support so that this book could be finished with pleasure in time.

Acknowledgments

We want to thank all who have contributed to this book, be it directly or indirectly.

First of all the reviewers with their detailed corrections and very useful comments as well as permission to use some of their own figures. The reviewers are:

Prof. Dr. Wim Brilman, University of Twente.
Prof. Dr. Sascha Kersten, University of Twente.
Dr. Jan Lerou, retired. Worked for Velocys and DuPont.
Dr. Michiel van de Stelt, University of Applied Sciences, Utrecht.

Second, we would like to thank those that provided comments and additional information during the manuscript writing and permission to use some of their own figures and tables. They are:

Dr. Wim Hesselink, retired. Worked for Shell.
Prof. Dr. Guy Marin, Ghent University.
Dr. Dominik Unruh, Shell, Amsterdam.

We also like to thank many (former) Shell colleagues and (former) managers. In particular, we thank the latter for René Bos' part-time professorship at Ghent University, which enabled the writing of this book.

About the authors

Ir. Jan Harmsen
Harmsen Consultancy BV
Hoofdweg Zuid 18, 2912ED, Nieuwerkerk a/d IJsel, The Netherlands
E-mail: jan@harmsenconsultancy.nl

Jan Harmsen is an independent consultant in sustainable process innovation. He provides advice and courses to industry and academia. After his graduation in chemical technology at the University of Twente in 1977, he joined Shell. There he held professional positions in process research, process development, reaction engineering, process concept design, process implementation at manufacturing sites in Rotterdam and South-Korea, and finally, a research position in process intensification till 2010. He was part-time Hoogewerff-Professor of Sustainable Chemical Technology in 1997, first at Delft University of Technology and later at Groningen University till 2013.

He is (co-)author of six books. Recent books are: Jan Harmsen, Maarten Verkerk, *Process Intensification: Breakthrough in Design, Industrial Innovation Practices, and Education*. De Gruyter, Berlin, 2020; Jan Harmsen, *Industrial Process Scale-up: A Practical Innovation Guide from Idea to Commercial Implementation*, 2nd revised edition, Elsevier, 2019, Jan Harmsen, André de Haan, Pieter Schwinkels, et. al., *Product and Process Design Driving Innovation*, De Gruyter, 2018.

Prof. Dr. Ir. René Bos
Shell Global Solutions International B.V.
Grasweg 31, 1031 HW Amsterdam, The Netherlands
E-mail: rene.bos@shell.com

Ghent University
Laboratory for Chemical Technology
Technologiepark-Zwijnaarde 125, B-9052 Gent, Belgium
E-mail: rene.bos@ugent.be

René Bos is Senior Principal Science Expert within the department Next Generation Breakthrough Research at Shell Projects & Technology, Amsterdam. Since June 2018 he is part-time (0.2 fte) seconded to Ghent University as Guest Professor of Industrial Reaction Engineering at the Laboratory of Chemical Technology (LCT).

He received his PhD in chemical engineering at the University of Twente on "Reactor and Catalyst Dynamics and Stability – The Selective Hydrogenation of Ethyne in Ethene" in the group of Prof. Roel Westerterp. He joined Shell in September 1991 where he has had a variety of roles in Amsterdam, Pernis, and Houston, mostly within R&D but also at manufacturing sites. In these roles he worked in wide variety of (reactor) technology fields, including ethylene oxide, DeNOx, MTO, SM/PO, Fischer-Tropsch, Gas-to-Chemicals, E-ODH, OCM, syngas technologies, and CO_2 conversions, spanning the whole range from ideation, proof of concept, development, and pilot plant testing to deployment.

In 2018 he was appointed Guest Professor at Ghent University. Overall, he has (co-)authored 41 scientific publications in the open literature (next to >100 Shell internal research reports) and 42 Patent Applications.

https://doi.org/10.1515/9783110713770-204

Part A: Multiphase reactors: chemical reaction engineering

1 Introduction

The purpose of this chapter is to introduce any reader to this book and into the subject of chemical reaction engineering (CRE), regardless of whether the reader is new to the subject, or whether the reader is experienced in reaction engineering.

1.1 Book introduction

There are many good and several truly excellent textbooks on multiphase reactors and CRE in general. This book, however, is different. It has more facets in its focus than the existing textbooks.

The first facet of this focus is on teaching students at universities (applied sciences and academic) on all levels: BSc, MSc, and PhD. Examples and exercises are provided at the end of each chapter, and Chapter 14 provides exercises for the reactor design. The second facet of this focus is on industrial practitioners in research development in process engineering and in operation. Industrial applications with their development trajectory are provided at that end. The third facet of the focus is on academic teachers and researchers. To that end, Part D with education guidelines and complete industrial cases for educational purposes are provided.

The approach is also different from other CRE textbooks. Although not an aim by itself, we wanted a book basically without equations beyond high school level. There is not a single differential equation to be found. The approach is on conceptual understanding, explaining the principles, the complex phenomena occurring in a reactor, and design and scale-up methodologies without resorting to mathematics. It goes without saying that we recognize and embrace the power of mathematics in translating (reaction engineering) concepts into clear and unambiguously defined models. We even have one chapter devoted to modeling.

However, in our experience, for quite a number of students and practitioners alike, the math actually forms a serious barrier. Even an ordinary differential equation can be such a barrier, let alone more lengthy derivations given in detail before ending at a final solution. The final solution may be simple enough and in fact what the practitioner will use for, for example, design or interpretation. What matters most from a CRE learning point of view is not the correctness of the math between model definition and that final equation, but a proper understanding of the concepts, their usefulness, as well as their limitations.

Furthermore, each theoretical subject is treated in such a way that industrial applications are at hand. A major portion of the book therefore comprises “real life” examples from the authors’ own direct experience to illustrate how the theory is applied in practice.

https://doi.org/10.1515/9783110713770-001

The scope is on all important reaction engineering phenomena, residence time distribution (RTD), mass transfer, heat transfer, which affect reactor performance parameters, conversion, selectivity, and product quality. Specific attention is paid to energy management for safe reactor design and operation.

This understanding of phenomena and their relation to reactor performance is then used for elaborate descriptions of reactor type selection, design, modeling, and experimental validation methods. Moreover, we treat in quite some detail the process of "stage-gated innovation." Different strategies and ways of working are required for the different stages between ideation, development, de-risking, detailed engineering, procuring and contracting, and ultimately start-up and deployment. We provide elaborate explanations for most subjects because we think those subjects are important. Some subjects we treat concisely. This is the case if the subject is less important, or when the subject is explained well in another textbook, to which we refer to.

All of the above led to the organization of this book, with four Parts A, B, C, and D, as shown in Figure 1.1.

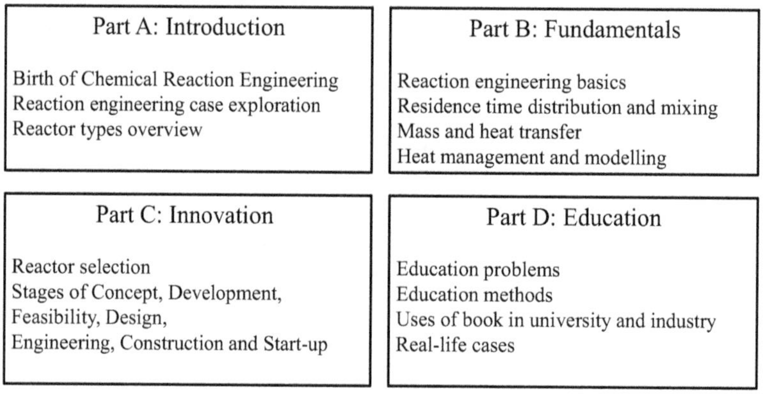

Part A: Introduction

Birth of Chemical Reaction Engineering
Reaction engineering case exploration
Reactor types overview

Part B: Fundamentals

Reaction engineering basics
Residence time distribution and mixing
Mass and heat transfer
Heat management and modelling

Part C: Innovation

Reactor selection
Stages of Concept, Development,
Feasibility, Design,
Engineering, Construction and Start-up

Part D: Education

Education problems
Education methods
Uses of book in university and industry
Real-life cases

Figure 1.1: Book content structure.

Part A Introduction: This provides a general introduction on chemical reactor and reaction engineering and shows all major reactor types with their major features as pictures. So readers get a feeling of how multiphase reactors look like. So here already some vocabulary is built up.

Part B Fundamentals: This describes the theory around major phenomena affecting reactor performances and how design and modeling play a role.

Part C Stage-gate innovation methods: This describes methods and guidelines for reactor type selection, reactor modeling, reactor design, and experimental validation of models and designs, for each innovation stage, from ideation up to deployment, including development. The other innovation stages such as engineering, procurement, and construction, start-up, normal operation, and demolition are briefly described.

Part D Education: This provides guidelines, methods, and "real-life" cases for education in academia and industry.

The remainder of this introduction section leads the reader to reaction engineering by conversations. We finish this first chapter with our potted history of the development of CRE as a separate discipline.

1.2 A reaction engineer meets an electronic engineer

Reaction engineering is a spectacular discipline. Jan Harmsen recalls:

> I still remember that I told my friend, an electronic engineer, about the effect of residence time distribution on reactor conversion. I told him: "Two reactors with the same feed composition, the same feed flow, the same temperature, the same pressure, and the same residence time of the fluid flowing through the reactor can have wildly different degrees of conversion, just due to different mixing pattern of the fluid."
>
> He said: "you must be kidding."
>
> I then explained him the effect of residence time distribution by just taking first a pipe reactor in which every fluid particle stays the same long time in the reactor reaching 99.999% conversion and then the other reactor, a mixed reactor, in which the average residence time of the fluid particles is the same as in the pipe reactor. In the mixed reactor however, some fluid particles directly swing to the outlet by the mixer blade, so without any conversion. The very deep conversion is not reached at all. "I see" he said.

1.3 Levenspiel's genius Problem 1.1

We treat here the wastewater treatment exercise case Problem 1.1 by the late Professor Octave Levenspiel to reveal a lot of important aspects of CRE and, in particular, of multiphase reaction engineering [1].

While still both working for Shell, we had designed from scratch a company internal course and delivered it many times in Amsterdam, Bangalore, Pernis, Bintulu, Doha, and Houston. It is called *Industrial Reaction Engineering and Conceptual Process Design*. The target audience is freshly recruited graduates with a master's degree or PhD in chemical engineering. The vast majority of this course comprises real-life case studies from the authors' own experiences. Deliberately, these case studies are quite different compared to "constructed" problems or examples typically practiced at universities.

Nevertheless, the course starts with Problem 1.1: the very first "constructed problem" from the classic *Chemical Reaction Engineering* textbook by Octave Levenspiel [1]. This is an easily underrated example of the educational genius that Levenspiel was. There are many lessons to be learned from this seemingly innocent student problem.

So we go step by step through the exercise. Prior to coming to our course, the participants are asked to read the short introduction of Levenspiel's book and then solve Problem 1.1 "Municipal wastewater treatment plant." This is their "tiny bit of homework." We start the course with compiling and discussing all the answers.

The problem stated seems simple enough:

Given

A stream of 32,000 m^3/day of wastewater flows through a reactor tank ("plant") with a mean residence time of 8 h; air is bubbled through and the microbes in the tank break down the organic material, i.e., the reaction, catalyzed by microbes, is:

$$\textbf{(organic waste)} + O_2 = CO_2 + H_2O$$

The BOD (biological oxygen demand) of the feed is 200 mg O_2/L, while the effluent has a negligible BOD.

Find the rate of reaction or decrease in BOD.

Most people in our course, chemists and chemical engineers alike, calculate the rate of reaction by simply taking the difference in biological oxygen demand (BOD) between inlet and outlet concentration and divide that by the 8 h mean residence time. So they calculate the reaction rate to be, for example, 2.17×10^{-4}. Some treat the given wastewater flow rate as superfluous information; others use it to determine the tank volume. Most folks – many of them are experienced chemists/engineers – mention not only a number but also a unit (dimension), for example, mol/s m^3 belonging to the number above, or they convert it into another unit for time or volume.

The first impression might be that this example is all about the importance of explicitly mentioning the units when communicating, for example, a rate of reaction to colleagues. Levenspiel's introduction before Problem 1.1 elaborates on it. This indeed is absolutely an important message. Examples of where this has led to errors in industrial practice are numerous. These go way beyond the obvious ones like per second versus per hour or joule versus BTU.

The genius of Levenspiel's Problem 1.1 is unveiled by elaborating on the enormous richness below the surface and on what actually goes wrong here. Many key aspects of CRE are implicitly addressed. We will make these explicit by means of eight different learning points from this Problem 1.1, which we observed when we started our industry course and interacting with the participants. We believe these were also points that Levenspiel intended to make. And we may even be incomplete here!

Learning point 1: Units are not so simple, especially for multiphase reactors.
When it pertains to multiphase reactors, getting the units fully correct is not as easy as it may seem. In Problem 1.1, people will typically have their rate *per cubic meter*.

When we ask: "to which *cubic meter* do you refer to?", the answer is nearly always "the volume of the tank."

We then ask them: "did you make a drawing of the reactor?" A remarkable thing always happens. Most did not draw anything at all. This is a major observation. Some people struggle to make an abstract drawing of something. Patience and providing hints such as "draw a box" can help, especially after suggesting to also draw a liquid level. This abstraction from a real artifact to a simplified abstract drawing is very important in reaction engineering. It can only be internalized into a habit by asking again and again to make a drawing.

Then we, the teachers, draw a tank on the white board and we also draw inside a wavy horizontal line as liquid level. After asking again "which liter to use?", the answer is changed unanimously into "the volume of the tank without the empty volume above the liquid level."

We then add to our drawing air bubbles in the tank and ask: "should we include or exclude the volume of the gas bubbles?"

And then it dawns on the course participants that even a thing like a volume is not simple in reactors and reaction engineering.

This exercise example illustrates why we always ask to draw a picture of the reactor. It also demonstrates the importance of being always very explicit about units and their precise object referred to, for instance, by using a subscript; in the earlier case, use L_{liquid}. Or even better write every unit in SI units, hence, m^3_{liquid}. A simple m^3 can mean many things as Levenspiel's Problem 1.1 already demonstrated. The same holds for m^2 and m^2/m^3 and other units as well. Even when a subscript is used, avoid ambiguities that often occur especially in multiphase reactors.

For example, when dealing with catalytic reactors with porous catalyst particles in a bed, unexpected ambiguities easily arise. For a catalyst expert, a catalyst density of 1 kg catalyst per liter of catalyst will likely mean "per liter including all the voids between the catalyst particles and including all the pores inside the catalyst." He or she would call this the compacted bulk density: the density of a large bag of catalyst. However, many reaction engineers would interpret a catalyst density of 1 kg/L as being "per liter catalyst *particle*, excluding the voids between the particles but including the pores inside the particle." A rationale would be that in the volume of the void, no catalytic reaction takes place. Note, however, that one can say the same for the pores inside the catalyst particle: the reaction only takes place at the catalytic pore surface!

In this fixed bed reactor example, the use of subscript "particle" would be much clearer than "catalyst" or even "solid phase."

Another brief example: if for a bed of such catalyst particles the superficial velocity is quoted as being 1 m/s, one should be alerted as well. It might be tempting to apply the subscript "reactor" here (or tube). However, the definition of superficial velocity is the volumetric fluid flow rate per cross-sectional area of the reactor. From this it follows that the unit "meter" in this velocity is in fact $m^3_{fluid}/m^2_{reactor}$. Another example here is the liquid-phase mass transfer coefficient k_L, where its unit is also not really m/s but $m^3_{fluid}/sm^2_{interface}$. In checking the correctness of mass and heat balances, using these very explicit units is very helpful and instructive indeed.

On writing reactor model equations and exact units

We always advise to write below the equations with all their parameter groups the same equation in the form of only the exact units of the variables and parameters. For example, for a superficial velocity appearing in an equation, write down $m^3_{fluid}/m^2_{reactor}$ s instead of ambiguous "m/s" or potentially incorrect $m_{reactor}/s$. For a liquid-phase concentration, write for example mol_i/m^3_{liquid}. For mol fractions, write for example mol_i/mol_{total}. Then inspect that all groups in your equations have exactly the same net unit by crossing out those that are exactly the same. Ensure you are not left with a seemingly dimensionless term such as m_x/m_y.

In our treatment of each of the various multiphase reactor systems, where there are many pitfalls especially with m, m^2, and m^3, we will provide similar specific examples and recommendations.

A phrase to remember here is: "I do not know what a cubic meter is!" Always use a subscript or make it unequivocally clear in another way.

Learning point 2: Averaged reaction rate values are different from local values.
Pointing again to the drawing of the reactor we ask: "you have all given us a reaction rate, but do you think the reaction rate near the inlet of the tank will be the same as the rate near the outlet?"

It then becomes clear to the course participants that the calculation, if anything, only would pertain to an averaged overall value. Speaking about "the" reaction rate is thus misleading.

Similar useful but potentially misleading approaches and concepts are widely used in industrial practice. Examples are the use of space time yield and work rate. These are averaged or "overall" productivities in kg of product per m^3 of reactor per hour. Within the reactor, the "local" values can vary widely. CRE teaches us to think in local terms. Along with other concepts, we then translate it to overall terms, that is, the overall reactor performance.

Learning point 3: RTD is important.
In solving Problem 1.1, the 8 h residence time was the basis to get a number. This is just an average residence time. We continue with our drawing and put in arbitrary streamlines. Some fluid elements may flow almost directly from inlet to outlet. Other elements may circulate in the tank for a long time.

We then ask the course participants: "Do you think that it would make a difference if, e.g., all fluid elements would flow like a plug from inlet to outlet or that, by putting in a stirrer, all fluid elements would nicely circulate, and the tank would be well mixed?"

Unanimously the chemical engineers in the participants group immediately agree that it definitely will have an effect. For the same mean residence time and same reaction kinetics, the conversion in a so-called plug flow reactor will be different than in a perfectly mixed reactor (except for zero-order reactions we should add). Even though this difference was known to them, it was not taken into account in the simple calculation of the rate. Some knowledge becomes dormant when not used often. Clearly if the RTD affects the overall conversion and mean reaction rate, then also the mean reaction rate or rate expression calculated from a conversion depends on the RTD. That latter is not provided in the problem statement by Levenspiel. Nevertheless, over the years, all ±300 participants of our course provided a number. Only some indeed expressed an explicit assumption like "perfectly mixed tank."

Learning point 4: It is a multiphase system and there is mass transfer too.
By pointing toward the bubbles of air in the drawing we ask: "In order to react, the oxygen must go from the gas phase to the liquid phase. Suppose the case that this transfer was very slow so that water was not saturated with oxygen. Are the calculated rates then not – at least partially – mass transfer rates instead of real reaction rates?"

Again the participants start to think and realize that a lot more is going on in this tank than imagined at the beginning of the exercise. Levenspiel here implicitly has introduced a very important aspect of reaction engineering: mass transfer.

Learning point 5: There is no kinetic information in the data, except a minimum value.
We then ask: "What would have to change in the problem statement if the rate of reaction would have been twice as high?"

This is another eye-opener for most people. For the unknown reaction kinetics in Problem 1.1, the BOD in the outlet was essentially 0 mg O_2/L_{liquid}. If the reaction rate would, for example, have been twice as high, then of course the outlet organic waste concentration (expressed as BOD) still remains 0 mg O_2/L_{liquid}. It could well be that also at a residence time of only 1 hour the BOD would still be 0. In other words, in the problem statement, there is no information from which reaction rates can be determined. At most, the calculated value might be seen as a minimum value. Even that determined minimum value is not necessarily a kinetic rate minimum value as the RTD, and the potential effects of mass transfer limitations are not considered. Finally, note that BOD of 0 in reality means "below the detection limit". Always be alerted when someone reports "100% conversion".

Learning point 6: All answers with a number are wrong; more info is needed.
We think Professor Levenspiel intended this, that is, any number given is wrong indeed, or at least only valid after adopting a number of speculative additional assumptions. We have no hard evidence, but it is remarkable that for most problems in his classic textbook the solutions have been published by Levenspiel himself in his "Omnibook" [2]. However, Problem 1.1 is missing there. We speculate that his answer on the question "Why is the answer to 1.1 not given?" might have been something like: "Before you give a number, study the book first."

Learning point 7: The reaction is not taking place in the water.
The waste conversion, that is, the chemical reaction, is taking place inside the microorganisms. There the organic components are broken down to sugars and are then converted with oxygen to carbon dioxide and water. In the water outside the microorganisms, no significant reactions take place. In this sense, there is a similarity with

heterogeneously catalyzed reactions, where the chemical reaction itself takes place on the surface of the catalyst (the active sites).

Learning point 8: The reaction heat production is also important.
The oxidation of organic material in the wastewater plant produces heat in large quantities. This heat is transferred to the water flowing out of the tank. Typically, this outlet water is thereby 5 °C higher in temperature than the inlet. This heat was until recently not noticed. Due to actions to find new renewable energy sources, it dawns on engineers that the energy content of the waste to the water treatment plant is enormous and the energy release by the biological conversion to the outflowing water can, for instance, be used in urban house heating. So always also consider the reaction heat in any reactor, not just as a nuisance requiring heat control, but as a useful side-product.

Take away learning points from wastewater treatment case
Learning point 1: Units are not so simple, especially for multiphase reactors.
Learning point 2: Averaged reaction rate values are different from local values.
Learning point 3: RTDs should not be forgotten.
Learning point 4: It is a multiphase system and there is mass transfer too.
Learning point 5: There is no reaction rate (kinetic) information in the data.
Learning point 6: All answers with a number are wrong; more info is needed.
Learning point 7: The reaction is not taking place in the water.
Learning point 8: Reaction heat production is also important.

1.4 A short history of chemical reaction engineering

1.4.1 Introduction

We present here a short history of CRE and its founding fathers. The purpose of this short history is to provide confidence to young chemical engineers that it is a well-founded field and also that it was from the beginning focused on solving practical reaction design problems. For the experienced CRE practitioner, this short history may create an *aha-erlebnis.* For many pieces of theory he uses, he now reads about the founding father who created this theory and how long ago this theory was generated and published. This may create a feel of well-being because the theory withstood the test of time and can therefore be used with confidence (if done with care and proper understanding).

1.4.2 Birth of chemical reaction engineering

Pieces of theory still used in CRE were generated in the first half of the twentieth century. Hatta, for instance, described gas–liquid reactions and defined the Hatta number as a ratio of reaction rate and mass transfer coefficient as early as 1928 [3]. Frank-

Kamenetskii [4] derived very neatly described dimensionless numbers for temperature excursions in reactors in 1942. Also Damköhler was certainly a pioneer here; for Professor Wicke he was even a "founding father of CRE"; see Section 1.1.5. In 1947, Hougen and Watson [5] published a book on kinetics and catalysis for process engineering, which was quickly used as a textbook in chemical engineering programs. In the USA, Dudukovic [6] describes these preludes to the CRE field as "voices in the wilderness".

These early remarkable works can be regarded as the embryonic phase of CRE. The real birth took place in the 1950s. Professor Kramers of Delft University of Technology invited Westerterp in 1958 to set up the discipline of CRE at Delft University [7]. This was supported by a group of young people at the Shell Amsterdam Research Center together with students from TU Delft who started CRE as a separate discipline. The originator of the term "chemical reaction engineering" is unknown, but it is attributed to Jo Vlugter [8]. He was also the organizer and chairman of the first European CRE symposium, held in Amsterdam in 1957 [6]. Vlugter became professor in 1963 at the new Twente University and immediately set up a CRE group. The textbook by Professor Dick Thoenes [9] mentions that during this symposium a definition of CRE was agreed upon. However, Thoenes attributes the coining of the term "CRE" to his then colleague at the DSM company: Professor Dirk van Krevelen (also known from the Mars–van Krevelen kinetics and van Krevelen–Hoftijzer diagrams).

Levenspiel expressed his surprise of the vision the Dutch expressed on CRE at the ISCRE in 1957 [10]. To quote Levenspiel:

> First came the remarkable book of Hougen and Watson (1947), which led to the American school of kinetics and catalysis, focusing primarily on petroleum processing, with its emphasis on high temperature reactors and mechanisms of surface catalysis.
>
> The second stage in the transition from descriptive technology to modern engineering came 10 years later at the First Symposium on Chemical Reaction Engineering (1957), where a number of earlier developments were brought together and synthesized into a discipline appropriately called Chemical Reaction Engineering, or CRE. As far as I know, it was here where the term CRE was coined. What an historic meeting this was with van Krevelen, Letort, Kramers, Danckwerts, Rietema, Denbigh and the many others mapping out the whole field. In going back and reading through the proceedings of this meeting I marvel at how far seeing were these participants. Since then we have filled in, extended here and there, pushed the methods of CRE into new fields of applications, however the grand scheme visualized so long ago (as measured by the progress of science and technology) is as sound today as it was then. CRE represented the crystallization of the European school of thought, and soon after it became adopted universally.

The definition of CRE was also established in the 1950s as "The key equation of CRE can be stated as Reactor Performance = f (input, kinetics, contacting). Product yield, or selectivity, or production rate can be taken as measures of performance. Feed and operating conditions constitute the input variables. Fluid mechanics of single or multiphase flows determines contacting, while kinetic descriptions relate reaction rate to pertinent intensive variables such as concentrations, temperature, pressure, catalyst activity" [6]. Levenspiel in his "The coming-of-age" of CRE paper (1980) simply introduced it as "the proper bringing to practice of chemical reactions" [10].

Within 12 years, CRE was established with textbooks and European and international CRE symposia. The very first symposium of the ISCRE (first international CRE symposium) series took place at our later workplace: KSLA, the Shell Laboratories, Amsterdam [7].

1.4.3 Founding fathers of CRE

Here, we provide a short list of founding fathers of CRE. The selection criterion: being the first to provide an important part of a theory that distinguished CRE as separate discipline in the period that CRE was established, so in the birth period 1950–1965.

Peter Danckwerts is the first founding father to mention. He introduced the element RTD of continuous flow systems in 1953 [11]. He also showed the difference in backmixing and plug flow on the reactor performance. Dudukovic states in 1987 that Danckwerts RTD concepts and their impact on CRE lasted several decades [6]. We can now conclude that Danckwerts concepts are still in use today, so it lasted at least seven decades.

Ernest Thiele is the second founding father we would like to put in the spotlight. He studied the effect of porous catalyst particle sizes on the reaction rate. To that end, he derived two dimensionless numbers for mass transfer plus reaction in porous catalyst. The first number is the Thiele modulus and the second is the effectiveness factor. He made beautiful graphs showing the effect of the Thiele modulus value on the porous catalyst effectiveness [12]. His method quickly entered into textbooks by Hougen and Watson [5], Levenspiel [1], and Kramers and Westerterp [13]. His theory is still used in textbooks. We will do so too by showing the original hand-drawn graphs by Thiele himself. Not many people know that this is the same Thiele known from the McCabe–Thiele diagrams for designing distillation columns.

Octave Levenspiel is the third founding father. He is the teacher. He published the first textbook for students on CRE in 1962 [1]. The book covers all major CRE elements, RTD, mass transfer and reaction in porous catalysts, mass transfer and reaction in gas–liquid and gas–liquid–solid reactors, and heat effects. He also treats some aspects of reactor type selection by taking these elements into account with the reaction kinetics.

At first, his book was mainly used in the USA. But now this book is used worldwide as a textbook in education. We found it to be the best book to be used in an industrial setting to introduce CRE to nonchemical engineers for three reasons. The writing style is easy to understand, which is illustrated with very clear drawings and graphs, and his exercises appeal because of their practical industrial setting.

Hans Kramers and Roel Westerterp are the twin fourth founding fathers. Of course, we are biased here. They published the book *Chemical Reactor Design and Operation* in 1963 [13]. The book evolved from a lecture course for senior chemical engineering students at TU Delft. The book is based on principles of Hougen and Watson,

Frank-Kamenetskii, and many others. Also, a lot of materials from the first two CRE symposia in 1957 and 1960 were used. And there had been frequent contacts with van Krevelen and Vlugter. Other contacts were with Bird, Danckwerts, LeGoff, van Heerden, Higbee, Hofmann, Hoftijzer, Horn, Letort, Messikommer, Pigfort, Rietema, Schoeneman, and Wicke. The Russians published a translation in 1967 [13]. The first edition of this book was mainly used in Europe but was also sold out quite rapidly. In 1984, a second completely revised and vastly expanded edition appeared, now with Beenackers and van Swaaij as coauthors besides Westerterp [14].

The last founding father to mention is Rutherford Aris. He is the founding father not only because of the content of his mathematical reactor modeling but also because of his lucid mathematical modeling style. In 1961, he published his book *The Optimal Design of Chemical Reactors: A Study in Dynamic Programming* [15].

Here, we end the short history of CRE, embracing the embryonic period till 1950 and its birth 1950–1963. It is interesting to note that the embryonic state lasted several decades while its birth only 13 years. The combination of textbooks and symposia appearing together for the first time accelerated this birth. Present new challenges to the worldwide society to contribute to the sustainable development goals also mean new challenges to the CRE practitioners. They have to come up with new methods and new reactors to produce basic people needs, which are environmentally benign and economically affordable to the poorest.

1.4.4 CRE as a language game

CRE has several features of a language game as defined by Wittgenstein [16]. This means that the members of a community who speak this language not only agree on their definitions but also agree on their judgments on the world as a whole. It could well be that CRE so quickly was established because of the combination of written articles, oral presentations, and by the lively discussions during the first three E/ISCRE's in 1957, 1960, and 1964. What may also have helped that these discussions were also written down in these early symposia proceedings. These ISCRE's are still held. They may be essential to keep the CRE community and its language alive and thereby maintain the existence of the field. Section D discusses the problem of educating CRE and elaborates on this language game aspect.

1.4.5 Dimensionless numbers in CRE: the persons behind the number

Before and after the birth of CRE as discipline, several important concepts and their dimensionless numbers named after often famous scientists have emerged. All these numbers have a very clear meaning, regardless of their specific application. Here we list the three most common ones which we believe all reaction engineers should know by heart. Further on in our book, all will appear at least once. We also include a short biography of the inventor of the number and his main achievement. We have used articles written by Wim Hesselink and published in NPT (*Netherlands Process Technology* journal) on biographies of the men behind a number of well-known dimensionless numbers, such as Reynolds, Sherwood, and Nusselt. We expanded this with Damköhler, who even has four numbers named after him and who probably was the first to define dimensionless numbers that involved both a physical and chemical timescale.

1.4.5.1 Reynolds: the person behind the number

This may easily be the best-known dimensionless number for chemical (reaction) engineers. The readers of this book need not to be explained: it is the ratio of inertial forces and viscous forces of fluids flowing in and around objects. Osborne Reynolds was born as a son of an Anglican minister in Belfast, Northern Ireland, in 1842. He died of influenza in 1912. Remarkably, he himself published his dimensionless number in 1883 [17] without giving it a name. It was only 25 years later that Arnold Sommerfeld gave this number its name in 1908 [18]. He wrote in German: ". . .eine reine Zahl, die wir die Reynolds'sche Zahl nennen wollen" [19].

Reynolds initially worked as civil engineer for an engineering contractor, but soon successfully applied for a professorship at Owens College (later called Queen's college), Manchester. He was only 25 years old. In his application, he wrote: "From my earliest recollection I have had an irresistible liking for mechanics and the physical laws on which mechanics as a science is based." He remained professor till his retirement in 1905.

Reynolds also contributed to teaching applied mathematics. He was a highly scholarly man who put the bar high. Maybe too high, because ironically, he was a poor teacher himself due to his tendency to dwell into sidetracks [19].

Looking back at Reynolds' life, it is interesting to note that at an early stage he already came into contact with mathematics and engineering, and these two fields kept his interest for the rest of his career both in industry and in university. He believed all engineering students should have a common basis in, among other things, mathematics. We also believe so. However, for reasons explained in the introduction, in our current book, we focus on understanding by explaining topics by pictures, metaphors, and written language and virtually completely abstain from mathematics.

1.4.5.2 Sherwood: the person behind the number

Chemical engineers know his number as the ratio of forced mass transfer over diffusion to (or from) an object. It is mostly quantified by using a correlation with the Reynolds number and the Schmidt number.

Thomas Kilgore Sherwood was born in Columbus, Ohio, USA, on 25 July 1903, but lived during his youth in Montreal, Canada. He received his BSc at McGill University in 1923. He then moved to Massachusetts Institute of Technology for his PhD. He collaborated there with Professor Warren K. Lewis, one of the founding fathers of modern chemical engineering. Sherwood obtained his PhD in 1929 with his thesis on "The Mechanism of the Drying of Solids."

This hooked him onto the subject of mass transfer in gases and liquids, which he never left for the rest of his career. He worked for a short while as assistant professor at Worcester Polytechnic Institute, but then returned to MIT in 1930. In 1941, he became professor, and in 1946–1952, he was dean of the engineering faculty. In 1969, he left MIT to become visiting professor of chemical engineering at the California University in Berkeley. He kept this position till he died in 1976.

Sherwood worked on many aspects of mass transfer, notably drying, absorption, extraction, and distillation. He established his name by his textbook *Absorption and Extraction*, published in 1937. After that he published several other textbooks for teaching purposes, where *Applied Mathematics in Chemical Engineering* (1939) had a great influence on students and practitioners because it contained many examples and practical applications. Other books were *Properties of Gases and Liquids*, *Course in Process Design*, and *The Role of Diffusion in Catalysis*.

During his career, he moved between academic research and industrial applications of chemical engineering theory for vulcanizing rubber, improvements of distillation trays, improvements of extraction processes, desalting of seawater, and the production of penicillin and vinyl acetate.

In a lecture at the end of his career, he stated his view on the role of computers in chemical engineering: "We have much more concern with complex physical phenomena, and we have not yet arrived at a point where all can be left to the computer. In a way I hope we never will, for chemical engineering is so much more fun when we do not know very much" [20].

Half a century later, we, the authors of this book, subscribe this statement.

1.4.5.3 Prandtl: the person behind the number

The dimensionless Prandtl number is the ratio of viscous impulse transfer over thermal energy transfer of a fluid. It is used in correlations of the Nusselt number with Reynolds and Prandtl.

Ludwig Prandtl was born in Freising, near Munich on 4 February 1875. His father, Alexander Prandtl, was professor at the "Landwirtschaftlichen Zentralschule" in Weihenstephan, at the dairy department. He developed a continuously operated milk centrifuge. He stimulated Ludwig's interest in physics and taught him working principles of machines and instruments.

In 1894, Ludwig started to study mechanical engineering at Technische Hochschule in München. In 1898, he graduated and started to work as assistant to Professor Föppl in the mechanical laboratory. He obtained his PhD with the thesis "Kipp-Erscheinungen, ein Fall von instabilem elastischem Gleichgewicht." After that he entered the

Maschinenfabrik Augsburg Nürnberg (MAN) in Nürnberg in 1900. Here Prandtl came for the first time in contact with rheology. There he made an improved design of a saw dust separator. Key points were the choice of a cyclone as gas-solid separator and optimizing the feed and outlet pipes.

At the age of 26, Prandtl was asked to apply for a professorship mechanics at Hannover University. He became the youngest professor in Prussia. He continued to explore his ideas about rheology. This led in 1904 to the boundary layer theory, which made him famous. Prandtl expanded his institute Göttingen. It took a front position in aerodynamics. Prandtl educated a large number of students, among them is Theodore von Kármán. He developed the aerodynamic wind flow theory around objects called the von Kármán gusts [21].

Looking back at his career, it is clear that he combined rheology theory development with engineering design thinking, leading to improved process equipment. He may be qualified as a reflective practitioner [21].

1.4.5.4 Damköhler: the person behind the number

In his 1985 paper, Professor Ewald Wicke [22] called Gerhard Damköhler "founder of chemical reaction engineering." Born on 16 March 1908 in Klingenmünster in Germany, at the age of 18, he started his studies in chemistry at the University of Munich. He graduated in 1931 summa cum laude as doctor of philosophy for a dissertation on osmosis. He remained at the university as a research fellow, during which time according to Inger [23]: "he began to display a marked ability for developing and applying analytical methods to practical problems in chemical engineering." Because of this, the director of Göttingen University's Institute of Physical Chemistry (Eucken) offered Damköhler a position as an assistant. Encouraged by Eucken, he started his work on a more systematic analysis of the relative timescale of reaction versus that of diffusion and convection.

Prandtl, of the Prandtl number, was teaching hydrodynamics at Göttingen at that time and Damköhler attended Prandtl's lectures. His "seminal" paper – in German of course – appeared in a German electrochemistry journal [24]. In that paper, he introduced what we now call the first and second Damköhler number: Da_I being the ratio of "flow time over reaction time" and Da_{II} the ratio of diffusion time over reaction time. After his so-called Habilitation, he had departed for Braunschweig, where he became an associate at the Aeronautical Research Establishment's Motors Research Institute. In that period, he wrote a few very influential surveys, including one for the "Handbuch der Chemie." His later work focuses on highly exothermic homogenous reactions like turbulent flames and internal combustion engines. From that work, the third and fourth Damköhler numbers Da_{III} and Da_{IV} surfaced.

At the age of 36, he tragically took his own life on 30 March 1944. Part of the reason appears rooted in a conflict that developed between him and the wartime national socialist government. Another contributing factor lies in Damköhler's personality and private life [22]: he has been consistently described as a highly intelligent man who, however, worked himself "25 hours a day" with little regard for the childless homelife he shared with his wife.

1.5 Reaction engineering as introduction to process design

The authors have noticed that reaction engineering is also a natural introduction to process design. There are many aspects of reaction engineering which are also useful for process design: notation of input and output streams with molar balances, mass balances, and heat balances; the importance of drawing reactor system boundaries separately from process system boundaries; and definitions of yield, selectivity, and by-product formation. All these reaction engineering items are also used in process design. Because of this, reaction engineering is treated in this book as distinct from process design, but where relevant, the process design context is provided.

1.6 Exercises

Exercise 1: Draw a picture of the wastewater reactor of Section 1.3
Draw the contours of the wastewater reactor
Draw all essential elements inside the reactor.
Draw all input and output streams as arrows.
Check whether all input and output streams are drawn.

Exercise 2: Wastewater process engineering of input–output table
Draw the wastewater treatment reactor and all its input and output streams and give each stream a label.
Make a table within the first column all components mentioned in the example.
Then make for each input and output stream a column within the first row its label.
Fill in for each component for each column its mass flow value.
Also make a row at the bottom of the table and add up for each column all mass flows.
You have now made an input–output mass balance table for each stream and each component.
These tables are a communication means of process engineers.

1.7 Takeaway learning points

Learning point 1
CRE is a well-founded field. From the beginning, it is focused on solving practical reaction design problems.

Learning point 2
Reactor performance = *f*(input, kinetics, contacting).

Learning point 3
Units (dimensions) are not so simple, especially for multiphase reactors.

References

[1] Levenspiel O, Chemical Reaction Engineering. Hoboken, NY, USA, J. Wiley, 1962.

[2] Levenspiel O, The Chemical Reactor Omnibook. Corvallis, OR, USA, OSU Book Stores, 1979.

[3] Hatta S, Absorption velocity of gases by liquids. I. Absorption of carbon dioxide by potassium hydroxide solution. Technology Reports of the Tohoku Imperial University, 1928, 8, 1–25.

[4] Frank-Kamenetskii AA, On the mathematical theory of thermal explosions. Acta Physico-chimica USSR, 1942, 16, 357.

[5] Hougen OA, Watson KM, Chemical Process Principles, Part 3: Kinetics and Catalysis. Hoboken NY, J. Wiley, 1947.

[6] Dudukovic MP, Chemical reaction engineering: Current status and future directions. Chemical Engineering Education, 1987, 21(4), 210–214.

[7] van den Akker HEA, Kleijn CR, van Kasteren J, Het gewicht van de Witte Olifant: 50 jaar Kramers Laboratorium voor Fysische Technologie, 1949–1999. Amsterdam, Spruijt, 1999.

[8] Bartleby, Chemical reaction engineering, 2021. Accessed November 9, 2021, at https://www.bartleby.com/subject/engineering/chemical-engineering/concepts/chemical-reaction-engineering

[9] Thoenes D, Chemical Reactor Development: From Laboratory Synthesis to Industrial Production. Berlin, Springer Science & Business Media, 2013.

[10] Levenspiel O, The coming-of-age of chemical reaction engineering. Chemical Engineering Science, 1980, 35, 1821–1839.

[11] Danckwerts PV, Continuous flow systems: Distribution of residence times. Chemical Engineering Science, 1953, 2(1), 1–13.

[12] Thiele EW, Relation between catalytic activity and size of particle. Industrial & Engineering Chemistry, 1939, 31(7), 916–920.

[13] Kramers H, Westerterp KR, Elements of Chemical Reactor Design and Operations. New York, USA, Acad. Press, 1963.

[14] Westerterp KR, van Swaaij WPM, Beenackers AACM, Chemical Reactor Design and Operation. New York, USA, Acad. Press, 1984.

[15] Rutherford A, The Optimal Design of Chemical Reactors: A Study in Dynamic Programming. New York USA, Academic Press, 1961.

[16] Wittgenstein L, Philosophical Investigations. Oxford, UK, Basil, 1953.

[17] Reynolds O, An experimental investigation of the circumstances which determine whether the motion of water in parallel channels shall be direct or sinuous and of the law of resistance in parallel channels. Philosophical Transactions of the Royal Society of London, 1883, 174, 935–982.

[18] Sommerfeld A, Ein Beitrag zur hydrodynamischen Erkläerung der turbulenten Flüssigkeits bewegüngen (A contribution to hydrodynamic explanation of turbulent fluid motions). International Congress of Mathematicians, 1908, 3, 116–12.
[19] Hesselink W, Het Reynolds G, et al. NPT Procestechnologie, 2013, 20ste jaargang februari, p. 8.
[20] Hesselink W, Het Sherwood G, et al. NPT Procestechnologie, 2013, 20ste jaargang, mei, p. 8.
[21] Hesselink W, Het Prandtl G, et al. NOPT Procestechnologie, 2013, 20ste jaargang, sept. p. 16.
[22] Wicke E, Gerhard Damköhler – Founder of chemical reaction engineering. International Chemical Engineering, 1985, 25(4), 770–773.
[23] Inger GR, Scaling nonequilibrium-reacting flows: The legacy of Gerhard Damkohler. Journal of Spacecraft and Rockets, 2001 Mar–Apr, 38(2), 185–190.
[24] Damköhler G, Einflüsse der Strömung, Diffusion und des Wärmeübergânges auf die Leistung von Reaktionsöfen. Zeitschrift für Electrochemie, 1936, 42, 846–862.

2 Overview of multiphase reactors

The purpose of this chapter is twofold. The first purpose is to give people an impression of the field of reactors. The second purpose is to provide main characteristics of the major reactors applied in the industry to facilitate a preselection of reactor type for a new innovation project.

2.1 Introduction

In this chapter, we discuss the most common industrial reactor types and their nomenclature. Industrial means are here applied at commercial scale. We spend most of the text on industrial reactor types that are in use for many different applications. These are sometimes called workhorses. Some new reactor types still under development – with the potential of becoming new workhorses – are also described in far less detail.

Detailed information about residence time distribution, mass transfer, heat transfer, and hydrodynamics is found in Chapters 4–8. Systematic reactor selection is treated in Chapter 10. Some industrial practitioners, however, wanting to have their own way of selecting a reactor type, are discussed in this chapter.

We use the term "reactor type" to indicate that it is about major elements constituting a reactor type. These elements concern:

- Which phases flow and which phases are fixed
- Which phase is continuous and which is dispersed
- Fluid flow feed points and flow directions
- How fluids are agitated
- How heat exchange can take place
- Feasible operation modes such as batch and continuous
- Main commercial-scale application features

The background reason for selecting these items to characterize reactor types is that phases that flow allow for continuous feed and withdrawal. That determines whether solids can be easily processed or not and that includes the facility of continuous catalyst replenishment or not. The phase holdup has a large effect on the residence time in the reactor. The continuous phase holdup is in general much larger than the dispersed holdup, which means a longer residence time. The way fluids are agitated determines to some extent mass transfer. Heat exchange is for reactor systems with a high heat production or consumption very important. For smaller capacities and for producing many different products in the same reactor, such as in fine chemicals and pharmaceuticals, batchwise operation is highly desired. Finally, industrial-scale applications indicate the maturity of the reactor technology.

https://doi.org/10.1515/9783110713770-002

Within a class of reactor types, there are still many variations in design details. A fixed bed reactor type, for instance, includes a reactor with several beds inside one shell. A fluid bed reactor type, for instance, includes a bubbling fluid bed, a turbulent fluid bed, a fast fluid bed, and a riser fluid bed.

A high-level classification is firstly applied by the number of phases in the reactor types.

First, the two-phase reactors are described, and then the three-phase reactors are described. A further subdivision for each of these two major classes is based on the condition of the solid phase, resulting in the following classification: fixed beds, fluid beds, and mechanically stirred reactors.

The phases present in the reactor are indicated, in which gas is indicated by G, liquid by L, and solids by S. So G-L-S means a reactor in which gas, liquid, and solid phases are present. A further detail is provided by describing the fluid phases as continuous or dispersed. Finally, for each reactor type, critical performance factors such as residence time distribution, mass transfer, and heat transfer [1] are qualitatively indicated.

The role of the solids phase can be heterogeneous catalysis but also it can be a feed component, such as in ore processing, or a product such as in polymerization. Where needed, the role is indicated. The role of gas and liquid phases is nearly always transport of feed and/or product components through the reactor.

Van Swaaij et al. [2] describe three families of catalytic reactors: fixed bed (including moving bed), fluidized bed, and barrier reactors. Van Swaaij represents the latter one as catalytic membrane reactors. He then shows for which applications these reactors are applied depending on the catalyst regeneration time as given in Figure 2.1.

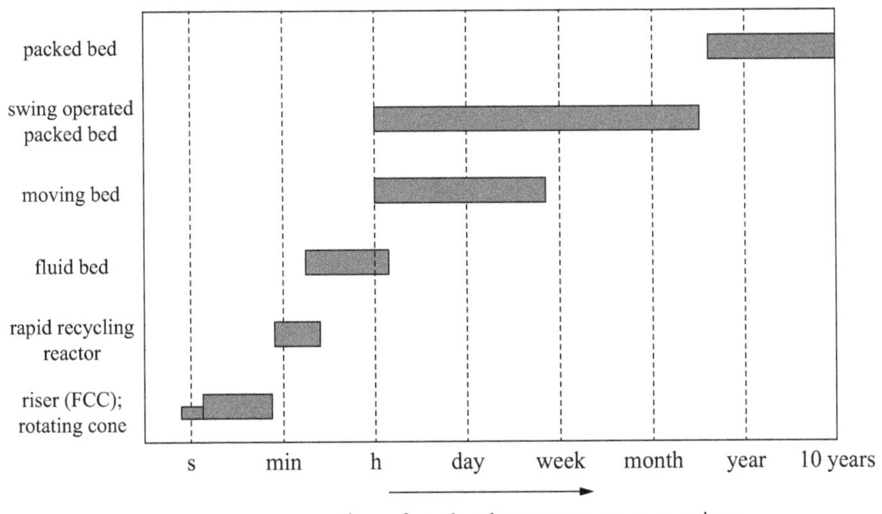

Figure 2.1: Reactor type applied versus on-stream time catalyst [2].

Van Swaaij made this figure for two-phase reactors, but he already hinted that it can also be applied to three-phase reactors. The fluid phases (gas and liquid) do not make the division but the catalytic solid phase does. Also catalyst can be withdrawn for slurry reactors and fed online to the reactor, and regeneration can take place outside the reactor in the same way as for a fluid bed reactor.

In the next sections, major reactor types of these main families are briefly described.

2.2 Two-phase G-S reactors

2.2.1 Fixed bed reactors

Two-phase fixed bed reactors, also commonly called packed bed reactors, can be regarded as the "workhorse" in the industry, especially in the petrochemical industry. Classic applications are the production of ethylene oxide, vinyl acetate, butadiene, cyclohexane, styrene, maleic anhydride, and phthalic anhydride. Recent examples of applications under development comprise oxidative coupling of methane and oxidative dehydrogenation of ethane to ethylene (EDHOX™). Some of these processes have also been developed using fluid bed technology. We will address this in the chapter on reactor selection.

Within the fixed bed reactor family, there is still a huge variety of specific reactor types. Professor Wim van Swaaij et al. [2] have made a nonexhaustive list of these reactors. Moreover, he very elegantly organized them in Figure 2.2 by using the analogy of the evolution of the mammals. The original archetype may be a small mammal which remained in its essential form but also gave rise to a wide radiation of species which were able to occupy niches of life in the environment. The direction of evolution is translated as a desire to show the analogy with the emergence of different fixed bed reactors stemming from the archetype fixed bed reactor, the basic adiabatic fixed bed.

2.2.1.1 Exercise for experienced chemical engineers

Already before the time that van Swaaij developed Figure 2.2.b, engineers from Shell managed to fix an underperforming parallel passage reactor (PPR) installed at a refinery in Germany. It underperformed in the sense that it had a lower conversion than expected from reaction kinetics and a plug flow gas phase model.

The PPR is reactor no. 10 as part of the low-pressure drop branch of Figure 2.2.b. In a PPR, conventional catalyst extrudates are placed in many parallel slabs formed by a set of wire gauzes. Flue gas is flowing through the empty channels between the slabs. The incredibly simple solution the engineers found – and that solved the too low conversion problem experienced in practice – was to alternatingly have closed and open

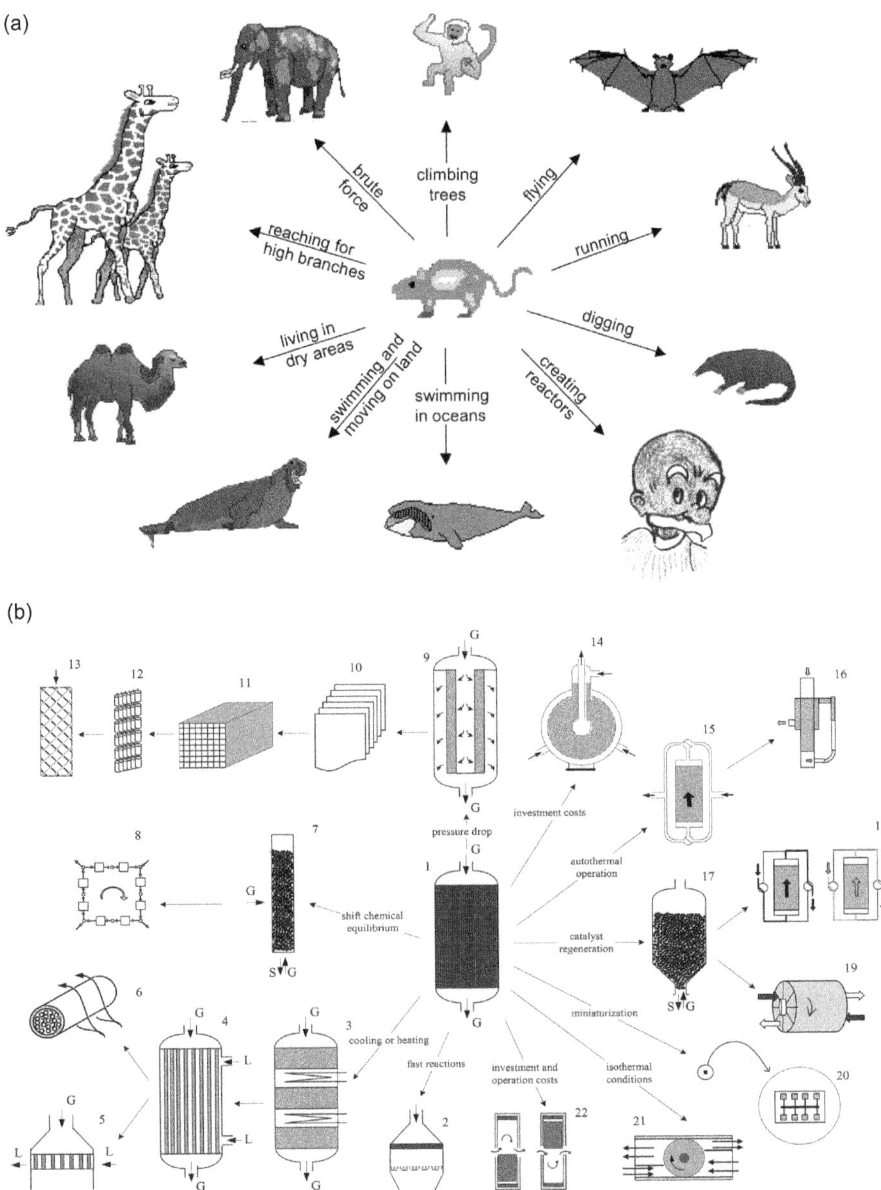

Figure 2.2: a) Radiation of mammal species from the archetype. Different mammal types evolved to fit into the various niches of life [2]. (b) The packed bed reactor family [2].
1, Adiabatic; 2, gauze layers; 3, adiabatic with intermediate cooling; 4, multitubular; 5, short bed with cooling/heating; 6, annular bed; 7, chromatographic; 8, simulated moving bed; 9, radial flow; 10, parallel passage; 11, monolith; 12, bead string; 13, polylith; 14, spherical; 15, reverse flow; 16, circulating loop; 17, moving bed; 18, coupling of endo- and exothermic reactions/simulated moving bed; 19, rotating fixed bed; 20, microreactor; 21, rotating disks; 22, pulsed compression.

channels on the side of each channel (see Figure 2.3). The reactor was named lateral flow reactor or LFR; see Van der Grift et al. [3] for the whole DeNOx system and Vandewalle et al. [4] for a detailed reactor engineering analysis of the LFR.

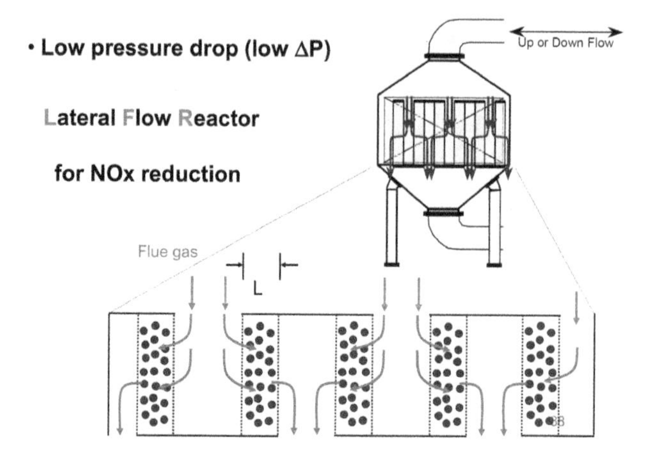

Figure 2.3: The lateral flow reactor for DeNOx [3, 4].

Questions

Q1: What would be the main reason for the lower-than-expected conversion of the PPR?

Q2: What would be the main reason why this simple modification dramatically improved the conversion?

Q3: What would you see as main disadvantage of the LFR over the PPR for flue gas DeNOx-ing?

In reality, it took one more iteration. The LFR only worked really well after increasing the empty(!) channel width.

Q4: What would be the main reason why increasing the empty channel width improved performance?

2.2.2 Gas–solid fluid bed reactors

2.2.2.1 Introduction

Fluidization transforms a fixed bed into fluid bed by simply increasing the gas flow to such a velocity that the particles are lifted. The friction between the particles reduced then enormously.

The resulting gas–solid mixture then behaves like a fluid, hence, a fluid bed. This is illustrated in Figure 2.4.

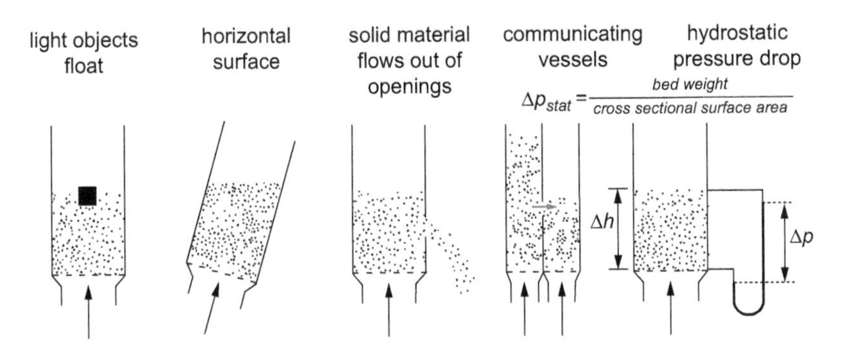

Figure 2.4: Illustration of the fluid like properties of a gas–solid fluid bed [5].

The fluid behavior of the solids creates all kinds of opportunities for the engineer. The mixture is now easily transported. The mixture can be rapidly mixed with other solids or liquid. Heat can be easily exchanged. The particles can be catalyst particles. These can be chosen to be very small, without resulting in high pressure drop, as would happen in a fixed bed.

One of the most spectacular applications of fluidization for reaction purposes is fluid catalytic cracking (FCC) of oil fractions in an oil refinery. It was developed and applied for the first time in 1942 by Exxon. The Second World War required a large production of petrol. Catalytic cracking could help to produce this petrol. However, the cracking catalyst deactivated within a minute by coke deposition. Fixed-bed reactor application would mean that enormous beds and many beds had to be constructed. To remove the coke from the catalyst, the fixed bed needed to be taken out of action. The coke would be needed to be removed by slow oxidation using nitrogen with about 1% oxygen in it, which typically takes hours. Hence, an enormous number of fixed-bed reactors, most of them in the regeneration mode, would be needed.

A team of chemical engineers and mechanical engineers in Exxon had found that fluidization using very fine particles in the range of 50–100 µm created a gas–solid mixture with a very low viscosity, similar to water. This mixture could flow downward by gravity through a sloped pipe at a velocity of several meters per second. The catalyst particles could also be transported vertically upward in a so-called riser reactor. The coke could be burned off in a fluid bed supplied by air.

So they designed an FCC process as shown in Figure 2.5. It consists of a riser where oil vapor was catalytically cracked. After separation of the spend catalyst particles from the cracked product vapor by cyclones, the coked catalyst was transported to a bubbling fluid bed using air as fluidization gas to burn the coke from the particles. This fluid bed combustor was at an elevated position. By a sloped pipe, called a standpipe, the hot catalyst particles moved as in fluidized state to the bottom of the riser reactor. This standpipe also acted as a seal so that no oil vapor would enter the combustor. At the bottom of the riser reactor, liquid oil was brought in contact with the hot catalyst particles. The oil gets evaporated. This vapor transported the catalyst particles upward, while cracking the oil molecules [6].

This FCC process had an enormous production capacity and could be run continuously, reliably, at very low cost. This FCC process was then used by many oil companies for many decades to produce petrol from heavier oil fractions.

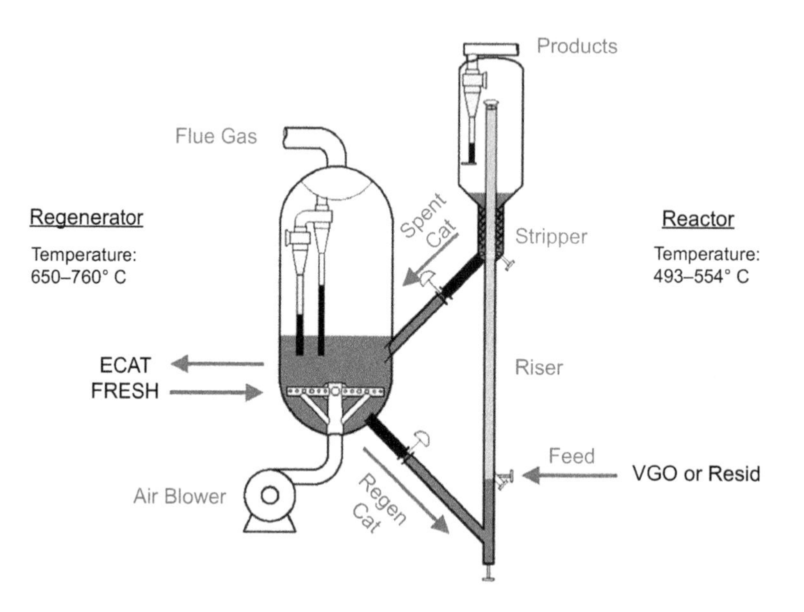

Figure 2.5: Fluid catalytic cracking process scheme [7].

Recently, the company BTG-BTL created a biomass to crude oil process also based on fluidization similar to the FCC process. The process scheme is shown in Figure 2.6.

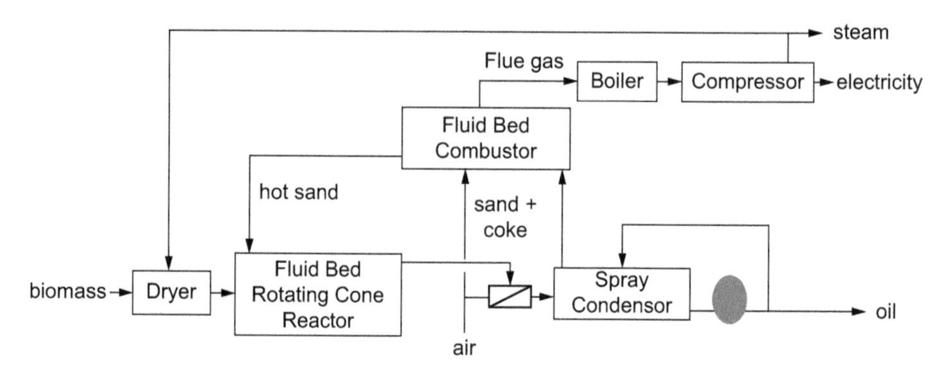

Figure 2.6: The BTG-BTL fluid bed process [8].

In the first fluid bed reactor, biomass is mixed with hot sand. The biomass is quickly cracked to oil vapor which acts as fluidization gas. Coke is deposited at the sand particles. This coked sand is moved to the second fluid bed, where the coke is burnt by air. The resulting hot sand is returned to the first fluid bed.

2.2.2.2 Fluidization behavior categories

This success created an enormous research activity into fluidization. One of the results was the classification behavior of the fluidization mixtures by Geldart into classes A, B, C, and D [9]. The classification is placed in a graph with the averaged particle size on the horizontal axis and the density difference between particles and gas on the vertical axis (see Figure 2.7).

Rietema pointed out that this classification is not physically correct. The attraction forces that cause the boundaries between class A and class B behavior and between class A and class C depend also on temperature and on molecules adhering to the particle surface. He proposed moreover a classification based on dimensionless numbers [10]. In his own group, he verified for some conditions his classification boundaries. Unfortunately, his theory so far has not been followed up by other researchers, so there is no validation yet over a wider range of conditions for his classification boundaries.

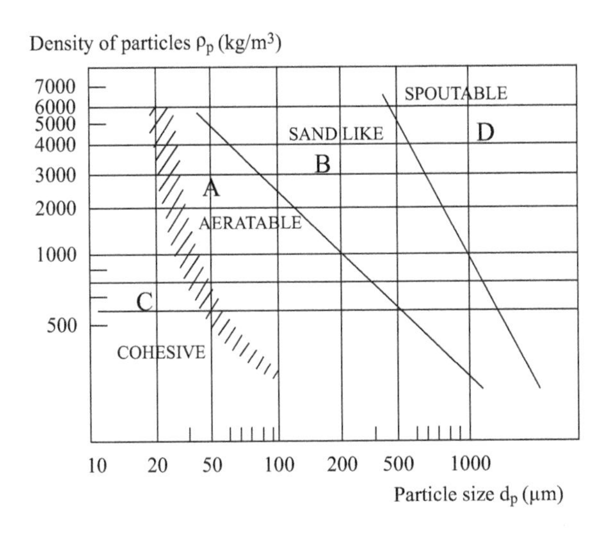

Figure 2.7: The Geldart fluidization powder diagram [11].

Here we describe the fluidization classes and their gas and solid residence time distribution characteristics.

2.2.2.3 Class A fluidization powder behavior

In this class, the dense phase is expanded a little beyond the packed bed density. This remarkable aspect is caused by attraction forces between the particles by which to a certain distance no bubbles are formed. This additional distance between the particles results in little friction between them, when moved relative to each other. So their dense phase viscosity is low and similar to water. These attraction forces can be increased by components from the gas phase adhering to the particles. This means that the boundary between classes A and B of Figure 2.7 is defined not only by parameters on the axes but also by these attraction forces (see [8]). Experimental proof of this phenomenon is provided by Cottaar [12].

Beyond the dense phase gas velocity bubbles are formed which move at a high velocity through the dense phase, thereby mixing the dense phase, while gas from the dense phase also flows through the bubbles upward.

2.2.2.4 Class B fluidization powder behavior

In class B fluidization, bubbles are immediately formed when the gas velocity is beyond the minimum fluidization velocity. The minimum fluidization velocity is reached when the pressure drop caused by the gas flow exceeds the packed bed gravity. For this class, the particles in the dense phase are still close to each other causing large friction forces. The dense-phase viscosity is much higher than for class A particles. Dense-phase circulation velocities are much lower than in class A fluid beds. For small diameter beds, the circulation velocity is very low. This class fluidization is hardly applied at industrial scale because of the larger particles involved and not having the benefit of class A powder.

Academic studies on fluidization have mostly been carried out with this class B powder. Little entrainment of particles with the gas leaving the fluid bed occurs, and entrained particles can be easily recovered by a cyclone.

2.2.2.5 Class C fluidization powder behavior

In class C fluidization, the very fine particles have strong cohesive behavior. Big lumps of particles separated by gas phase cracks are nearly stagnant. At very high gas velocities beyond 1 m/s, the lumps break down to smaller lumps and irregular gas bubbles flow through the bed. This class C powder is applied in polyolefin fluid bed reactors. This application is described in Chapter 14.

The solid residence time distribution is hard to predict. No models are available. It should be remarked that the boundary between classes C and A is less sharp than indicated in the graph. The cohesive forces are governed not only by particle size but also by double-layer attraction forces. For those particle systems, the boundary between A and C moves to the right-hand side. Polymer particles show these cohesive forces.

Reliable residence time distribution models for the solid phase and the gas phase are not available. A major part of the gas will rapidly flow via cracks in between the lumps, causing short cutting flow behavior. The particle-phase residence time distribution will also be erratic and wide.

2.2.2.6 Class D fluidization powder behavior

In class D, the gas velocity through the dense phase powder is higher than the bubble velocity. The heavy and large particle powder shows a higher viscosity and less circulation behavior than for classes A and B.

2.2.2.7 Fluid bed reactor types

There are many fluid bed reactor types. Here, we only describe in some detail the types that are mostly applied in industry.

2.2.2.8 Bubbling fluid reactors

Bubbling fluid bed reactors are the most commonly used types in industry. The applications are in oil refineries, mineral processing, polyolefin production, and combustion of coal and biomass for power generation.

Bubbling fluid beds with class A fluidization behave similar to a bubble column. For commercial-scale fluid beds, the vertical circulation velocities of the dense phase containing the majority of particles are in the order of 1 m/s, so circulation times will be less than 100 s. This means that for solids, the average residence time is more than 1,000 s and the solid residence time distribution will be close to an exponential one corresponding to fully backmixed, like a CISTR, but "segregated" (see Section 4.7).

The gas phase residence time distribution will be hard to describe. A large part of the gas will quickly move in bubbles through the central core section with limited exchange of gas with the dense phase. Gas residence time distribution will therefore show short-cutting behavior. Moreover, reliable prediction of the gas phase residence time for large commercial-scale fluid beds is not available. Hence, deep conversion of the gas phase reactant is not practically achievable in bubbling fluid beds, and its prediction is unreliable.

Often the designer choses to go for a limited single-pass gas conversion with an external gas recycle by which a fully backmixed (CISTR) gas phase residence time distribution is reliably obtained. Chapter 14 describes a design case on fluid bed polyolefin reactors.

More information on bubbling fluid bed class A powders regarding residence time distribution is found in Section 4.9.2.

Mass transfer is briefly described in Section 6.3. Heat transfer is briefly described in Section 7.3.

The detailed designer of fluid beds is referred to textbooks on fluid beds.

2.2.2.9 Bubbling fluid beds of classes B and D

Bubbling fluid beds of B and D are applied in coal combustion to generate steam. The heat exchanger is often placed in the beds as well as in the gas stream leaving the bed.

For the solid phase, no reliable residence time distribution model could be found. In one case study at one scale, the solid residence time could be modeled with a plug flow with a dead zone and some CSTRs in series (see [13]). In other words, the solid-phase residence time distribution has an erratic wide shape.

The gas phase residence time distribution is also not characterized well by the available models, and so far no reliable information could be found. The gas phase residence time distribution is therefore also not simple to describe; other than that it is wide for most cases and is also likely to depend on the column diameter.

2.2.2.10 Circulating fluid bed

This reactor type is also called riser fluid bed, or vertical pneumatic transport fluid bed.

In this reactor type, the gas velocity is so high that particles are transported upward through the riser pipe. At the top, they are separated from the gas phase by a cyclone, as shown in Figure 2.8. When the particles are transported downward by gravity and then moved by a sloped standpipe to the bottom of the riser, then the whole system is called a circulating fluid bed.

If the particles move to a different process section, then the reactor is called a riser reactor, or a pneumatic transport fluid bed reactor.

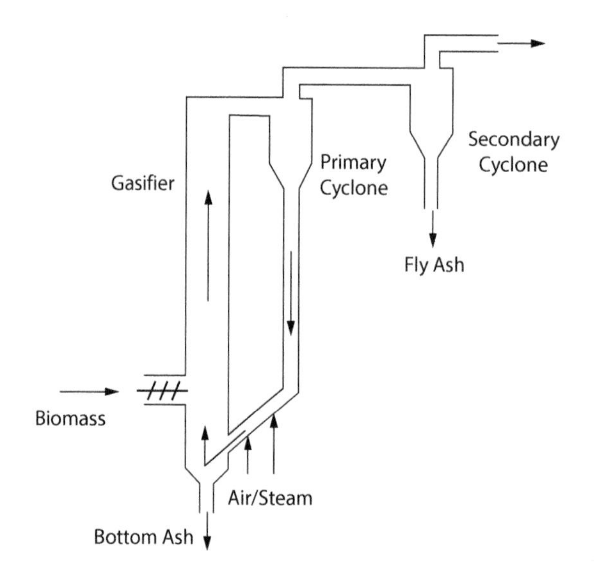

Figure 2.8: Circulating fluidized bed [11].

This type of reactor is often used for coal and biomass combustion to generate steam for electricity production. Then the sand acts as a heat carrier. The higher gas velocity compared to bubbling fluid bed means that the energy dissipation is higher causing more turbulence, which results in more intense mixing of coal (or biomass) particles with the sand particles. The resulting higher heat transfer between sand particles and coal means a more rapid combustion. The conversion of fine gas entrained particles is also deeper because of the more staged gas flow residence time distribution.

Clearly, there are many more types of fluid bed reactors. Rather than to make an exhaustive list, we show the elegant way that Professor Wim van Swaaij organized the different fluid bed as one family of reactor, just as he had done for the family of fixed beds, see Figure 2.9.

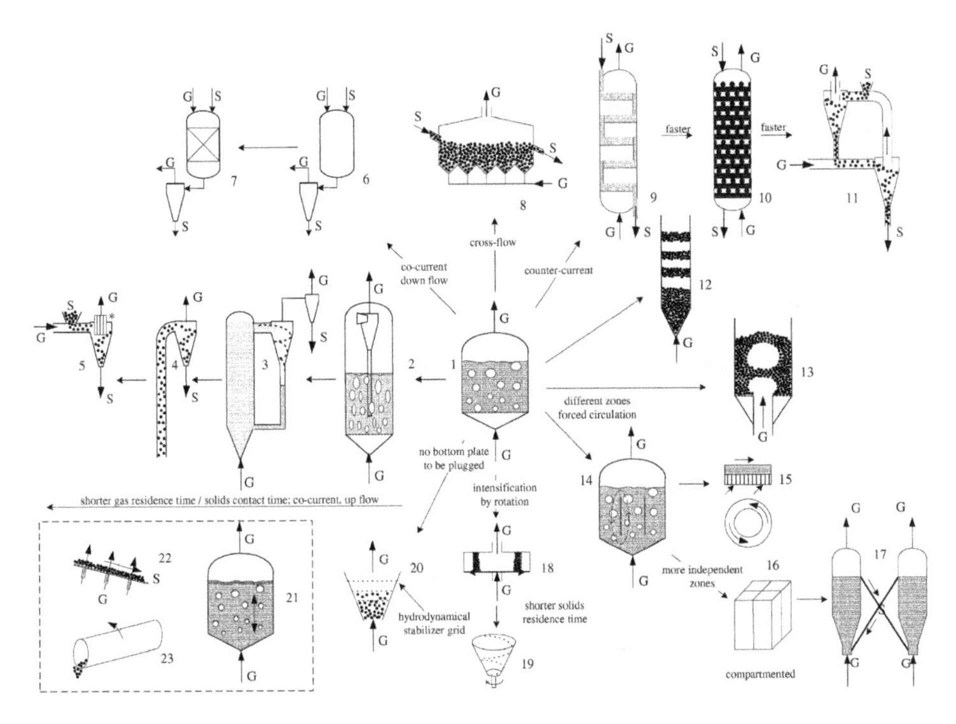

Figure 2.9: The fluid bed reactor family [2].

2.3 Two-phase G-L and L-L reactors

2.3.1 Gas–liquid bubble column reactors

2.3.1.1 Main characteristics

In gas–liquid bubble columns, the continuous phase is the liquid and the dispersed phase is gas.

Gas is fed at the bottom of the reactor via a distributor. Liquid can be fed in many ways to the column. It can be fed at the top or at the bottom but also it can be fed at multiple points along the height. Figure 2.10 is therefore just a schematic picture.

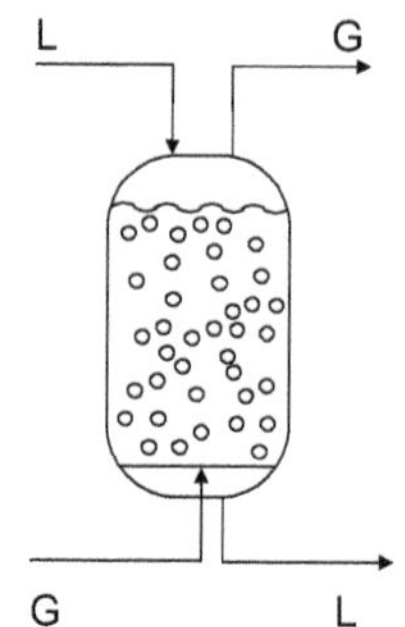

Figure 2.10: Schematic picture of a basic bubble column reactor.

The liquid phase circulates inside the vessel and is agitated by the gas flowing as bubbles through the reactor. Due to this liquid circulation pattern, the liquid feed location may be not critical. If, however, rapid mixing of the feed with the bulk liquid is needed, then the liquid feed may be near the gas distributor.

The gas is often separated inside the reactor by having a liquid outlet below the top of the reactor. In the disengagement zone above this liquid, the outlet gas is separated from the liquid by gravity. In this way, gas can leave the reactor at the top while liquid can leave at a lower location.

In most cases, some of the reaction components will be supplied in the gas phase to the reactor. By mass transfer, the components are transferred to the liquid phase and react there with the other reaction components, fed as a liquid feed.

The reactor can also be designed and operated as a three-phase reactor. Solids are then present in the liquid phase. The solids can be a heterogeneous catalyst. The solids can move out of the reactor with the liquid outflow. The solids (catalyst) may also be kept inside the reactor by a special liquid–solid separation device.

The bubble column can be configured as a horizontal vessel in which the liquid flows horizontally through the vessel and the gas mainly vertically. In this configuration, liquid backmixing can be limited by having baffles (or vertical walls with specially designed openings) so that the liquid can flow horizontally through the reactor. The baffles require a special configuration to avoid shortcut flows [14]. A picture of

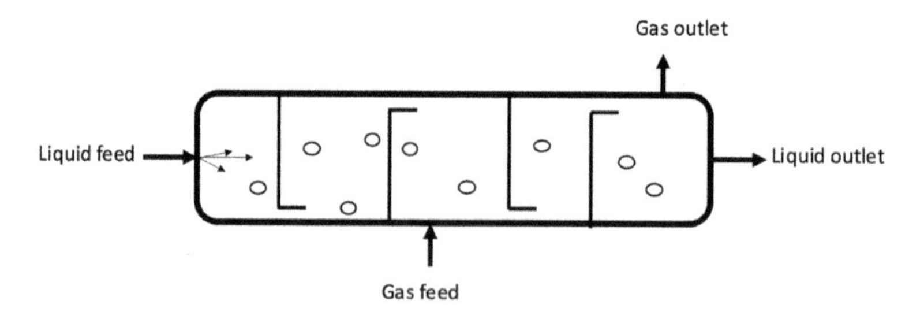

Figure 2.11: A horizontal bubble column reactor.

such a horizontal cross-flow bubble column is shown in Figure 2.11. Our industrial case in Section 13.3 provides more details.

The reactor can be designed and operated as an adiabatic reactor when the adiabatic temperature rise is low. It can also be designed as a reactor with an internal heat exchanger.

2.3.2 Continuous and batchwise operation

For the liquid phase, the bubble column reactor can be operated continuously or batchwise. For the batch operation, heating up to the reaction temperature and cooling the reactor can be done by installing a heat exchanger. For the latter, also evaporation of the reaction medium can be applied. Outside the reactor or in the top of the reactor, a condenser can be installed to remove the reaction heat and recycle the liquid back to the reactor. However, we have not seen such a batchwise-operated bubble column in industry. Most bubble columns are operated continuously and for large production capacities.

2.3.2.1 Main characteristics

The liquid can be uniformly distributed to the column by a distributor or by a simple nozzle. The gas will be uniformly fed by a sparger.

The liquid-phase residence time distribution for vertical columns will show considerable backmixing, and this is highly sensitive to the column diameter size. The subject is treated in detail in Chapter 5.

For horizontal columns, the liquid residence time distribution can be staged using vertical walls. Then each chamber in between the vertical walls can be considered as backmixed.

The other main hydrodynamic characteristics of bubble column reactors are summarized:

- Mixing: Liquid mixing can be obtained inside the column at a reasonable rate. Gas phase mixing will be low and will depend on the bubble flow regime. In the small bubble regime, mixing will be absent.
- Shear rate distribution: The shear rate distribution will be nonuniform and highly dependent on the superficial gas velocity and column dimensions.
- Mass transfer: The mass transfer rate in bubble columns is low. More information is found in Chapter 6.
- Heat transfer: Heat transfer in bubble columns with a heat exchanger inside is moderate. More information is provided in Chapter 7.
- Impulse transfer: Impulse transfer of the liquid on column internals can be considerable, when the column diameter is beyond 1 m, and liquid circulation velocities are several meters per second. Care should be taken that the gas sparger does not break down by the liquid impulse forces.

2.3.2.2 Industrial applications

The main feature of a bubble column reactor is that it facilitates contact between a reacting component available in the gas phase and a reacting component available in the liquid phase. It is very suited for slow reactions with typical reaction times of 10–100 min. The gas phase average residence time is in the order of 1 min. The slow reaction also means that the gas–liquid mass transfer is often not limiting the overall reactor rate. So the reactor scale-up can be reliably executed.

There are numerous commercial-scale applications in many process industries up to very large production scales. It has been applied for over a century. What distillation is to separations is the bubble column reactor to reactions. It is a true workhorse. Because the reactor vessel is a simple cylinder, the investment cost and maintenance cost are low.

2.3.3 Mechanically stirred gas–liquid reactors

2.3.3.1 Main characteristics

The main characteristics of mechanically stirred gas–liquid reactor are that the liquid is the continuous phase and gas is the dispersed phase. The liquid feed point can be chosen near the location where the stirrer blades move around in order to create maximum local shear rate. This can be important for micromixing with reaction as explained in Chapter 4. The gas will be supplied via a gas sparger often placed near the bottom of the reactor.

Due to the turbulent field created by the stirrer, bubbles may be broken up near the impeller blades and may coalesce far away from the impeller blade.

Heat exchange area can be created by the wall and also by placing tubes inside the vessel. Often, however, heat is removed by evaporating part of the liquid. The vapor is then cooled in a condenser and the cooled liquid is returned to the reactor. The condenser is often placed in the reactor top dome.

The reactor can be operated batchwise or fed-batch wise. The latter is often applied to limit the reaction heat rate so that the temperature can be controlled.

2.3.3.2 Commercial-scale applications

This reactor type is applied at least since the sixteenth century [15]. The main reason is that the mechanically stirring action allows for an enormous freedom to the designer to create what he/she wants in contacting components for reactions. Nowadays, this reactor type is mainly used in fine chemicals and pharmaceuticals. In those industry branches, it is probably the main reactor type used. Therefore, it is a workhorse.

In bulk chemicals and oil refining, this reactor type is virtually absent. The main reason for this absence is that scale-up is limited to a certain size. Beyond this the forces on axis and bearings become too large. Reliability is also lower than for reactors with no rotating equipment parts. A bubble column does not have these negative items so this is often preferred.

2.3.4 Gas–liquid spray tower reactor and Venturi washer

2.3.4.1 Main characteristics

In spray tower reactors, gas is the continuous phase and liquid is the dispersed phase. The liquid is sprayed from the top of the reactor. By gravity, the droplets fall down. The gas phase can flow co-currently or countercurrently. The reaction components may all be present in the liquid phase and the gas is only used as a heat exchange medium.

A variant of the basic spray tower is the Venturi washer. Here also the gas is the continuous phase, but rather than using a spray nozzle to make fine droplets, a nozzle throat is used. The liquid atomizes in tiny droplets due to a high velocity. These subsequently partially evaporate due to a big decrease in pressure in this converging section, according to Bernoulli's law. In the subsequent diverging section, the pressure rises again and the vapor condenses. The big advantage being that solid contaminants in the gas phase feed G serve as nuclei and are separated from the gas via liquid at bottom. Figure 2.12 shows schematic pictures of the reactors [16].

Feed distribution: Liquid via a distributor will be nearly uniform. Gas feed distribution is nearly uniform but requires special detailed design attention.

The residence time distribution of the liquid will be near-plug flow if the liquid distributor is properly designed and does not create small (satellite) droplets which can move around with the gas phase.

The residence time distribution of the gas phase, however, will in general be wide due to gas entrained by the falling droplets moving down, causing a gas phase circulation pattern, similar to circulation patterns in bubble columns, where the liquid is entrained by the rising gas bubbles. Thanks to, among other things, to the fine droplets, and some typical characteristics are:

- Mass transfer rates between the gas and liquid phases will be high.
- Heat transfer rates between the gas and liquid phases will be high.
- Shear rate distribution is nearly uniform on the droplets.
- Impulse transfer between the gas and liquid phases is low.
- Pressure drop of the gas phase is low.

More information is provided in Chapter 6.

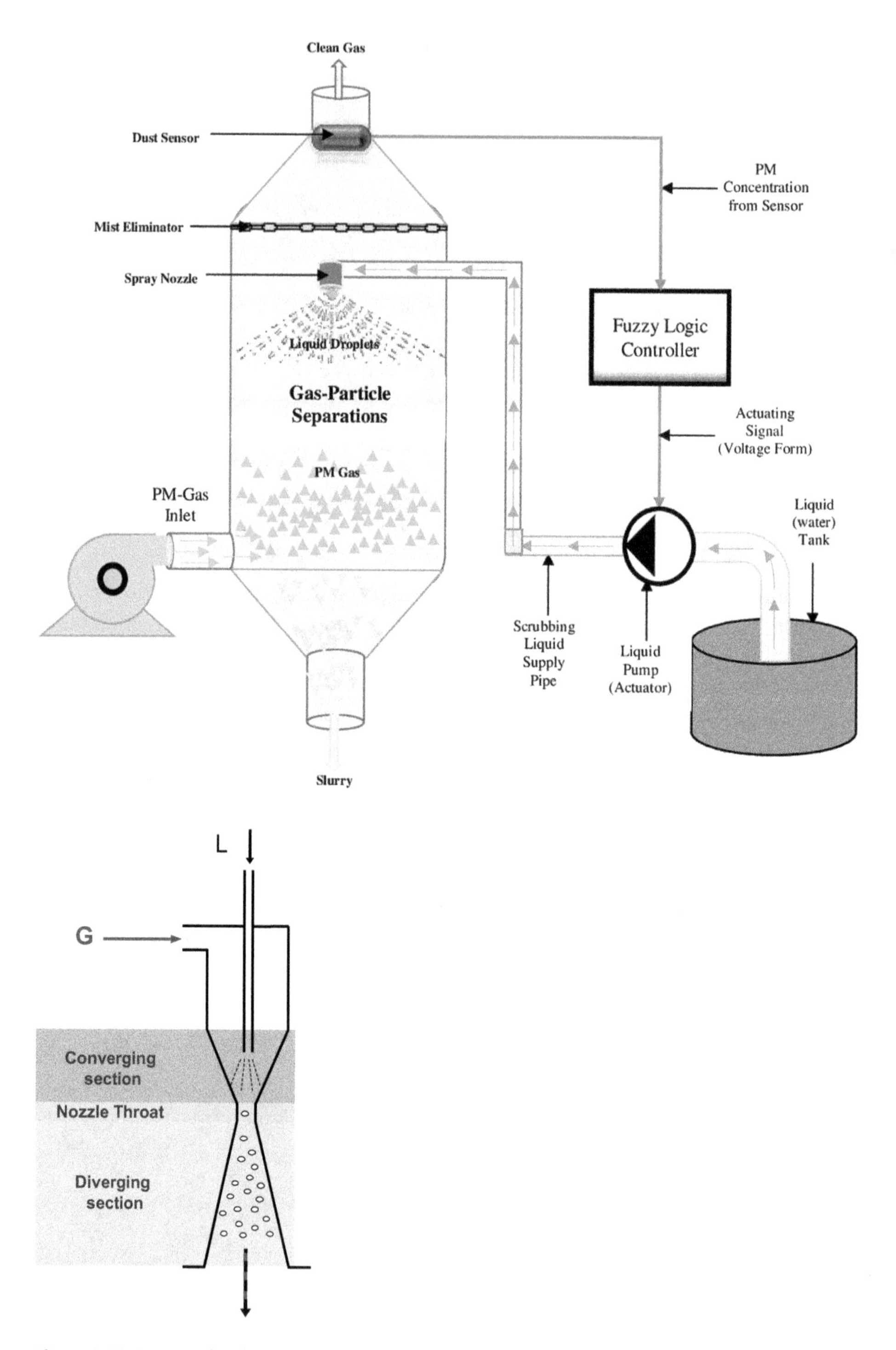

Figure 2.12: An example of a spay column reactor (top) and of a Venturi washer (bottom) [16].

2.3.4.2 Commercial-scale applications

Spray tower reactors are mainly used in the industry to remove trace components from a gas stream to avoid atmospheric pollution, instead of a packed column when fouling of the packing could be an issue [16].

The low liquid residence time limits the applications to very fast reactions only.

Gas–liquid spray towers are, however, used a lot in various process industries for various other applications such as cooling the liquid and drying a liquid to form dry solids (spray drying). A lot of knowledge is therefore available from spray towers used as a dryer (milk powder production from milk) or as a solidifier (also called a prilling tower).

On the term "spraytower"

A warning for finding more information: the term "spraytower" is sometimes used wrongly or confusingly. For example, sometimes a gas–liquid co-currently flowing packed bed (a trickle bed) is called a spraytower.

2.3.5 Gas–liquid packed bed reactor

In the gas–liquid structured packed bed reactor, liquid flows down over an inert packing. The packing can be Raschig rings or Pall rings or any internal structure similar to internals to distillation columns. Gas in general flows upward through the column. Components in the gas phase are transferred to the liquid and react there in general by a homogeneous reaction.

It is, for instance, used in amine treatment columns to remove H_2S from a gas stream. The H_2S reacts with the amine in the aqueous phase. In a separate unit, the reverse reaction runs at a higher temperature and the amine is recycled. For those cases, the reactor is designed countercurrently.

It is, however, mostly applied to remove trace components from gas streams to be released into the atmosphere. In most industrial applications, it is not called a reactor, but a gas washer, a polishing column, an absorber, a scrubber, or even a spray tower. In many of these industry cases, the design gets little attention.

2.3.5.1 Warning on column design for effluent gas treatment

For effluent gas treatment where trace components have to be removed by a liquid, often the column height is determined from mass transfer correlations applied in oil refineries for sizing distillations (height of a transfer unit is used as a fixed value), where the mass transfer limitation is on the gas phase side, while for the design case, the mass transfer limitation is often on the liquid side. Consequently, the column removes the trace component not deep enough. The process designer is often not aware of the limitations of the height of transfer unit used for his/her design and designs a far too short column height. When it is reported to him/her that the column does not perform according to the design specification of the degree of tracer removal, he/she has no clue why this is the case. We have seen this problem more than once.

The solution is to use the liquid-side mass transfer coefficients as well as gas-side mass transfer coefficients to determine the overall mass transfer coefficient. See Chapters 5 and 6 on this subject.

2.3.6 Two-phase L-L reactors

2.3.6.1 Introduction to liquid–liquid reactors

The authors could not find many publications about many industrial-scale applications of liquid–liquid reactors. So it cannot be defined as a workhorse reactor. For that reason, we leave out detailed descriptions of this reactor type but write a short text of the main features holding for most liquid–liquid reactors.

Reactor types mentioned in literature are:

- packed column reactor,
- baffled pulsating flow reactor,
- rotating packed bed reactor,
- rotor-stator (spinning disk) reactor, and
- multistage mechanically stirred reactor.

2.3.6.2 Phases and flows and residence time distribution

In liquid–liquid reactors, one liquid phase will be continuous and one phase will be dispersed. The reaction phase can be in either of the two phases. Because the density difference between the two phases is small, the countercurrent operation has only a small operating window. However, for reactions, the countercurrent flow is for most cases not needed, as the components react away.

The two feeds can enter at the same side of the reactor. The dispersed phase will need a specially designed distributor to create fine droplets. This holds in particular for packed columns, where the shear rate and energy dissipation are low. The initially formed droplets may not coalesce, so the distributor determines the mass transfer performance of these columns. In general, the fluid flow is staged by internals. Typically, 5–10 stages are easily obtained.

2.3.6.3 Fluid agitation and mass transfer

Except for the packed column reactor, all other reactors mentioned here have intense shear applied on the phases to create high mass transfer. This shear is created by either an intense turbulent field such as in the baffled pulsating reactor and the multistage mechanically stirred reactor. Or it is created by a high gravity field in combination with small channels for the fluids to flow through.

2.3.6.4 Heat transfer

Heat transfer can take place in the baffled pulsating reactor via the wall. The turbulent field creates high heat transfer coefficients. The heat exchange area is however limited to the reactor wall. The rotor-stator reactor has very high heat transfer coefficients and also a very high heat exchange area relative to the reactor volume. This specific surface area is a factor of 100 larger than in other reactors.

2.3.6.5 Batchwise and continuous operation options

No commercial-scale batchwise operation is reported. However, a mechanically stirred reactor could be operated batchwise or fed-batch wise. In the latter operation one liquid feed is loaded into the reactor and heated to the desired temperature, while stirring. Then the other liquid is fed slowly to the reactor while stirring. The second liquid will form the dispersed phase. Deep conversion can be easily obtained due to plug-flow behavior of the batchwise operation.

All other described reactors can be artificially operated batchwise by installing holdup tanks for the product streams and operating the reactor continuously.

2.3.6.6 Commercial-scale application features

Descriptions of commercial-scale applications for these reactors are hard to find. Harmsen published about a commercial-scale application of a liquid–liquid multistage mechanically stirred reactor where a conversion of over 99% was reached. Here is a short description of the application.

The reaction takes place in the continuous organic phase. One of the reacting components, an organic chloride, is in this phase. An alkaline component is fed as an aqueous solution to the reactor and is in the dispersed phase. By mass transfer that component moves to the other phase. There it reacts to form an organic molecule and a salt. The salt transfers back to the aqueous phase.

The reactor is a vertical column with a central vertical axis on which several stirrers are mounted at equal distance. The column, furthermore, is divided into chambers by horizontal plates with a small hole in the center through which the axis goes. A small slid next to the axis allows the two phases to flow concurrently from one chamber to the next chamber.

The residence time distribution was measured using a salt pulse injection and conductivity measurements at the inlet and outlet. It appeared that the residence time distribution closely resembled CISTR's in series with the number of CISTR being equal to the number of chambers. No distinction was made between either of the two phases [17].

2.4 Three-phase gas–liquid–solid reactors

2.4.1 Slurry reactors (liquid is continuous phase)

There are many three-phase slurry reactors. We describe here only the main reactor types: three-phase bubble column, three-phase fluid bed (liquid feed up flow causes particles to be fluidized), mechanically stirred slurry reactor, and Venturi loop reactor.

2.4.1.1 Three-phase bubble column and fluid bed reactors

In a three-phase bubble column reactor, gas flows upward through the column and agitates the slurry (liquid with the solids in suspension). In a three-phase fluidized bed, liquid flows upward through the column at such a velocity that the solids are fluidized. Gas is sparged at the bottom and also flows upward through the column. These two types have so many features in common that we treat them here together.

The characteristics of this reactor type are similar to a two-phase bubble column. The solid particles are inside the liquid phase and, in general, have the same backmixed residence time distribution as the liquid phase of a bubble column.

The gas phase also has a residence time distribution similar to bubble column reactors. Mass transfer rates for the gas–liquid interface are also similar to bubble columns.

The reactor may have internal heat exchanger so that it can be operated as nearly isothermal.

This reactor type is applied for Fischer–Tropsch synthesis: reactions of syngas (CO and H_2) to hydrocarbons.

An ebullated bed is a special version of this reactor type. Here the catalyst particles are so large that they tend to settle in a quiet zone near the conical bottom wall and then are whirled upward by the gas and liquid in the central section. This type is applied in the LC-FINING™ process developed by Lummus for hydro-upgrading of residual oil fractions. This description is present in Section 13.5.

Another special version is a plated column slurry reactor. This is described in Section 13.7.

2.4.1.2 Three-phase mechanically stirred reactors

Three-phase mechanically stirred tank reactors are very similar to gas–liquid mechanically stirred reactors.

The gas phase residence time distribution will be wide. The liquid residence time distribution will be very narrow if the reactor is operated in batch mode. In continuous operation mode, the distribution will be backmixed.

The solid-phase residence time distribution will be narrow if operated batchwise. In continuous mode, it will be backmixed.

If the solids phase is made, such as microorganism growth in fermentation, then the residence time distribution cannot be simply stated.

Mass transfer across the gas–liquid interface will be similar to bubble columns unless the power dissipation is much higher than in a bubble column.

Heat transfer can be obtained via the wall and heat exchange tubes placed in the reactor. Also overhead vapor cooling can be applied. This means that part of the liquid is evaporated, condensed in the overhead cooler, and the condensed liquid drops back into the reactor.

2.4.1.3 Three-phase Venturi jet loop reactor

A special three-phase slurry reactor is the Venturi jet shown in Section 2.3.3 but then a vessel is placed below the Venturi and a pump, and a recycle pipe feeds the slurry back to the Venturi.

The Venturi sucks in the gas and creates very fine bubbles for a high mass transfer. The jet from the Venturi device enters a vessel where additional mass transfer occurs. The system also has an external liquid recycle in which a heat exchanger may be placed. It is sometimes called the Buss loop reactor, after the technology provider Buss.

Critical performance factors of the Venturi jet loop reactor are:

- Feed distribution: The liquid is fed to the main vessel as a three-phase jet. The gas is well dispersed in the Venturi and then by the jet enters the main vessel.
- Residence time distribution: For a continuous operation, the relevant residence time distribution can be considered close to plug flow if the reaction is so fast that a single-pass conversion occurs in the Venturi plus jet. If the reaction is slow than the considerable backmixing in the liquid phase will have to be accounted for.
- Mixing: Mixing in the liquid phase will be fast. Gas mixing is complex and will generally be unknown.
- Shear rate distribution: The shear rate distribution is very different in each part of the system.
- Mass transfer: Mass transfer is high.
- Heat transfer: Heat transfer in the external loop heat exchanger will be medium.
- Impulse transfer: Impulse transfer on the venture device will be high.

It can be operated batchwise and continuously.

Its main applications are for specialty and fine chemicals production, often for hydrogenation using a heterogeneous catalyst.

2.4.2 Trickle-bed three-phase reactor with gas as the continuous phase

2.4.2.1 Main characteristics

The liquid flows downward as a thin film over a packing in trickle flow. So the liquid can be seen as a continuous phase The packing can be an inert or catalytic. The liquid distribution is critical to the performance; hence, a dedicated liquid distributor with many nozzles over the whole cross-sectional area is applied; see Figure 2.13 for an artist's impression of the importance of a good distribution.

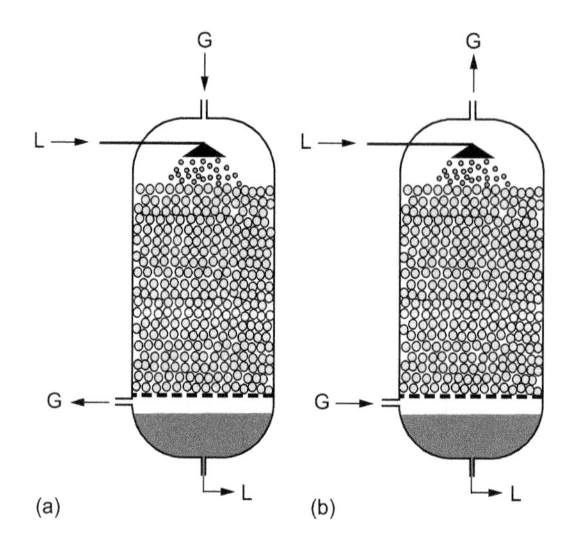

Figure 2.13: Co-current (a) and countercurrent (b) trickle-bed reactor. (adapted from Prof. Guy Marin, Ghent University).

The gas flows in most applications downward, hence, co-currently with the liquid. In the majority of applications, the gas also is a continuous phase flowing in between the liquid films around the particles. This flow pattern is called trickle flow. The reactors are called trickle-bed reactors.

For trickle flow, the gas feed distribution at the top can be a simple gas chamber with a gas inlet device to avoid a gas jet hitting the bed. For water treatment applications, the gas is air, which is entrained by the liquid spray flowing downward onto the packed bed. For hydrogenation applications, in particular, the gas is forced by pressure through the bed. The gas flows in some applications upwards through the column countercurrent to the liquid flow. The gas is then fed at the bottom via a dedicated designed distributor to ensure an even gas flow.

Concurrent downflow of gas and liquid is mostly applied for heterogeneous catalytic reactors at commercial scale. For those commercial-scale reactors, the residence time distribution of both gas and liquid can be approximated to be in plug flow.

Over the cross-sectional bed area, gas and liquid mixing is virtually absent. Gas and liquid flow as segregated streams through the packing. This phenomenon stresses the importance of a good liquid and gas feed distribution.

Gas/liquid and liquid/solid mass transfer in commercial-scale reactors is in general high compared to the reaction rate. So mass transfer limitations are absent.

Academic research has spent a large effort on stagnant liquid pockets in the neck of particle contact points. The results are summarized by Westerterp et al. [20].

These pockets, however, have a characteristic length which is smaller than the particle diameter. This means that mass transfer by diffusion is still higher than diffusion inside the particle. By proper design methods, provided in Chapter 6, mass transfer limitations can be avoided for commercial-scale applications.

Heat exchange is absent when the trickle-bed reactors are designed adiabatically. The design can be such that the temperature rise of the gas and liquid over the entire bed is only a few degrees. This is often obtained by having a large surplus of the gas phase reactant fed to the reactor.

Packed bed reactors can also be designed as multitubular reactors with heat exchange via the tube walls. This is described in some detail in Chapter 7.

2.4.2.2 Industrial applications

This reactor is presently used probably more than any other reactor in all kinds of process industry branches, such as oil and gas, refineries, bulk chemicals, and wastewater treatment. In general, a gas phase component reacts with a liquid-phase component. Often the reaction is fast (reaction time of seconds or lower) and deep conversion of the gas phase component is desired.

A special application is in biological water treatment. Air and wastewater flow down over the packing. The packing contains a film of microorganisms oxidizing the organic components. This reactor is applied at large commercial scale for urban wastewater treatment but also on a small scale in small communities in developing countries to convert rainwater into safe drinking water.

The reaction heat rate per unit volume is low due to the low concentration of the gas phase component. A counter-current packed bed column is fed with liquid from the top and with gas from the bottom. The high mass transfer and near-plug-flow behavior of the gas and the liquid facilitate a deep conversion.

2.5 Reactors with heat control

2.5.1 Introduction of reactors with heat control

Nearly all chemical reactions have a nonzero reaction enthalpy. So nearly all reactors need heat control. If the heat production is low, meaning that the adiabatic temperature rise of the fluids inside the reactor is low, then the reactor can be designed as an adiabatic reactor, and the reactor has still to be temperature controlled. But the reactor does not need a heat exchanger as part of the reactor system.

When the reactor system needs to be heat controlled, then there are several options available to do so. For the workhorse reactors of the previous sections, heat control options are indicated. Here we provide general ways for reactor heat control.

2.5.2 Adiabatic heat control

Adiabatic heat control is often applied in industry. If the adiabatic temperature rise of the reaction system would be too high for adiabatic reactor operation. The following options for still having an adiabatic reactor are available:

Option 1: Surplus feed with external cooling and recycle
Operate with a surplus of one component so that the adiabatic temperature rise drops to an acceptable level. Separate the surplus. Cool it in an external heat exchanger and recycle the component back to the reactor. This design is applied for hydrogenation reactions. A very large surplus of hydrogen (typically a factor of 10 is more than needed for a stoichiometric feed) is fed to the reactor, by which the adiabatic temperature rise is in the order of 10 °C. The hydrogen is separated from the product stream, cooled in a heat exchanger, and then fed back to the reactor. To do so also a recycle compressor is needed.

The alternative multitubular cooled fixed bed is far more expensive for these high-pressure reactors (typical pressure of 150 bar), as the multitubular reactor volume would increase by more than a factor of 2.

Option 2: Medium feed with external cooling and recycle
An inert medium feed rather than a surplus of a reacting component can also be applied to reduce the adiabatic temperature rise. The medium is then separated from the product stream, cooled, and recycled. This option, however, is inferior than option 1, as an additional component is introduced to the reactor and the process. This requires a research effort to determine its effect on the reaction and the product quality, on safety measures, and the cost of additional storage. The use of media (solvents) in bulk chemicals is therefore not used for heat control.

2.5.3 Multitubular fixed bed reactor

Fixed bed reactors, both two-phase and three-phase, can be designed as multitubular reactors. A large number of tubes filled with catalyst is surrounded by a cooling medium. The cooling medium may be in the boiling mode, so the amount of cooling fluid to be supplied is greatly reduced, as the heat is mainly used for boiling the cooling liquid.

A large-scale industrial application is described in Section 13.2.

2.5.4 Wall (jacket) heat exchange

Wall heat exchange is used in mechanically stirred reactors. The heat transfer wall area is small relative to the reactor volume and moreover decreases relative to the reactor volume with increasing reactor diameter. So its applicability is very limited. It is mainly used in small-scale batch reactors.

2.5.5 Heat transfer by evaporation and a condenser

Reactor heat transfer can also be obtained by evaporating a reaction component or a solvent in the reactor and then condensing the vapor in a condenser placed at the top of the reactor or outside the reactor. This is commonly a commercial practice design in emulsion polymerization, fine chemicals, and pharmaceuticals. In most cases, a solvent is needed anyway to facilitate the liquid-phase reaction. So using the solvent also as a cooling agent does not cause additional concerns. A condensing vapor has a high heat transfer coefficient, so the condenser costs are low.

2.5.6 Heat transfer by coils inside the reactor

Heat control by placing heat exchange coils inside a reactor is a feasible option for reactors where only fluids are inside the reactor. So these can be (and are) applied in two-phase and three-phase bubble columns and fluid beds.

2.5.7 Microwaves heating

Heating by radiation is mainly applied in the food industry. Most technologies have a moving belt with a radiation source on top for continuous operation. Most applications are used for food drying. However, rapid microwave heating is also used for reactors [8]. For instance, they are used to convert some of the proteins in the food so that they no longer act as a food poison. Many technology providers provide information and often also have a test facility. A web search using the keywords “industrial” and “microwave” will provide results from many companies with their technologies.

2.5.8 Electrical heating

Electrical heating is an emerging technology for endothermic chemical reactions such as alkane dehydrogenation. There are a number of different methods for electrical heating, including indirect and direct heating, resistive and inductive. Especially for equilibrium-limited endothermic reactions, the very high temperatures created can be beneficial to obtain a high product yield; hence, one of the front-runners for commercial-scale application is steam methane reforming. This is under development by Haldor-Topsoe. The subject is, however, beyond the scope of this book.

A short introduction to this subject (and several other reactor types) is provided by de Haan and Padding [18] and a more extensive review from Eurokin [19].

2.6 Exercises

2.6.1 Industrial exercise 1: reactor types for PVC depolymerization start-up company

Given: A young chemical engineer enters into a techno start-up company. The company has obtained a patent to convert waste PVC into VC monomer. The waste PVC is dissolved in a solvent and a homogeneous catalyst depolymerizes the reaction. The monomer evaporates from the solvent. The reaction is endothermic, so heat has to be supplied. A proposal has to be made to the bank to obtain investment money for the concept stage and the development stage. When the development stage is finished, the business model is to obtain investment money for the commercial-scale implementation by issuing shares. To show the credibility of the invention to the bank, a shortlist of reactor types which can be applied has to be part of the proposal to the bank.

Problem to solve:
Which phases are involved in the reaction system?
Which reactor type options are potentially applicable?

2.6.2 Industrial exercise 2: reactor type options for precipitation reaction

Given: A chemical engineer is added to an innovation project in the feasibility stage. The project is about a gas–liquid reaction producing the product in the liquid state and a salt by-product that precipitates in the reactor as solid particles. In the concept stage, a mechanically stirred tank reactor in continuous operation, cooled by wall heat exchange, is chosen. But now the managers have asked to provide an overview of all potential reactor types for the commercial scale, so all types are evaluated and the best reactor type can be chosen.

Task: Make a short list of applicable reactor types for this reaction.

2.7 Takeaway learning points

Learning point 1
There are dozens of commercially applied reactor types for two- and three-phase reaction systems available. Only within Shell, we had counted in 2010 close to 60 different ones.

Learning point 2
A preselection of reactor types can be made based on the phases present.

Learning point 3
A major distinction between multiphase reactors is fixed bed versus fluidized bed and slurry reactors. If a large amount of solid particles have to be processed, such as in catalytic cracking, then reactor types where particles behave as a fluid such as in fluidized beds and slurry reactors are preferred.

References

[1] Harmsen J, Industrial Process Scale-up: A Practical Innovation Guide from Idea to Commercial Implementation. Amsterdam, Elsevier, 2019.
[2] Van Swaaij WPM, van der Ham AG, Kronberg AE, Evolution patterns and family relations in G–S reactors. Chemical Engineering Journal, 2002, 90(1–2), 25–45.
[3] Van der Grift CJG, Woldhuis AF, Maaskant OL, The Shell DENOX system for low temperature Nox removal. Catalysis Today, 1996, 27(1–2), 23–27.
[4] Vandewalle LA, De S, Van Geem KM, Marin GB, Bos R, Reactor engineering aspects of the lateral flow reactor. Industrial & Engineering Chemistry Research, 2020, 59(24), 11157–11169.
[5] Kunii D, Levenspiel O, Fluidization Engineering. Boston MA, Butterworth-Heinemann, 1991.
[6] Vogt ET, Weckhuysen BM, Fluid catalytic cracking: Recent developments on the grand old lady of zeolite catalysis. Chemical Society Reviews, 2015, 44(20), 7342–7370.
[7] Vasconcelos PD, Mesquita LA, Gas-solid flow applications for powder handling in industrial furnaces operations. In: Ahsan A, ed. Heat Analysis and Thermodynamics Effects. Rijeka, Intech open access publisher. 2011, Chapter 10.
[8] Harmsen J, Verkerk M, Process Intensification – Breakthrough in Design, Industrial Innovation Practices, and Education. Berlin, De Gruyter, 2020.
[9] Geldart D, Gas Fluidization Technology. Chichester, J. Wiley & Sons, 1986.
[10] Rietema K, Powders, what are they. Powder Technology, 1984, 37(5), 5–23.
[11] Karmakar MK, Loha C, De S, Chatterjee PK, eds, Hydrodynamics of circulating fluidized bed systems. In: Coal and Biomass Gasification, Singapore, Springer, 2018, 93–114.
[12] Cottaar EJE, The influence of interstitial gas on powder handling. PhD thesis, Eindhoven. TU/e, 1985.
[13] Kiadehi AD, Development of a new technique for determining the RTD of a dispersed solid phase and its application in a deep fluidized bed, Doctoral dissertation, Compiègne France, Université de Technologie de Compiègne, 2019.
[14] Harmsen GJ, Rots AWT, Process for the preparation of alkylene glycol. Patent EC C07C29/12, 2009, WO2008068243, 2008.

[15] Agricola G, De Re Metallica Libri XII. Basel, J. Froben & N. Episopius, 1556, as referred by Stankiewicz A, Moulijn JA, Re-engineering the Chemical Processing Plant – Process Intensification. New York, USA, Marcel Dekker, 2019.
[16] EPA, Air Pollution Control Technology Fact Sheet EPA-452/F-03-016, Accessed October 29 October 2021 by, https://www3.epa.gov/ttnchie1/mkb/documents/fsprytwr.pdf
[17] Harmsen GJ, Product quality at lowest cost: Robust processes by parameter and tolerance design using integration tools. Proc. Sec. Int. Conf. Computer Integrated Manufacturing in the Process Industries, Eindhoven 1996, 293–297.
[18] de Haan AB, Padding JT, Process Technology – An Introduction, 2nd ed. Berlin, de Gruyter, 2022.
[19] Zheng L, Ambrosetti M, Tronconi E, Review on approaches and materials for the electrification of catalytic processes. Eurokin Task B16d, 2022, https://eurokin.org/?cat=42
[20] Westerterp KR, van Swaaij WPM, Beenackers AACM, Chemical Reactor Design and Operation. Chichester UK, John Wiley & Sons, 1984.

Part B: **Fundamentals**

3 Scale-independent basics relevant for all reactors

The purpose of this chapter is to provide essential basic chemical reaction engineering theory, independent of reactor type and scale. It is a grounding chapter for all other chapters. Important definitions and terms for chemical reaction engineering practices are explained.

3.1 Reaction stoichiometry and kinetics

3.1.1 Introduction

Reaction kinetics and connected stoichiometric equations are an essential part of chemical reaction engineering. Theories on residence time distribution, mass transfer, and heat transfer can only be applied when reaction kinetics are available. At the beginning of a project that involves new chemistry, the kinetic rate expressions (kinetics, in short) can be simple. Often, it is sufficient to have kinetics of the main reaction described in a simple way. In that stage, only a choice of reactor family type may be made, so accuracy is not needed. The kinetics and their stoichiometric equations, however, should be correctly defined. Wrong definitions can result in hidden errors, which only appear much later. Therefore, we write in detail, here, on correctly formulating stoichiometry and kinetics.

Before we delve into the details of chemical reaction engineering, it is worth mentioning that "chemical" in the term "chemical reaction engineering" refers only to (a change in) a chemical molecule composition. The term and its applications do not refer to the chemical engineering industry; it refers to molecular changes. Chemical reaction engineering is relevant for any molecular reaction. The reaction can be in food, ceramics, minerals, biomass, water treatment, or any other process industry.

3.1.2 Reaction stoichiometry

Reaction stoichiometry is about the molecules and the ratio of molecules involved in a reaction. Once known and written down, it clarifies immediately about which products will be formed by the reaction in what fixed ratios and which reacting (feed) components in fixed ratios are needed for the products.

This information is essential for reaction kinetics. It also gives direct information on what separations are needed, anyway, in the process design. In our experience, when asked for advice about reaction engineering design or reactor trouble shooting, often, the stoichiometric equation(s) were not known. After some minutes of discussion, the stoichiometric equation of the main reaction was written down. In some

https://doi.org/10.1515/9783110713770-003

cases, the researcher or the process engineer went silent. Clearly, he or she was thinking about the problem in a new way.

The general stoichiometric equation format of a reaction is:

$$aA + bB = pP + qQ$$

The composition of the molecules *A, B, P,* and *Q* are described in their simple chemical formula with subscripts indicating the number of each of the different atoms present in each molecule. The stoichiometric coefficients of the equation are: *a, b, p,* and *q*.

The correctness of the stoichiometric equation with its stoichiometric coefficients is easily determined by a simple atom count.

If the experimental results cannot be described by the single stoichiometric equation, then a different reaction may (also) take place, for which a separate stoichiometric equation needs to be established.

An example of a stoichiometric equation is the oxidation of acetic acid to carbon dioxide and water:

$$C_2H_4O_2 + 2\,O_2 = 2\,CO_2 + 2\,H_2O$$

In some textbooks [1, 2] the stoichiometric equation is not written with an "=" but with a "→", especially when highlighting the forward reaction. When it involves an equilibrium reaction, often, the "⇌" sign is used. However, whether the reaction is an equilibrium reaction or not may not be known upfront. (The reaction rate and the equilibrium are treated in the reaction kinetics part). We, therefore, advocate the use of the "=". It also stresses the point that it is an equation. It directly shows the fixed ratios of molecules involved in the reaction. However, it does not state anything about the reaction direction or equilibrium involvement.

In most reactors, more than one reaction takes place. This is easily observable by analyzing the reactor feed and product composition with the stoichiometric equation. When the observed molecular ratios are not in agreement with the stoichiometric equation, then, an additional reaction is involved. By running several experiments and by hypothesizing what additional reaction is involved, the stoichiometry of the additional reaction can be discovered.

In certain industrial branches such as in oil and gas refining and biofuels, hundreds to thousands of reactions may be involved. Stoichiometric equations can then still be useful by lumping molecules similar in chemical composition, such as, for instance, olefinic (unsaturated) molecules or cellulosic molecules. For reactions such as a hydrogenation with olefins containing unsaturated chemical bonds, a stoichiometric equation can still be written down by considering the unsaturated bonds as a "molecule." An industrial example is found in Chapter 14, where two lumped reactions are considered – the polymerization reaction of monomer to a polymer chain and the hydrogenation reaction of an unsaturated group in the polymer. Hence, all monomer additions are lumped together and all hydrogenation reactions are lumped together.

In the ideation stage, when, for instance, several feed types for a certain product have to be evaluated, the stoichiometric equation for each feed type gives insight into the maximum yield of product on the different feeds achievable and also the number of separations minimally needed, in each process concept.

This is even more the case when intermediate reaction steps are involved. A block flow diagram showing the reaction blocks, separation blocks, and recycle streams are then, often, sufficient to make a preselection of most promising chemical routes, by taking into account the investment cost and energy cost for separations, regardless of the reaction rate and the reactor volume required.

Also, in existing processes in commercial operation, a stoichiometric analysis of feed and product composition can reveal that an unknown reaction is also present in the process, as the known reaction with its stoichiometry is not in agreement with the product composition. Further investigation with samples from the reactor outlet and the product outlet stream may indicate that the reaction takes place outside the reactor. Often, the undesired reaction takes place at the bottom of a distillation column. A remedial action can then be taken.

3.1.3 Definition of reaction rate and kinetics

Reaction rate is, first of all, about the speed at which a reaction takes place and, second, about variables and parameters that influence this rate. Connecting the observed reaction rate with the variables and parameters is, in general, called reaction kinetics determination.

Reaction rates and reaction kinetics play pivotal role in reaction engineering. It helps select a reactor type. This is essential in designing the reactor concept and in optimizing the concept. Kinetics is also a key element of reaction engineering modeling and, thereby, predicts the commercial scale reactor behavior; in that way, it can speed up the development of a new reactor. Hence, we will give a lot of attention to reaction kinetics.

Reaction kinetics is first of all about description of the reaction rate r. This reaction rate r is defined as the rate at which a species is formed or disappears by a chemical reaction per unit fluid volume. In S.I. units, it has the dimension mol/s m^3.

The component on which r is specified is indicated in lowercase; so for a reaction of A to B, the reaction rate for disappearance of A is described as r_A and for the formation of B as r_B. Furthermore, the reaction rate belongs to the stoichiometric equation of the reaction. That stoichiometric equation is often numbered. If it has the number 1, then the reaction rate for that stoichiometric equation can then be indicated as r_{1A}.

The reaction rate is an intensive variable, like temperature and pressure are intensive variables. The rate depends on the local (point) conditions in a system. It is independent of the type of system itself. Of course, different systems (reactors), in general, will have different local conditions in time and space.

When a component, i reacts away, then by convention, r has a negative value. When a component is formed, then r has a positive value.

Let us now take an example to further clarify the reaction rate. The reaction rate is specified for the reaction of one stoichiometric equation and for one of its components, i. Let us now take the reaction of: molecules A react with molecules B to form molecules P and Q. The general stoichiometric equation belonging to this reaction is provided:

$$aA + bB = pP + qQ$$

And let us specify the reaction rate on component A, so we have r_A.

Then, it follows immediately from the stoichiometry equation that the reaction rates of B, P, and Q for this reaction are simply connected by

$$-r_A/a = -r_B/b = r_P/p = r_Q/q$$

Hence, for this reaction, we are free to select one of the components in the stoichiometric equation by which we define the reaction rate. The others are then easily obtained by the stoichiometric equation coefficients.

If there are more reactions taking place, more stoichiometric equations are involved; then, it is important to number the reactions and also indicate the reaction number in the definition of the reaction rate, so the indication then becomes r_{1A} for the above reaction. If component A is involved in more than one reaction, then the subscript number indicates to which reaction (and stoichiometry) the reaction rate belongs.

On the definition of the rate of reaction

A common misconception of the reaction rate is to use the definition, $r_A = dC_A/dt$, in which C_A is the concentration of A, the number of moles per unit volume. In fact, this is not a definition of the reaction rate at all. If you read it from right to left, it is clear that it is a mole balance for a specific control volume. Even if this control volume is infinitely small, it is just a mole balance, valid only with additional assumptions, the most obvious one being that there is no flow into or out of this control volume. A second assumption needed is that there is no density or volume change due to the reaction.

Proper ways of writing mole balances for reactors and also for local phenomena inside reactors such as mass transfer and reaction are described in other chapters of this book.

For heterogeneously catalyzed reactions, the control volume V can be replaced by the catalyst particle volume V_p or by the catalyst surface area S_{cat}. Therefore, any kinetic rate r should always be clearly defined for a specified reaction volume type or catalyst surface type in which, or on which, the reaction takes place.

The reaction rate can vary enormously between different reaction systems. It can easily be different by a factor of 10^8 [1], for reactions in a space rocket and for a biological wastewater treatment plant. But even within a given reaction system the reaction rate can easily differ by 10^6, just by increasing the temperature by some 200 °C, so there is no yard stick for reaction rates. Reaction rates have to be determined experimentally and also the dependency on temperature, concentrations of species in the reaction system, and catalyst types have to be determined experimentally.

3.1.4 Reaction rate equations

The reaction rate r in general depends on concentrations of components, and if catalyzed, on the catalyst involved. It also depends on the temperature. Dedicated experiments are needed to determine the dependency of r on these variables. In the majority of cases, the dependency is provided by an equation called the rate equation. Sometimes, it is called the rate law. The word "law," however, suggests that it is founded on basic theory, like the gravity law of Newton. However, it does not have that sound base. It is just an empirical expression linking the local concentrations (or for gases partial pressures or even fugacities) to the reaction rate.

A simple dependency of the reaction rate on the concentration for reaction stoichiometry $A = P$ can be

$$r_A = -k\ C_A$$

This is called a first-order reaction rate: the reaction rate depends on the C_A to the power 1.

Note the negative sign in this expression. r_A has a negative value as component A reacts away.

If the reaction rate is not dependent on the concentration of A or P, then the reactor order is called zero order. If the reaction rate depends on the square of C_A, then the reaction is called second order in A.

The temperature dependency of the reaction rate is put into the reaction parameter k. It is often given by the Arrhenius law expression:

$$k = k_0 e^{-E_a/RT}$$

where k_0 is the pre-exponential constant, in units depending on the kinetics, E_a is the activation energy in J/mol, T is the temperature of the reaction control volume in Kelvin, and R is the universal gas constant with its value and dimensions 8.314 J/mol K [3].

The reader is however warned that different dimensions are still used as well. It is particularly important that the activation energy E_a applied has the same S.I. dimensions as R.

This way of describing the temperature dependency using the reaction rate parameter is very elegant. The researcher can first focus on finding the effect of concentrations by performing experiments at one concentration. Later, she can focus on the temperature effect and determine parameters in the Arrhenius law expression.

More complex expressions for the temperature dependency of the reaction rate constant k are also available but are beyond the scope of this book.

On the pre-exponential factor

From fundamental kinetic theory, k_0 is not really a constant, but is somewhat dependent on temperature. For chemical engineering purposes, this can be neglected nearly always, in particular, because the temperature dependency is only reliable inside the experimental temperature range.

The background of the Arrhenius law expression is that temperature is a macroscopic averaged parameter accounting for the reality that each molecule has a different temperature (different energy content) and that the molecules follow the Boltzmann exponential distribution around the average temperature. Only molecules beyond a certain energy level (called activation energy) react. When the average temperature of all molecules increases, more molecules reach this minimum energy level, according to the Boltzmann distribution. This explains the remarkable shape of the Arrhenius law expression. More information can be found in physical chemistry textbooks.

3.1.4.1 Reaction order

If the reaction rate expression is $r_A = -k\ C_{A,}^n$ then, the reaction kinetics is called an nth-order reaction in reactant A. However, the reaction rate expression in general is far more complex than a simple exponent expression. In fact, usually a simple nth-order rate expression with n not equal to a "rational" number like 1, 0, 0.5, or 2 suggests complex underlying kinetics. The use of an "average n" of 0.83 may be valid only in a very limited range of application. Any reaction engineer should be alerted here when applying such "broken" nth-order kinetics for reactor design.

The underlying detailed kinetic descriptions are called elementary kinetics. From that elementary description, overall, all kinetic expressions can be derived using, for instance, the steady state approximation. The reader is referred to other books for this method, such as Levenspiel [1]. Here, we use nonelementary kinetics with only observable concentrations in the expressions.

Table 3.1 shows most common reaction rate expressions with their names and applications. A short description for most of these is given further.

Table 3.1: Common reaction rate expressions. (For brevity, species concentrations are denoted by e.g. A instead of C_A)

Stoichiometry	Reaction rate equation	Catalyst	Kinetic name	Application
$A = P$	$r_A = -k\ A$	–	First order	Any
$A = P$	$r_A = -k\ A/(K{+}A)$	Enzyme	Michaelis–Menten	Biotechnology
$A = C + P$	$r_C = k\ C\ A/(K{+}A)$	Microbe	Monod	Biotechnology
$A = P$	$r_A = -k\ K\ A/(1{+}K\ A)$	Heterogeneous	Langmuir–Hinshelwood	Chemicals
$A{=}P$	$r_A{=}\ -k_1\ A\ A,\ r_P = -k_2 P$	–	Equilibrium reactions	Chemicals
$A{+}B = P$	$r_A = -k\ A\ B$	–	First order in A and B	Chemicals

3.1.4.2 Michaelis–Menten kinetics

The presented expression is the simplest version. The assumption is that the enzymatic concentration has a fixed value for the chosen reaction system. The enzyme concentration effect is not revealed in this expression but hidden in the constant K. Hence, the K value obtained holds only for the given enzyme concentration.

This kinetic expression shows the following behavior: for low concentrations of A, where $A \ll K$, the denominator goes to its limit value of K. The reaction then becomes first order in A. For very high concentrations of A, where $A \gg K$, the denominator goes to the value of A, so the overall reaction becomes zero order in A.

The physical reason for this latter behavior is that all enzymes are occupied with component A; so higher concentration of A does not increase the reaction rate anymore. The enzymes are saturated with A, so to say.

3.1.4.3 Monod kinetics

Monod kinetics hold good for microbial reactions where the microbe reaction rate, called growth and the product reaction rate are directly coupled via the stoichiometric equation. The feed component A is often called substrate.

Note that the reaction rate depends both on the concentration of the feed component A and on the concentration of the product component P. This makes the reactor behavior very sensitive to the residence time distribution and to the feed concentration of P. This is further explained in the section on RTD.

The presented kinetic expression is the simplest version. In reality, the feed (substrate) will contain many components needed for the microbial growth and a component for the microbial growth and product formation. If the components other than A are in excess, thus, not limiting the growth, and if the product is not poisoning the microbe, then this expression can be used.

For most practical cases, product P will poison the growth. The reaction rate should include this poison effect of the P concentration on the reaction rate, for those cases. There are also microbiological reactions in which the production rate is not coupled to the growth rate.

3.1.4.4 Langmuir–Hinshelwood kinetics

Langmuir–Hinshelwood kinetics may hold for heterogeneously catalyzed reactions. The basic principle is the same as for the Michaelis–Menten kinetics for enzymes. A molecule attaches to the catalytic surface. It changes thereby to an active state. It then reacts to a product molecule. That molecule leaves the active site, and the catalytic active place becomes available for the next molecule A. At low concentration of A, most active sites are available and the reaction behaves like a first order in A. For a high concentration of A, all active sites are occupied. Further increasing the concentration of A does not increase the reaction rate, so the reaction becomes zero order in A.

For certain type of Langmuir–Hinshelwood kinetics, the order in A can even become negative: the higher the concentration of A, the lower the rate of reaction. This can occur for bi-molecular reactions with competative adsorption where, in the denominator of the reaction rate equation, the nett exponent for A is higher than in the nominator.

Equilibrium reactions

Equilibrium reaction kinetics means that there is not only a forward reaction but also a backward reaction taking place. At equilibrium, the forward and the backward reaction will have equal reaction rates. For a simple reaction A=B, and rate expressions first order in A and B concentration, this means that

$$K = k_1/k_2$$

The equilibrium constant K is a thermodynamic parameter. It only depends on the temperature. This means that only one of the two kinetic constants can be chosen to fit experimental results. The other follows directly from the value of K and the determined kinetic constant.

3.1.4.5 First order in *A* and *B*

Many reactions involving two feed components are first order in A and B. An example is shown in Chapter 14. However, this is not always the case, especially for catalytic reactions. It could well be that the reaction rate expression only contains one of the feed components. It all depends on the rate-limiting step. If, for instance, the rate-limiting step only involves component A, then the kinetic expression only contains the concentration of A. This can also be the case when B is in large excess. Then, this reaction is said to be pseudo-zero-order in For catalytic reactions involving dissociative adsorption the square root of a concentration may arise.

3.1.4.6 Reactor performance and reaction rate expressions

The conversion and selectivity of reactors depend not only on kinetics but also on macro-, micromixing, and mass and heat transfers. Chapters 4–7 explain this behavior.

Nowadays, many software packages are available for solving the differential equations for given kinetics and residence time distribution. Due to this, no analytical solutions for conversion as function of time for given reaction kinetics in ideal reactors are presented here, but the reader can easily find them. For some simple cases such as single-phase plug-flow and fully back mixed reactors, analytical solutions are provided by Levenspiel [1].

3.1.5 Experimental kinetics determination

The reaction rate expression and its parameters have to be determined experimentally in dedicated laboratory setups where the temperature is controlled, and where the control volume is kept uniform, and where the concentrations are uniform over the control volume.

Ideally, reaction rates are measured in such a so-called "gradientless" reactor. A perfectly back mixed continuously operated reactor is the prime example here, shown in Figure 3.1. In such a reactor, the concentration and temperature are uniform throughout the reactor, provided the system is also ideally "micro-mixed"; see Chapter 5. Hence, the whole reactor volume is the control volume. Moreover, the reactor outlet concentrations are the same as the concentrations inside the reactor, so these concentrations can be measured easily by taking samples from the outlet. Also, many samples can be taken in the same conditions, so that the experimental error can be determined and catalyst decay can be easily observed.

In such a gradientless reactor, the reaction rate r is obtained directly from the mole balance over input and output flows and the reaction fluid volume:

$$-r = (F_{in} - F_{out})/V$$

where r is the reaction rate of a defined component in mol/s m^3, F_{in} is the molar feed flow of the component in mol/s, F_{out} is the molar outlet flow of the component in mol/s, and V is the reaction fluid volume in which the reaction takes place, m^3.

For heterogeneously catalyzed reactions, the reaction control volume V can be the catalyst volume, provided this is well defined; see Section 1.3.

For heterogeneous catalysts, kinetics can be determined for two-phase systems with a specially designed spinning basket or stationary basket reactors; the latter are widely known as Berty-type reactors [4]. For three-phase systems, the Robinson–Mahoney reactor has been successfully developed [5].

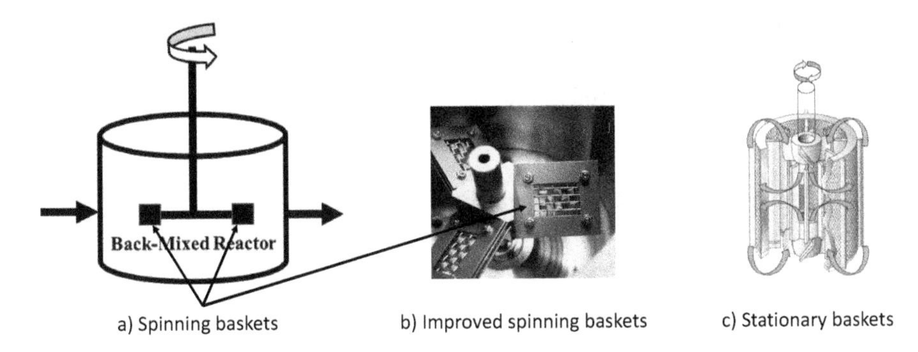

Figure 3.1: Gradientless reactors for heterogeneous kinetic studies: (a) spinning basket-type concept; (b) close-up from an improved design [4]; and (c) stationary type: Robinson-Mahoney [5].

The experimental reaction rate data determined for different concentrations of the reacting components and for different temperatures are then used to obtain kinetic rate expressions and the parameter values for the reaction orders and the activation energy. It is important that the experimental range for concentrations and temperature is beyond the range of the actual reactor design. It is also important that the experimental design of concentration values and temperature are independent of each other. For the concentration range, it is important to measure also for low feed concentrations, so that the kinetics also holds for deep conversions in the reactor design. An experimental design setup could go for log (concentration) at equal distance to avoid bias for high concentrations.

For determining the effect of temperature on the reaction rate constant k, it is important that the $1/T$ values are at equal distance to avoid bias to high temperature results.

Furthermore analyzing experimental results to obtain k_0 and E_a values is not a reliable way of obtaining parameter values. The value of k_0 obtained will have a large uncertainty range as it is the reaction rate constant k at T = infinite. The experimental temperature range will be far away from this temperature. Any small change in (estimated) E_a implies a relatively big change in k_0 to maintain the same rate at the reaction conditions. It is, therefore, far better to use a k_{ref} value at an otherwise arbitrary reference temperature within the experimental range. That k_{ref} is simply the rate constant at that reference temperature. The expression for k as a function of temperature is then rewritten as

$$\ln(k_0/k_{ref}) = (E_a/R)\ (1/T_{ref} - 1/T)$$

Hence, at the reference temperature any change in E_a has no effect at all on the k value; so now, the two parameters are decoupled.

This expression can be used to determine E_a and k_{ref}. In other words: use k_{ref} as the fit parameter. After the optimization, the traditional k_0 can simply be recalculated from the same expression. This is again to avoid a bias to the single experimental result at T_{ref}.

It should furthermore be stressed that the reaction rate is a locally specified parameter inside the reactor. Chapters 4–7 show how the reaction rate kinetics is linked to other reactor phenomena to obtain the reactor design performance.

3.2 Reactor performance definitions

3.2.1 Process and reactor boundaries

Before we can treat performance terms such as conversion and yield, we have to determine the boundaries of the reactor and the boundaries of the process.

Figure 3.2 shows these boundaries.

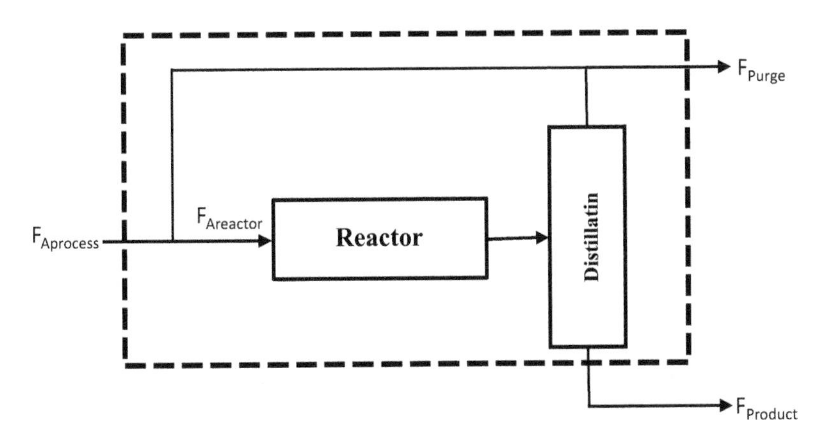

Figure 3.2: Process and reactor boundaries with input and output streams.

Virtually, a purge flow is always included in such a process scheme. The purpose of a purge flow is to prevent buildup of any components in the recycle loop. If that purge flow is extracted from the recycle stream, then that purge stream will also contain some reactant A. There will also be minor amounts of product P as well, because the separation will never be 100% selective. This is then a loss of component A and product P. The feed to the process has then to increase to compensate for this loss of A.

For simplicity, we consider here no purge flow and 100% separation efficiency to keep component balances simple.

For each definition, we will indicate whether it is about the reactor with its input and output streams or the process with its input and output streams.

3.2.2 Reactor conversion

Conversion is the term used for the fraction of feed components that is converted in a reactor. Let us take the case that component A is converted. The conversion X_A for a continuously operated *single-phase* reactor can then be determined from

$$X_A = (F_{A\text{in}} - F_{A\text{out}})/F_{A\text{in}} = (\phi_{\text{in}} C_{A\text{in}} - \phi_{\text{out}} C_{A\text{out}})/\phi_{\text{in}} C_{A\text{in}} \tag{3.1}$$

where $F_{A\text{in}}$ is the molar inlet flow rate of A in mol_A/s, $F_{A\text{out}}$ is the molar inlet flow rate of A in mol_A/s, ϕ_{in} is the volumetric feed flow rate to the reactor in m^3/s, $C_{A\text{in}}$ is the concentration of A in the reactor feed in mol_A/m^3, $\phi_{V\text{out}}$ is the volumetric outlet flow rate of the reactor in m^3/s, and $C_{A\text{out}}$ is the concentration of A in the reactor outlet in mol_A/m^3.

Thus, the reactor conversion of component A is the ratio of moles of component A converted over the moles of A fed to the reactor. (This only holds if there is a no purge stream out of the process containing component A.)

This way of calculating the conversion makes sure that any volume changes in the flow through the reactor are properly taken into account in the conversion determination.

For a batch reactor, the conversion can be calculated by replacing the flow rate in eq. (3.1) by the reaction volume at the start and replacing the flow rate out by the volume at the end of the reaction.

On moles versus mass

Some practitioners prefer to use base conversion on the mass of A. It suffices to note that the values for X_A remain the same when the conversion of A is determined from the mass balances over the reactor. The numerator and the denominator of expression (3.1) are just multiplied by the molecular mass of A to obtain the mass balance-based expression. For selectivity, it is not so simple!

The conversion dimension is a fraction. It has no dimension (or unit) attached to its value. It is also related to a specified molecule. Sometimes, however, practitioners talk about conversion as the amount converted. For instance, they state: The conversion is 10 ton/h. By stating the dimension (ton/h), it is clear that they talk about something other than conversion in a chemical reaction engineering sense.

If the reaction system has several reactors in series, then the conversion of each reactor is defined by eq. (3.1), with the input and output specified for each individual reactor. If each reactor has the same residence time and also all other conditions such as temperature are the same, then still the conversion values will, for most cases, be different, due to the fact that a lower amount of component A is fed to each subsequent reactor.

Reactor conversion per pass

If there is a recycle flow from the reactor outlet directly to the reactor inlet, such as shown in Figure 3.2, then it has to be specified whether the reactor system is about a so-called single pass or about the whole reactor system, including the direct recycle flow. In the section on residence time distribution with a recycle flow, this will be explained in detail in Chapter 4.

If the recycle flow is after a separation step and the recycle is only containing feed components, then the total feed to the reactor should be taken as in the input stream to the reactor.

3.2.3 Integral versus differential reactor selectivity

Selectivity (σ_P) is a term used when more than one reaction is involved. It reflects that part of the converted reactant(s) may not have formed the desired product P.

Westerterp et al. [2] properly define selectivity as follows: the ratio between the amount of desired product P obtained and amount of key reactant A converted. If defined on a molar basis, it is essential to take the reaction stoichiometry into account in order to get selectivity numbers that make sense. If the main reaction stoichiometry of the desired reaction is $aA = pP$, then the selectivity, or rather the *integral* selectivity can be calculated from [2]

$$\sigma_P = \frac{(\text{moles}\,P\,\text{formed})\ (a)}{(\text{moles}\,A\ \text{converted})(p)} \tag{3.2}$$

Realizing that this selectivity is equal to the amount of A converted to P per total amount of A converted, it becomes clear that this selectivity always lies between 0 and 1. This selectivity σ_P is dimensionless, but note that the unit can best be written as mol_A/mol_A.

This selectivity, σ_P is called "integral selectivity" because it is an average value resulting from all the conditions in time and space that the reaction has proceeded inside the reactor.

However, in an industrial environment, it has become customary to express it on a mass basis, for example, mass P formed/mass A converted. This can easily lead to miscommunication as the numbers on a mass basis are very often (very) different than those on molar basis.

The *differential* selectivity is the reaction rate at which the desired product is formed over the reaction rate at which the reactant is consumed at a specific condition. For known kinetics, it can be calculated for any condition by dividing the rate expression(s) of the reaction(s) yielding P by the sum of all reaction rate expressions for A reacting to any product. The concept of differential selectivity is very useful because, unlike the integral selectivity, it is not an average over time or space. It is generally more insightful than the integral one for the effects of temperature and concentrations. Nevertheless, most nonreaction engineers will only talk about integral selectivity. When you hear a "selectivity" number, nearly always, this will be the integral one. But always ask what is meant by selectivity.

Of course, ultimately, indeed it is the integral selectivity that one is interested in. It is used to determine the best reactor concept and the reactor design for highest selectivity. It is, however, difficult to determine directly from standard experiments.

Yield

Yield is the term used to express the amount of product formed over the amount of feed components to the reactor. The formal definition of yield, or integral yield, is simply

$$Y_P = \sigma_P X_A$$

It is the conversion of A multiplied by the integral selectivity of P from the reaction A to P. If these latter two are properly expressed on a molar basis, taking into account the stoichiometry, then the yield is always between 0 and 1.

However, just like for selectivity, in industrial practice, the yield is often expressed as mass of product per total mass of key feed component. These yields can easily be well over 1. Hence, always ask for the definition used for yield.

3.2.4 Reactor production capacity

The reactor production capacity is, in general, expressed as the amount of product produced per unit time. In reactor design, it is often an input design variable provided as the required amount of product to be produced per year, for the new process. It may be obtained, for instance, from a market survey and a management decision to go for a new plant with a certain capacity. The relation with the reaction rate expressions and the reactor volume is in general complex, as it depends on the reactor type with its residence time distribution and reactor conditions. This is dealt with in a separate chapter.

The reactor production capacity is defined as the product flow rate F_P in mol/s. In the petro-chemical industry, it is usually expressed in tons/h or even ktons/yr.

Reactor production capacity is often used as an input variable for reactor design, as it is the required production rate of the product. Note that the reactor capacity typically will have to be slightly higher than the plant capacity due to minor but inevitable losses in the downstream part of the plant, as discussed below.

3.2.5 Process conversion and yield

If the process consists of a reactor and a separator, as shown in Figure 3.3, where the unconverted feed is separated and recycled back to the reactor, then it is obvious that the process conversion will be close to 1 (100%), while the reactor conversion per pass is much lower. So reactor conversion is different in definition from process conversion and often has different values.

The process input flow will, for most cases, also be different from the reactor input flow. Referring to Figure 3.2, the process conversion and process yield are respectively:

$$X_{A\text{process}} = \left(F_{A\text{process}} - F_{A,\text{process, out}}\right) / F_{A\text{process}}$$
$$Y_{P,\text{process}} = F_{\text{product}} / F_{A\text{process}}$$

The reactor output flow of product (mol/s) is essentially the same as the process output flow. This, of course, only holds if product P is only made in the treated reactor and 100% of the product is recovered.

3.2.6 Definitions of terms: space velocity, GHSV, WHSV, LHSV

Space velocity (SV) is a term mostly used in manufacturing companies. It is defined as the total volumetric feed flow rate (m^3/s) to the reactor divided by the relevant reactor volume (m^3). Note that for a constant density single phase system, this would simply be the inverse of the average residence time.

In the case of a heterogeneous catalyst, the reactor volume is usually replaced by the volume of the catalyst. As mentioned in our Levenspiel example of Chapter 1 with "volume of catalyst" the voids in the catalyst bed are included. For a slurry reactor, it is the compacted bulk volume of the catalyst prior to loading, that is, again including the voids.

Here are some specific space velocity definitions:

- LHSV: If the feed is a liquid, then the term "liquid hourly space velocity" (LHSV) is sometimes used. The feed flow rate is then expressed in m^3/h and the reactor or catalyst bed volume in m^3.
- GHSV: If the feed is a gas, then the term "gas hourly space velocity" (GHSV) is used. The feed flow rate is then expressed in m^3/h and the reactor volume or catalyst bed volume in m^3. Moreover, in industry, nearly always, the actual volumetric gas feed flow rate at reactor pressure is not used but the gas flow rate at ambient temperature and pressure, that is, Nm^3/m^3 h; so when the GHSV term is used, then always find out how it is defined.
- WHSV: This is the ratio of feed mass flow rate over the mass of catalyst in the reactor, so the dimension is kg_{feed} / kg_{catalyst} h.

On catalyst volume

With respect to "volume of catalyst" an even more subtle error can occur, especially at lab-scale, when the volume is calculated on the basis of the "compacted bulk density" or CBD of the catalyst. For small reactors with diameters less than 20 times the catalyst particle size, the packing density in the reactor may deviate substantially from that CBD. Also, the way the catalyst is packed will determine the CBD, both on the lab scale and on the commercial scale. Hence, using packed bed volume as a base for reaction engineering is asking for trouble.

The process engineer, having to design a reactor for a new project but with the same catalyst and all other conditions, will use this term by stating: The reactor is designed for the same SV. Thus, if for the new project, the process capacity is twice as large as the previous design, then he will just increase the feed flow rate and the reactor volume by a factor 2, and thereby keep SV the same. It is, of course, essential that he also keeps all other parameters relevant for the reactor performance the same. Put differently, this way of scale-up implies that there would be no further scale-up effects!

The SV term also often causes confusion and even wrong designs. Some think that SV is just the inverse of the residence time and use the term in a reactor performance expression, in that way. For all cases where the reaction medium changes volume, this is a wrong way of using the term. For details on residence time determination, see Chapter 4.

Sometimes the SV term is also used in the scale-up from a pilot plant to a commercial scale as follows. The process development engineer may say: "We don't have a scale-up problem, because we scale-up with constant Space Velocity." By just keeping SV the same, actual velocities of the fluids can change a lot in scale-up. Thus, such a statement only creates a false sense of security. And a false sense of security in scale-up is one of the surest ways to failure.

If a new project is started to improve a previous design (that has been commercially implemented), then using the SV parameter has a very limited meaning. SV is not uniquely related to the reactor performance or to overall economics, so it cannot be used for reactor design and optimizations. If it is still used, it can cause confusion and debates that are hard to detangle. Hence, the free advice is not to use this term.

Work rate (WR) and space time yield (STY)

Reactor work rate (WR) is a term used in manufacturing companies. It is defined as the reactor production capacity expressed (kg/h) or (kton/year, kta) divided by the reactor or catalyst bed volume (m^3). This is also called space time yield (STY). In the academic world, STY is often expressed in moles rather than kilograms and per second rather than hours.

For a new project, the process engineer may choose to go for the same WR or STY as the previous project. Again, this is a proper way of designing only if nothing has changed with respect to the reactor parameters or in any other area of the process or of the economics. If there are improvements desired in the reactor performance, the reactor should be redesigned using reaction engineering methods.

The WR may be determined after all the optimization has taken place. It is very likely that, then, it appears that the WR has increased or decreased. The optimal WR is typically also dependent on the economics, for example, balancing the lifetime of the catalyst and cost of replacement of the catalyst or balancing the productivity with product selectivity. A higher price of the catalyst or the product will typically shift the optimum WR more towards lower WRs that typically give higher lifetimes and higher selectivity.

3.2.7 Residence time and space time

Residence time in a reactor is often seen as a key parameter governing the performance of the reactor and as a clear and simple concept. Apart the topic of residence distribution that we will deal with some detail in the next chapter, even the average residence time τ is often ill-defined, especially in multiphase reactors. For a constant density single phase liquid flow in a liquid full reactor, it would simply be the volume of the reactor divided by the volumetric flow rate. But in general, the density is not constant and for multiphase reactors, one has to face the question: "which volumetric flow rate to take?" Hence, we prefer not to use the term residence time in this book for reactor design purposes.

Having worked with Professor Guy Marin of Ghent University, we like to make a point here that from a chemical reaction engineering point of view, the concept of "space time" is much more powerful than that of residence time. Space time is defined as the relevant volume[1] V divided by the molar feed flow rate of the (limiting) reactant A. In words, it is a measure for the "time" available for reaction of reactant A. In a general form of a reactor performance equation, it is space time rather than residence time that is the parameter to look at. Intuitively, it is clear that having to react to a feed of 100 ppm of reactant at the reactor inlet is a much easier challenge than to convert it at an inlet concentration or rather mole fraction of 500,000 (50%). Unlike residence time, space time takes this inlet fraction into account and, hence, is a much better measure for the "time" available for reaction. The main disadvantage is that it is not in seconds or hours but in m^3 s/mol. However, we also will not use that term either for reactor design purposes in this book, as it can cause confusion.

3.2.8 Limiting reactant

The term "limiting reactant" (also stated above) is the reactant that is limiting the formation of product, P, because of its limiting feed flow rate relative to the other reactants. The term is used when more than one reactant is involved and when the other reactant is fed in excess to limiting reactant. That other reactant is called "excess reactant," so that reactant cannot be fully converted in the reactor. In most cases, it will be separated from the product and the excess amount will be recycled to the reactor inlet, together with fresh reactant feed.

1 The use of space time instead of residence time does not solve the "volume" issue.

On the misuse of definitions and dimensions

Often, in industry, other words or terms are used for the terms treated in the previous sections. When that is the case, just finding out what the meaning of the terms used is sufficient to be able to use the theory provided in these sections.

However, in some cases, the same terms are used but their meaning is different in the particular company. Sometimes, conversion is used, but it appears to have a different meaning. Sometimes, the term process conversion is used as if it is the same as the reactor conversion. Sometimes, the term yield means overall plant yield and not reactor yield. These two can be very different, indeed. Take for example, a process with 90% reactor selectivity and 50% conversion in the reactor, the "conversion per pass". The reactor yield is 45% (on a molar basis). Usually, in the back end of the plant, there is a separation section. The unreacted feedstock is separated and recycled to the reactor. If the separation is perfect, all of the feed will be converted with 90% selectivity. The plant yield can then be as high as 90%. If the latter is expressed as tons of product per ton of main feedstock, the plant yield becomes 90% of the ratio of molar weights of the product and feed component. Hence, this could be 140%.

The general advice is to always ask the *exact* meaning of the terms used. That way, a lot of confusion or mistakes can be avoided.

3.3 Physical properties

3.3.1 Reaction medium density modeling

Reaction medium density modeling is treated here separately from all other physical properties, because for modeling any reactor performance, a reaction medium model is needed. This follows directly from the reaction rate definition in which the reaction control volume is an essential part. Any reaction occurring will change the molecular composition of the control volume and, in most cases, will thereby change the density. For gas phase reactions where the number of product molecules in the stoichiometric equation differs from the number of feed molecules, this will be obvious. But for liquid phase reactions too, the liquid density will change due to the molecular composition changes. For modeling purposes, a density model is needed as a function of molecular compositions over the full range of potential compositions for which the reactor model will be applied.

These density models are, in general, provided by the physical models of the flow sheet program used for the reactor performance modeling. An experimental check on the model predictions is, however, needed. Even for liquids, the density can change by 20% or more, due to the different molecular composition of the resulting reaction medium. If the density change is not correctly modeled, the reactor can be sized wrongly. As there is no design safety factor that can be applied (a reactor that is too large can mean more by-product formation, while a very small reactor can mean too little conversion) reactor medium density models need to be accurate.

3.3.2 Physical transport properties

Physical transport properties for fluid flow, heat transfer, and mass transfer of the phases involved need to be known. For most components and also for mixtures, these data are readily available from book, and from physical data bases in flow sheet packages. Some physical properties need special attention.

First of all, this is the interface tension between gas and liquid. The real interface tension can change widely from the model or listed value. Trace components from the feed or formed during reaction can preferentially adhere to the interface, lowering or increasing the interface tension. Prediction of the bubble size using the tabled interface tension can then be widely different from the real bubble size. The mass transfer rate in the reactor or in the distillation can then be much lower than then the predicted values. The adherence of components to the interface can also cause frothing or foaming, resulting in poor gas–liquid separation inside the reactor or distillation, thereby disturbing the operation. A hot mini-plant in glass or with view spots will be needed to show this behavior.

The second is viscosity. The viscosity of highly viscous solutions of large molecules in a solvent may be different from the predicted value using correlations, due to the three-dimensional configuration of the macromolecules.

The third is phase equilibria. Knowing gas–liquid phase equilibria for the conditions in the multiphase reaction system is very important for the gas–liquid mass transfer modeling and prediction. For many components, data bases and physical property equilibrium models are available.

When physical data are lacking, dedicated institutes can be contracted to obtain the required information. Sometimes, engineers rely on the physical models inside their flow sheet package, without checking whether the model has been validated by experiments for the reactor system conditions. This lack of reliability will show up in the integrated mini-plant tests or in the pilot plant tests, but then, problem analysis and experiments at the institute to determine the phase equilibria will take considerable time, and hence, the development stage will be much longer than anticipated.

3.4 Reaction enthalpy

Reaction enthalpy ΔH_r, also known as "heat of reaction", is the enthalpy change due to a chemical reaction at constant pressure and temperature [3, p. 64]. For an exothermic reaction, ΔH_r is negative. Values of reaction enthalpy are tabled at standard conditions. For most tables, the standard conditions are $T = 298$ K and $P = 1$ bar. However, the standard conditions of the table should always be checked. Some report the data at $T = 273$ K.

It is common misunderstanding that the reaction enthalpy is a constant and does not change with temperature and pressure. But it does. The reaction enthalpy and conditions are different from the standard conditions. The temperature effect is given by Levenspiel [1, pp. 208–209]:

$$\Delta H\ (T_2) = \Delta H\ (T_{\text{standard}}) + \text{integral}\,(\nabla C_p\, dT)$$

in which $\nabla C_p = rC_{pR} + sC_{pS} - a\ C_{pA}$, belonging to the reaction stoichiometry $aA = rR + sS$.

The molar specific heats term ∇C_p also depend on temperature, as each individual C_p value depends on temperature. The dependence for A is often given by $C_{pA} = \alpha_A + \beta_A T + \gamma_A T^2$. The values of α, β, γ may be found for each component.

The values of the reaction enthalpy for each reaction taking place in the reactor needs to be determined accurately in the envisaged reaction temperature range, for several reasons. The first reason is that the reaction enthalpies are used for selecting the basics of the reactor design. If it is an adiabatic reactor – hence, with no heat exchange, or a reactor with heat-exchange. Second, the reaction heat is likely to be heat-integrated in the plantwide process.

Because of the required accuracy, the reaction enthalpies will have to be determined experimentally. Nowadays, engineers rely often on reaction enthalpy data obtained from flow sheeter data bases. However, these data often stem from theoretical calculations and not from experiments. A 10% difference with the actual data is not uncommon. This 10% difference can mean that the production capacity of the plant at startup appears to be 10% lower, due to heat transfer limitations or due to plantwide heat integration.

Experimental determination of reaction enthalpies can be carried out in in-house laboratory equipment or carried out by outside specialized companies. For multiphase reactions, special precaution is to be taken to account for phase changes with accompanying heat effects.

Accurate reaction enthalpy values are not only important for reactor design, nowadays with optimized process heat integration, it is important for the process design as a whole. A 10% higher value in reaction enthalpy used in process design can mean that the whole process, once started up, does not reach its design capacity.

3.5 Reaction runaway behavior

Reaction runaway means that the reaction rate of at least one chemical reaction increases so rapidly that the reaction is no longer controlled but runs away. Runaway reactions as typically studied in the field of "Reactive Hazards Analysis" which focuses on advent undesired exothermic side reactions upon an increase in temperature beyond the normal conditions. This phenomenon is not to be confused with "reactor runaway" or "reactor stability". Those topics will be discussed in Chapter 7.

Reaction runaway behavior is caused by two phenomena: heat produced by reactions and reaction rates increase with temperature. The theory of reaction runaway is relatively simple and involves the coupling of reaction rates, described as functions of temperature, with a heat balance involving reaction enthalpy and heat capacities of the reaction media. Reaction runaway is preferably studied experimentally in special equipment.

On reactor runaway

Reactor runaway will be dealt with in Chapter 7. There, the focus is on the balance between the local heat generation rate and the heat removal rate. This is highly dependent on the reactor design. Due to the exponential dependency of the reaction rate, even for a simple single exothermic reaction, this balance can lead to complex phenomena such as multiple steady states, ignition/extinction, hysteresis, parametric sensitivity, and even oscillations.

The reason for having the subject discussed in this chapter too is that, unlike reactor runaway, reaction runaway behavior is mainly dependent on the reaction system itself and not on the reactor design. Reaction runaway depends on all reactions that can take place, including reactions of the catalyst. It is best to study it experimentally early in the process development, so that when the reaction system appears to have strong runaway behavior considerations, one can start early to change the reaction system.

Temperature runaway behavior has to be known reliably for the reaction system to be able to design and operate the reactors safely. During a runaway, not only do the known reactions play a role, but also reactions not known, and not expected. These reactions may occur at temperatures outside the normal operation range. The reactions may also occur inside the catalyst or with the catalyst and the reaction components. Kummer [6] mentions in his review this unknown occurrence, but does not mention at all the need for experimental determination of runaway behavior to discover this unknown behavior. Instead, the whole review is about reactor design to avoid runaway – but based on models. However, dedicated temperature runaway experimental systems are available; see, for instance, the article by Fauske [7].

3.5.1 Example from Jan's experience

When auditing a novel catalytic reactor development inside a large chemical company I observed that the runaway analysis was solely based on their reaction kinetics and on a model for the reaction heat. The developers were surprised that I mentioned that dedicated runaway experiments would be needed with the actual catalyst and with the actual reactor feed to obtain reliable conclusions on a safe design, with runaway prevention measures. When I explained that outside the normal operation window other reactions could occur and also that the actual feed could contain trace components creating different reactions, they agreed to perform dedicated runaway experiments.

Hence, papers and textbooks on temperature runaway should not just be based on models but also stress the limitations of models and on the need for experimental runaway behavior determination.

3.6 Exercises

Exercise 1: Kinetic order determination
Given: The stoichiometric equation $A + B = P + Q$

Question: What is the order of the reaction rate kinetics in A and B?

Exercise 2: Central heating boiler
Given: Experimenters found in the exhaust gas of a natural gas fired heating boiler, methane, carbon monoxide, carbon dioxide, oxygen, and water.

Problem: Determine the stoichiometric expressions for a central heating boiler fired with natural gas and air.

Exercise 3: Selectivity profiles in a batch reactor
In a reaction scheme, a product is formed out of component A and reacts further to component X. Hence, the reaction scheme is $A = P = X$. The reactions are carried out in a perfectly mixed batch reactor: Assume simple first-order kinetic rate expressions for both reactions and sketch both the integral and differential selectivity of the desired product P as a function of the batch time.

Exercise 4: Reaction heat
Given: A group of chemical engineering students visits a biological wastewater treatment plant, where air bubbles through the wastewater to facilitate oxidation of the organic waste. A student asks the operator: "What is the reaction heat?" The answer is: "Zero, we need no cooler in the process."

Then, they visit a brick manufacturing plant, where clay is heated to 1,300 °C, and converted to solid bricks. The same student, who had just taken a course in chemical reaction engineering, again asks the question: "What is the reaction heat." The answer given is: "Very high, you can feel the heat."

Question 1: Is the wastewater answer correct?
Question 2: Is the brick answer correct?
Question 3: Is the wastewater reaction endothermic or exothermic?

Exercise 5: Increase reaction rate

Given: A chemist has carried out a reaction and found a 1% conversion. He wants to have a much higher conversion.

Task: Generate at least five ideas to increase the conversion.

3.7 Takeaway learning points

The first eight learning points stem from Chapter 1 are reformulated here, based on the provided theory.

Learning point 1

Proper units (dimensions) regarding reaction rates are not easy to define, especially for multiphase reactors. This chapter provides guidance.

Learning point 2

Reaction rate is a local variable and rate equations are about local conditions.

Learning point 3

Care should be taken with respect to the exact meaning when people refer to commonly used terms such as space velocity, space time yield, work rate, yield, selectivity, and residence time.

Learning point 4

From a reaction engineering point of view, the concept of space time is more useful than the more commonly used residence time.

Learning point 5

Reactor input and output stream composition information, as such, do not provide meaningful or reliable kinetic data. Additional information will be needed.

Learning point 6

Reactor performance factors such as conversion and yield are different from process performance factors.

References

[1] Levenspiel O, Chemical Reaction Engineering, 3rd ed. New York, J. Wiley, 1999.

[2] Westerterp KR, van Swaaij WPM, Beenackers AACM, Chemical Reactor Design and Operation. Chicester UK, John Wiley & Sons, 1984.

[3] James AM, A Dictionary of Thermodynamics. London, UK, Macmillan Press, 1976.

[4] Borman PC, Bos ANR, Westerterp KR, A novel reactor for the determination of kinetics for solid catalyzed gas reactions. American Institute of Chemical Engineers, Journal, 1994, 3, 862–869.

[5] Lauwaert J, Raghuveer CS, Thybaut JW, A three-phase Robinson-Mahoney reactor as a tool for intrinsic kinetic measurements: determination of gas-liquid hold up and volumetric mass transfer coefficient. Chemical Engineering Science, 2017, 170, 694–704.
[6] Kummer A, Varga T, What do we know already about reactor runaway? – A review. Process Safety & Environmental Protection, 2021, 147, 460–476.
[7] Fauske HK, Leung JC, New experimental techniques for characterizing runaway chemical reactions. Chemical Engineering Progress, 1985, 81(8), 39–46.

List of symbols

C_p	Heat capacity of a component	J/mol K
C_{Ain}	Concentration of *A* in the reactor feed	$\mathrm{mol/m^3}$
C_{Aout}	Concentration of *A* in the reactor outlet	$\mathrm{mol/m^3}$
E_a	Activation energy	J/mol
F_{in}	Molar feed flow of the component	mol/s
$F_{A\mathrm{in}}$	Reactor molar feed flow rate of component *A*	mol/s
$F_{A\mathrm{out}}$	Reactor molar output flow rate of component *A*	mol/s
F_{out}	Reactor molar outlet flow rate of the component	mol/s
F_{Process}	Reactor production capacity	mol/s
GHSV	Gas hourly space velocity	$\mathrm{Nm^3/m^3\ h}$
k	Reaction rate constant, *n*th-order reaction	$\mathrm{m^{3(n-1)}\ mol^{1-n}/s}$
k_0	Pre-exponential reaction rate constant, *n*th-order reaction	$\mathrm{m^{3(n-1)}\ mol^{1-n}/s}$
LHSV	Liquid hourly space velocity	$\mathrm{m^3/m^3\ h}$
r	Reaction rate of a component	$\mathrm{mol/s\ m^3}$
R	Universal gas constant	J/mol K
STY	Space time yield	$\mathrm{kg/s\ m^3}$
T	Temperature of the reaction control volume	K
V	Reaction volume in which the reaction takes place	$\mathrm{m^3}$
WHSV	Weight hourly space velocity	kg/kg h
WR	Reactor work rate	$\mathrm{kg/m^3\ h}$
X_A	Conversion of component *A*	$\mathrm{mol_A/mol_A}$
Yp	Yield of product *P*, fraction of reactant converted to product *P*	$\mathrm{mol_A/mol_A}$
ΔH_r	Reaction enthalpy	J/mol
∇C_p	Difference in heat capacity between products and reactants	J/mol K
$\phi_{V\mathrm{in}}$	Volumetric feed flow rate of the reactor	$\mathrm{m^3/s}$
$\phi_{V\mathrm{out}}$	Volumetric outlet flow rate of the reactor	$\mathrm{m^3/s}$
σ_P	Integral selectivity, eq. (3.2)	$\mathrm{mol_A/mol_A}$

4 Residence time distribution and mixing theory

The purpose of this chapter is to explain residence time distribution (RTD) theory, its importance for reactor design, and the RTD of major reactor types applied in industry.

4.1 Residence time distribution theory

RTD is a very important phenomenon in continuously operated reactors as it affects their performance in conversion and selectivity. It is about the phenomenon that some of the fluid parts entering the equipment have a shorter residence time than the mathematical mean residence time, while other fluid parts have a longer one. Hence, the fluid residence time has a distribution.

An analogy is the distribution of the salary of household income distribution of people in a country; see Figure 4.1 [1].

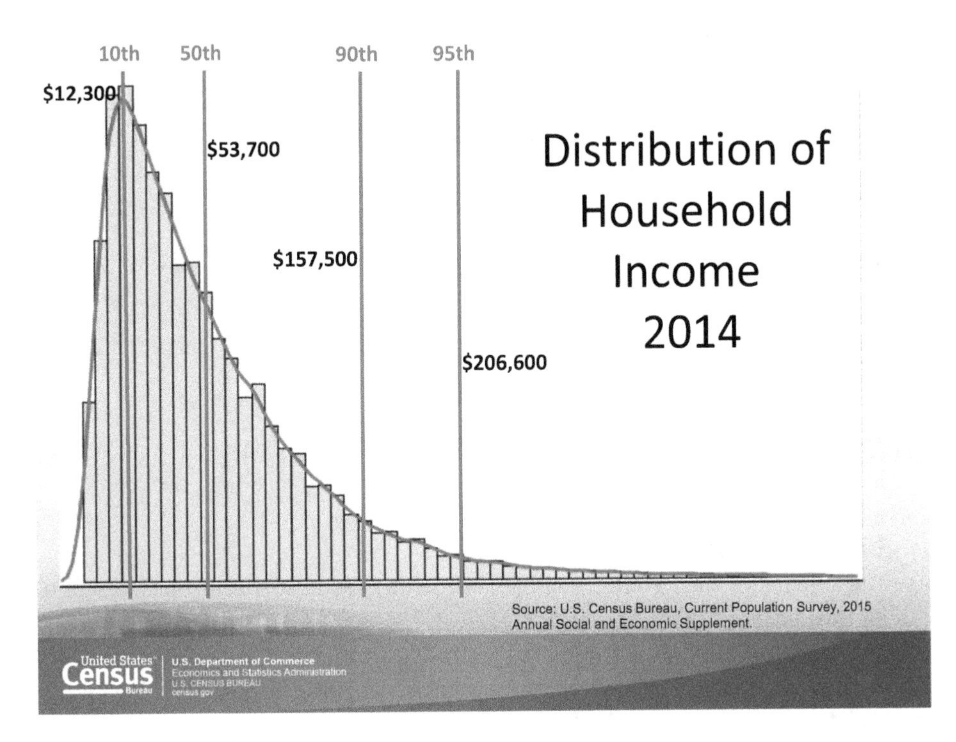

Figure 4.1: Household income distribution.

In such an income distribution graph, one takes certain brackets of incomes, for example, \$10–15k, \$15–20k, \$20–25k. Then, for each bracket, the number of people that has an income in that range is simply counted. It is then expressed as a percentage or

https://doi.org/10.1515/9783110713770-004

fraction of the total population. We then get all the blue vertical bars, for example, 3% has an income between $10 and 15k, 5% between $15 and 20k, and 0.2% between $300 and 305k. Since the total population is, of course, 100%, the sum of all the fractions must add up to exactly 1.

This is still a discrete graph. By making the brackets smaller and smaller, one ultimately would get a continuous distribution like the red curve in the figure.

However, in the basic form, the y-axis is the number of people or fraction of the total population in the bracket of income. Unit = (no. of people)/(no. of people in population). Clearly, if one would make the income bracket narrower, the number of people in that bracket and, therefore, the y-axis scale would also decrease.

Analogical with how chemical engineers work with RTD, we divide the y-axis by the width of the income brackets. Let us call that the $P(i)$ function, i being the annual income here, in, for example, dollars. The fraction of people that have an income between i and $i+\Delta i$ is then $P(i)\Delta i$. The unit of the function $P(i)$ is the *number of people per dollar* or, more conveniently, the *fraction of people per dollar*.

Since the sum of all people must add up to the total population, the sum of $P(i)\Delta i$ must be 1. Mathematically, the full integral becomes 1. Graphically, the area below the $P(i)$ curve becomes 1.

From the $P(i)$ curve, we can also calculate the average income. If 1,000 people would have an income between $9.9 and 10.1k, the combined income of that cohort is approximately $1,000 × 10k. If we add up the combined incomes of all the cohorts, this becomes the total income of whole population. Dividing that by the total number of people, we obtain the average income.

The same simple concepts are applied for RTD:

- Income becoming residence time
- Number of people or fraction of people becoming number or fraction of "fluid elements"
- Average income becoming mean residence time
- The $P(i)$ function becoming the $E(t)$ function – the RTD function
- $E(t)dt$ being the fraction of "fluid elements" with residence time between t and $t + dt$
- The unit of $E(t)$ is 1/s
- The area below the RTD function being equal to 1 (because this area is the sum of all fractions)

The concept of RTD was introduced to the reaction engineering field by Danckwerts in 1953 [2].

An extensive explanation of the concept of RTD theory, including the formal mathematics, can now be found in many reaction engineering textbooks, see, for example, Froment and Bischoff [3] or Westerterp et al. [4]. They also discuss in detail, how an RTD can be determined for any piece of equipment. Here, we focus on conceptual understanding and guidelines on how to apply it in practice for multiphase reactors.

First of all, it is important to realize that RTD only relates to phenomena on the scale of the reactor, that is, to macromixing. It is unrelated to micromixing effects! We will discuss micromixing in some detail in Section 4.6, but Figure 4.2 illustrates the high-level difference between macromixing and micromixing for a stirred tank. The blue parts in that figure relate to macromixing and the orange parts to micromixing. The RTD only pertains to macroscale. Effects at microscale are treated under "micromixing."

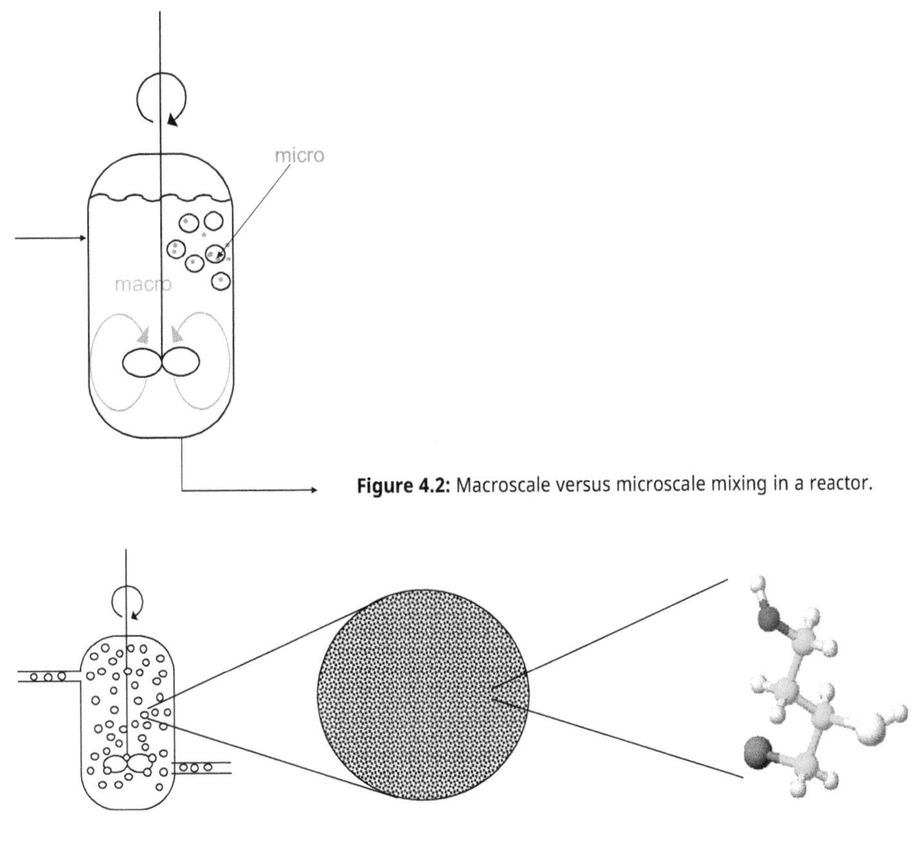

Figure 4.2: Macroscale versus microscale mixing in a reactor.

Figure 4.3: The concept of "fluid element" .
(adapted from Guy Marin, Ghent University).

In RTD theory, the concept of "fluid elements" is introduced. Figure 4.3 illustrates this concept. Here, a fluid element is, on the one hand, very much smaller than the scale of the reactor, but also much larger than the scale of a single molecule. For this introduction of RTD, we will assume this is a *micro*-fluid and not a *macro*-fluid, that is, it has a uniform concentration up to almost the molecular scale and is not (partially) segregated. Such a typical element is then considered when it enters the reactor (or any other piece of equipment for that matter).

The RTD function $E(t)$, as introduced earlier, then describes the probability that that fluid element leaves the reactor at a certain time, t, or rather between t and $t + dt$.

It is often convenient to render the time and the E-function dimensionless, using the average residence time, τ_0. This dimensionless time is then $\Theta' = t/\tau_0$. The area under the dimensionless $E(\Theta')$ curve will also always be 1: the sum of all fractions.

The RTD of a real reactor or vessel in general can take many shapes. One of the strengths of classic reaction engineering is to adopt a few simple and idealized concepts representing the extremes and combinations or variations thereof.

4.2 The plug flow reactor concept: PFR

A first simple concept of RTD is (ideal) plug flow, meaning that the residence time of each fluid element is exactly the same. The fluid flows as a plug through the equipment without any mixing (see Figure 4.4). This RTD concept is often treated as a reactor. It is then referred to as "PFR": **Plug Flow Reactor**. It should be realized that it is not a real reactor; it is just an "idealized" concept.

Sometimes it is called an "ideal reactor." However, this would be misleading. It suggests that it is always a very good reactor. But that is not necessarily true. In many cases, a close to plug flow RTD is not the best choice. It is also not a real reactor because "plug flow" only refers to the RTD and not to other important aspects such as micromixing and mass transfer.

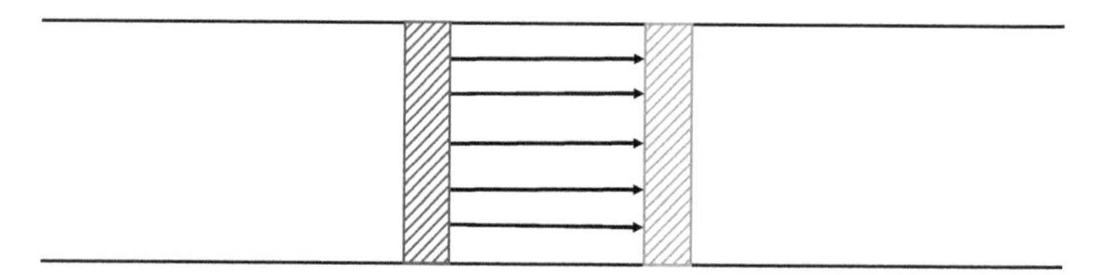

Figure 4.4: The concept of plug flow.

The RTD function $E(t)$ of such a PFR is a vertical line at exactly the average residence time. As the area below the dimensionless $E(\Theta')$ curve (the sum of all fractions) must be equal to 1, that vertical line is infinitely thin and infinitely high (a so-called Dirac function). Figure 4.5 shows the dimensionless RTD for the PFR.

Because in a PFR, all fluid elements have the same residence time, conceptually, the PFR is just like a perfectly mixed *batch* reactor. What is clock time for a batch reactor is residence time in a PFR – a kind of space-time analogy.

In practice, the flow in a real reactor or tube will never be perfectly in plug flow, even for turbulent flow in long tube with a small diameter; see Figure 4.6. Then the vertical line of the PFR becomes more like a narrow Gaussian distribution around the

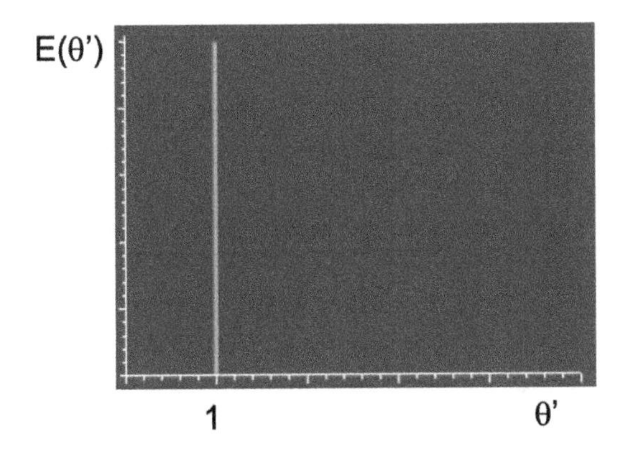

Figure 4.5: Dimensionless residence time distribution for a PFR.

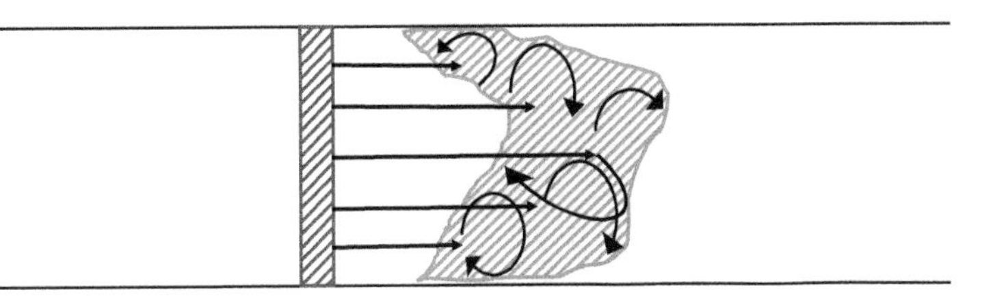

Figure 4.6: Nonideal plug flow.

average residence time. Whether or not a seemingly small or large deviation from plug flow is really significant for the performance of the reactor can only be assessed by taking into account the reaction kinetics and the desired degree of conversion. We will elaborate on this later in the section "nonideal flow reactors."

4.3 The perfectly backmixed reactor: CSTR or CISTR

Another well-known *idealized* RTD *concept* is that of the "perfectly backmixed" system. Also this concept is often misleadingly treated as a real reactor. The classic textbook by Octave Levenspiel uses the term *Mixed Flow Reactor* (MFR). But by far, the most commonly used name is *CSTR*. This stands for: **C**ontinuous **S**tirred-**T**ank Reactor. However, Westerterp, van Swaaij, and Beenackers prefer the name *CISTR*: Continuous **I**deally Stirred Tank Reactor. This indeed stresses the importance of the assumption of ideal mixing.

A common misuse and misunderstanding of the term CSTR is that it is used for a mechanically stirred tank reactor, operated *batchwise*. CSTR then would stand for "Continuously Stirred Tank Reactor". It refers, however, to the continuous stirring and not the mode of operation of the process with continuous flow into and out of the reactor. Actually, a perfectly mixed batch reactor behaves in RTD terms as a plug flow reactor, so the opposite of a CISTR.

Moreover, instead of CSTR or CISTR, we would have preferred one of two other names sometimes seen in papers or textbooks: the backmixed reactor (BMR) or ideally backmixed reactor (IBMR). This removes any suggestion that the perfect mixing is to be achieved via mechanically stirring the reactor. As IBMR is not commonly used at all, we will follow our own teachers, Westerterp et al., and will mostly use CISTR throughout this book. By using CISTR, rather than CSTR, it is always immediately clear this is a hypothetical "ideally" backmixed reactor operated continuously rather than a real continuous(ly) stirred tank reactor (of not defined operation mode).

In a CISTR, all fluid elements entering the reactor are assumed to be instantaneously and uniformly mixed with the fluid elements already in the vessel. This implies an exponentially decaying RTD curve, starting at a value equal to 1 divided by the average residence time for residence time $t = 0$, see Figure 4.7.

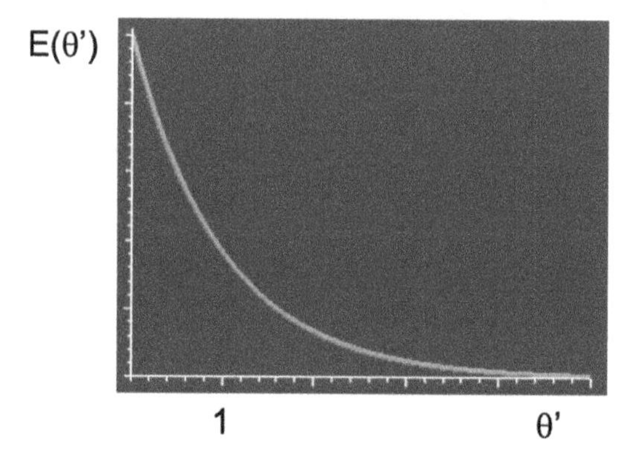

Figure 4.7: The RTD function of the perfectly backmixed reactor (CISTR).

In our experience, for many, it can be baffling at first sight that for a CISTR, the highest value of $E(t)$ is for $t = 0$. So the highest fraction of fluid elements is for a residence time of 0 s!

This can be understood by realizing first that a CISTR is an idealized concept: the mixing is assumed to be perfect and instantaneous. All fluid elements that enter the reactor at $t = 0$ are assumed to be instantaneously and uniformly mixed throughout the whole reactor. Hence, the chance that a fluid element leaves the reactor is equal for all elements inside the reactor. Assume that at time $t = 0$, say 1,000 fluid elements are introduced in the reactor. Then, at $t = 0$ there are 1,000 of the original elements present in the reactor, all with the same chance of exiting. Exited elements will be replaced by new elements from the feed stream. The chance that any of the original 1,000 elements will be leaving at, for example, $t = 1$ s will now be lower. This is because there are now less than 1,000 of the original elements present and all elements ("old and new") still have the same chance of leaving.

A simple analogue is injection of a droplet of red dye (reflecting the 1,000 fluid elements) in a perfectly backmixed flow reactor, operated just with clear water. In such a CISTR, all the red dye will be instantaneously and uniformly distributed in the reactor, so the whole reactor will turn red. From then on, red water is leaving the reactor, while no more red dye is entering, just clear water. So the color in the reactor will turn lighter and lighter; its concentration decreases exponentially, like in Figure 4.7. The reactor is most red at $t = 0$: the highest fraction of fluid elements is for residence time $t = 0$.

4.4 Intermediate macromixing

The PFR and CISTR are two very useful concepts, even when no real reactor will be a perfect PFR or CISTR. For "small" deviations from either two, these deviations may be neglected, resulting in simple reactor models and relations between, for example, conversion as a function of average residence time for given reaction kinetics. However, for larger deviations, the concepts can be modified to better reflect the intermediate degree of macromixing and the corresponding intermediate form of the RTD. Moreover, such concepts are needed to assess whether or not the RTD deviations are truly "small" or "large," when it comes to the impact on the performance of the reactor, that is, conversion and selectivity.

4.4.1 Tanks-in-series concept

The simplest and widely used concept for this is known as tanks-in-series. The "tank" here is the perfectly backmixed reactor (CISTR). By placing, hypothetically, any number n of such ideal CISTRs in series, a whole range of RTDs can be described, ranging from the exponentially decaying function found the for the single CISTR to more and more Gaussian types of RTDs up to the infinitely narrow and high RTD found for the PFR. The latter is obtained when an infinite number of CISTRs are put in series. Figure 4.8 shows a whole range of these, that is, the RTD for $n = 1, 2, 3, 5, 10$, and 20.

Even though this concept can cover the range from CISTR to PFR, in reality, the RTD of a reactor may have very different shapes. In other words, not all types of RTDs found in practice can be described by the tanks-in-series concept.

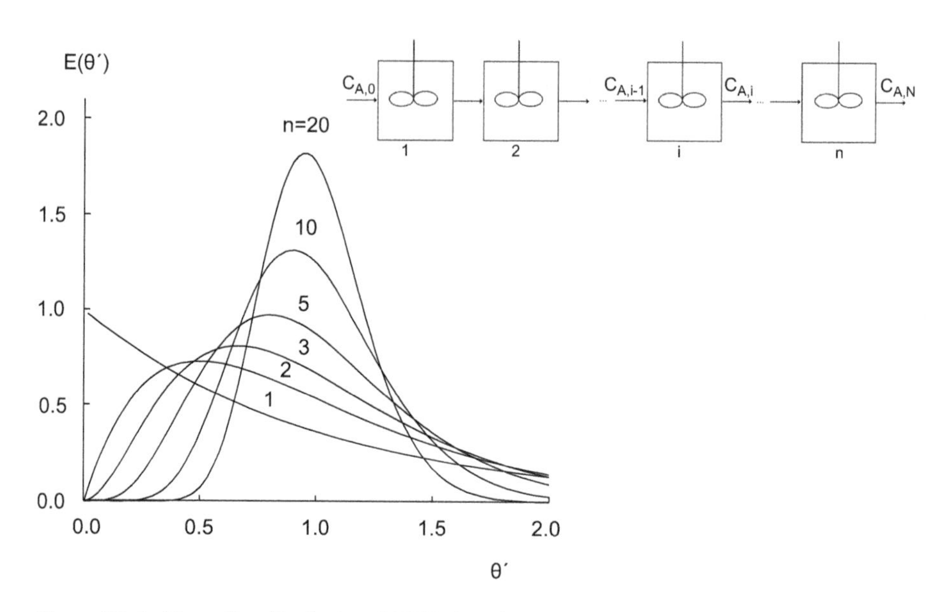

Figure 4.8: Residence time distribution of CISTRs in series.

4.4.2 Axial dispersion concept

The other commonly applied concept, especially in academic literature, is that of *plug flow + axial dispersion*. As should always be kept in mind, ideal plug flow is only a theoretical concept. There are no 100% PFRs in real life.

One of the founding fathers of CRE, Peter Danckwerts, has become famous, among other things, for his work on a natural "correction" to this ideal plug flow. Figure 4.9 shows the essence.

In this concept, dispersion is superimposed on the convective ideal plug flow. This is analogous to Fickian diffusion. The actual mechanism is not only molecular diffusion as the "eddies" in Figure 4.9 illustrate. Therefore, it is not called diffusion, but dispersion. If, for whatever reason, there is a concentration gradient of component A in the axial direction, then besides the convective flow component F_A there is a flux J_A. By analogy, this flux is assumed to be proportional to this gradient. For pure ideal binary diffusion, this is like in Fick's law, where the proportionality constant is the (Fick) diffusion coefficient D. In the axial dispersion model the same concept is used, albeit that the mechanism for the flux comprises not only diffusion but also any other reasons for "dispersion." Hence, in the model, it is called the axial dispersion coefficient D_{ax}. The main reasons for this dispersion are already indicated in Figure 4.9 but listed more explicitly:

- Eddies due to turbulence
- Radial velocity profiles (e.g., a parabolic profile for pure laminar flow)
- Catalyst particles

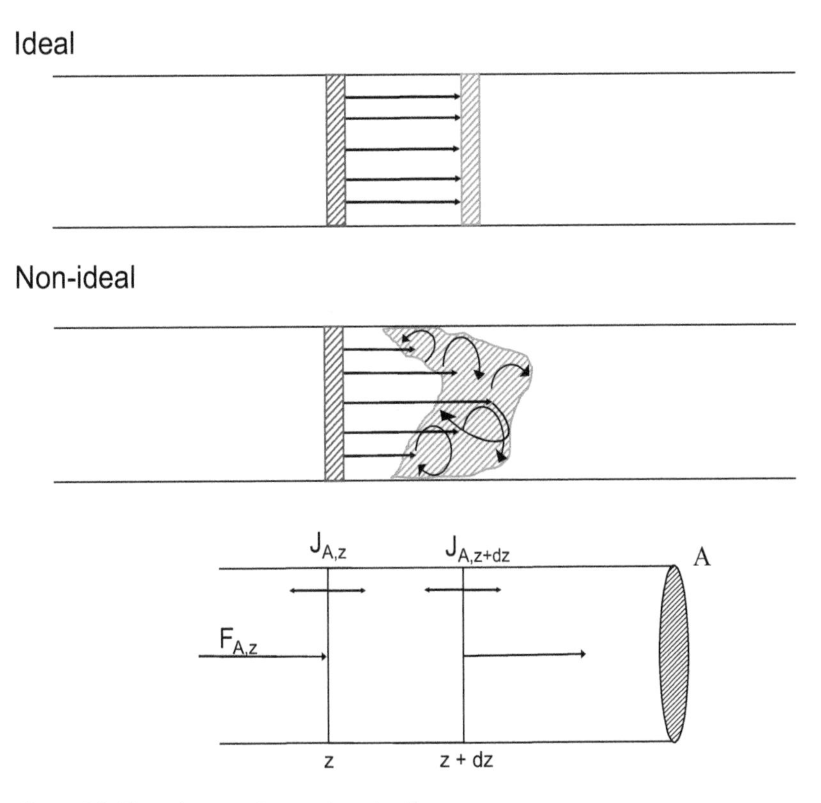

Figure 4.9: Dispersion superimposed on plug flow.

Clearly, the more dispersion (higher D_{ax}) the more "backmixing" there is. Here, another famous dimensionless number arises: the Péclet number, named after the French physicist, Jean Claude Eugene Péclet. In the general physical definition, a Péclet number is the ratio of advective over diffusive transport rates. For our application to axial dispersion in chemical reactor, it is, in simple words, the ratio of convection over dispersion. To be precise:

$$\mathrm{Pe} = uL/D_{ax}$$

where u is the fluid velocity, L is the reactor length, and D_{ax} is the aforementioned axial dispersion coefficient, lumping all effects leading to deviation from ideal plug flow into just one model parameter.

So a high Péclet number corresponds to small deviations from plug flow. What "high" and "small" are, really depends on several factors, especially also on the degree of targeted conversion and reaction kinetics (for simplicity, reaction order), see also the next section. We will not discuss that here, in any detail.

Suffice to say, in this chapter, that simply by looking at the standard deviation of the RTD predicted by the tanks-in-series model, it can be derived that the Péclet number becomes approximately equal to two times the number of tanks-in-series N:

$$\mathrm{Pe} \approx 2N \ (\text{for}\ \mathrm{Pe} \gg 1)$$

In other words, unless there is a huge amount of backmixing or that the flow-pattern really is very different from convection plus backmixing, the very different concepts actually will predict similar results. And quantitatively, the two models, in general, even give numerically very close results, if one substitutes $\mathrm{Pe} = 2N$ or $N = \mathrm{Pe}/2$ in case the backmixing is not really severe.

On Levenspiel's remark on the name Péclet number for dispersion

In the third and latest edition of Levenspiel's classic textbook, he deliberately avoids the use of the Péclet number when introducing the axial dispersion model and prefers to call it the "vessel dispersion number". Hidden at literally the very last page of the very last Appendix, he explains: "This [D_{ax}/uL] is a new and different type of dimensionless group introduced by workers in chemical reaction engineering. Unfortunately. someone started calling the reciprocal of this group the Péclet number – this is wrong. It is neither the Péclet number nor its mass transfer analog, which is widely called the Bodenstein number." As explained above, the difference, indeed, lies in referring to diffusion versus dispersion. However, Levenspiel's preference here never got traction, and the "wrong" use of both the names Bodenstein and Péclet are still the norm.

A common misunderstanding is that plug flow and fully backmixed flow are the two extremes of RTD. This is totally wrong. An RTD can be far more extremely distributed than a backmixed flow. For instance, a considerable fraction can show short-cutting flow, causing less deep conversion than expected from backmixed flow. A bubbling fluid bed reactor or a horizontal bubbling bed reactor, are just two of many examples. The industrial scale-up case 13.5 describes the latter behavior.

4.5 Residence time distribution effects on conversion/selectivity

The RTD of reactors can have enormous effects on the concentration profiles inside reactors. Figure 4.10 shows this for two cases; the PFR and the backmixed reactor (CISTR). This means that it also has, in general, quite significant effects on reaction conversion and selectivity. For a proper comparison of the two reactor types, the conversion can be made equal by adjusting the reactor volume. However, for selectivity, such a simple adjustment usually cannot be made.

The RTD concepts provided in section 4.6 allow for rapid evaluations of the effect of RTD on conversion, selectivity, and the reactor volume needed for a given design capacity, so that reactor types can be compared, sized, and compared. Subsequently, reactor conditions, such as temperature and pressure can also be optimized.

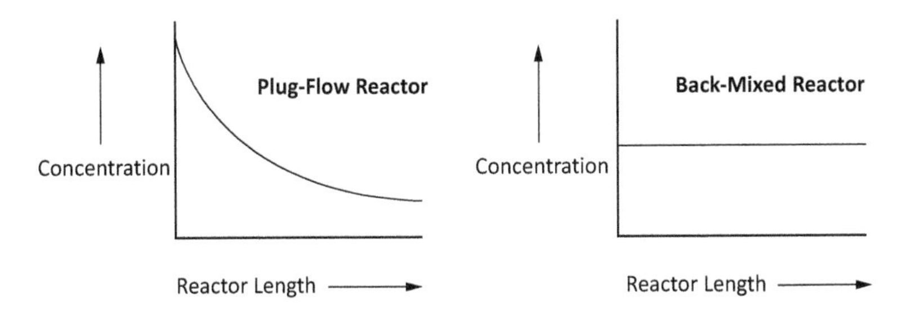

Figure 4.10: Concentration profiles of a plug flow reactor and a fully backmixed reactor. (same volume and conditions).

For the idealized concepts of PFR and CISTR, relatively simple relations can be derived for the conversion as a function of the average residence time. For a variety of kinetic rate expressions, these can be found in many textbooks, often along with their formal derivations. Here, we limit it to listing just these for first, zero, and *n*th-order reactions in a PFR and CISTR (see Table 4.1).

Table 4.1: Performance expressions for plug flow and backmixed (CISTR) in case of isothermal single phase and no volume change.

Reaction order n	Plug flow X_A	Backmixed X_A	Plug flow $k\tau$	Backmixed $k\tau$
0	$k\tau/C_{A0}$	$k\tau/C_{A0}$	$C_{A0}\,X_A$	$C_{A0}\,X_A$
1	$1-e^{-k\tau}$	$k\tau/(1+k\tau)$	$\ln(1/(1-X_A))$	$1/(1-X_A)$
2	$f_{\text{pfr}}\,(k\tau, C_{A0})$	$f_{\text{CISTR}}\,(k\tau, C_{A0})$	$X_A/(C_{A0}\,(1-X_A))$	$X_A/(C_{A0}((1-X_A)^2$

Reaction rate expression $r_A = -k\,C_A^{\,n}$ (mol/s m^3).
Residence time $\tau = V/\phi_0$ with V = reactor volume (m^3), ϕ_0 = volumetric feed flow rate (m^3/s)
C_{A0} = feed concentration component A (mol/m^3)
X_A = conversion of reactant A

Note that:
- In all expressions, the rate constant k and averaged residence time τ appear as product, $k\tau$.
- For first-order reactions, the conversion is independent of the inlet concentration.
- For orders $n > 1$, the conversion increases with inlet concentration.
- For orders $n < 1$, the opposite is true.
- The expressions for the averaged residence time only hold for no density changes between inlet and outlet flows.

Comparisons of the outcome of these relations for both idealized concepts show that, in general, the RTD has very significant influence on the reactor volume needed to obtain the same conversion. Figure 4.11 shows a basic comparison of a first-order reaction carried out in a PFR (a) or a CISTR (b). This graph already demonstrates three important points, *valid for first-order reactions*:

- For any average residence time t_0 the PFR has a higher conversion than a CISTR.
- For low residence time, or to be more precise, for low conversion the difference is quite small.
- For high conversion, the difference can become very large.

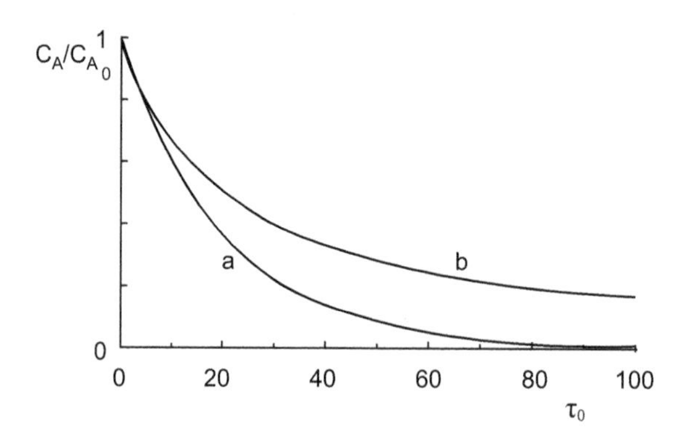

Figure 4.11: Reactor outlet/inlet concentration ratio versus averaged residence time. Curve a: plug flow reactor, Curve b: CISTR.

The idealized concepts, in combination with some basic insights into the reaction system at hand, can also be used to quickly make a preselection of reactor types in the concept innovation stage. We will deal with reaction systems A, B and C, here, to illustrate this reactor pre-selection.

Example Case A: Reaction rate increases with feed component concentration increase

If the reaction rate increases with increasing feed component concentration, then the reactor volume is required to reach a certain conversion, and hence a PFR reactor will require less volume than a CISTR. The stronger the effect of the feed concentration, the larger is the difference in volume required between the two ideal reactors.

A second effect is that deeper the conversions of the feed component, larger the difference in reactor volumes required will become.

Table 4.2 shows that for deep conversion the V_{CISTR}/V_{PFR} ratio becomes very large, which was already hinted at in Figure 4.11. There, we were primarily looking at the difference in conversion for a given τ_0, and hence comparing the curves in the vertical direction. In Table 4.2, we highlight the differences in the horizontal direction of Figure 4.11.

Table 4.2 shows that for a 10% conversion, the required CISTR volume is only 5% higher than for a PFR. For a conversion of 99.99%, however, the required volume for

Table 4.2: Reactor volume ratio V_{CISTR}/V_{PFR}.

Kinetics: first order in feed component; no volume changes	
Conversion %	V_{CISTR}/V_{PFR} –
10	1.05
50	1.44
90	3.90
99	21.4
99.9	144.6
99.99	1,085.8

a CISTR is larger than for a PFR by a factor of 1,000. The PFR is then clearly the preferred option.

The simple explanation for this striking difference in behavior is that, in a CISTR, the feed is immediately diluted with the product P and for positive-order reactions, this will give a lower rate of reaction and hence, a larger reactor is needed for the same degree of conversion.

Yet another way of explaining this: consider a CISTR having the same conversion as the PFR. Everywhere in the CISTR, the reaction proceeds at a rate corresponding to the outlet concentration. In a PFR, everywhere before the outlet, the concentration is higher than this outlet concentration. Therefore, for a positive-order reaction, the average rate of reaction will be higher than that of the CISTR, and hence the PFR can be smaller in volume to reach the same conversion.

For higher order kinetics, the ratio increases even more with conversion, so the conclusion holds for reactions where the reaction rate increases with the concentration of feed components.

The reactor volume ratios are also independent of the reaction rate constant. Hence, this conclusion is also robust to uncertainties in the reaction rate constant.

For negative-order reactions where the rate of reaction decreases with increasing concentration, the CISTR will not be larger but in fact smaller than a PFR. Negative-order reactions are quite common in heterogeneous catalysis, for example, proceeding via a so-called Langmuir–Hinshelwood mechanism. There, for example, reactants A and B have to adsorb on the catalyst surface to react to P. An increasing concentration of A can "self-poison" the reaction by competitive adsorption with B, that is, the rate decreases because the surface concentration of B decreases.

Example Case B: The reaction rate is dependent on the product concentration

The case where the reaction rate depends on the product concentration (and not on the feed component concentration) is relevant for microbial growth systems such as single cell protein production and for autocatalytic reactions, where the product catalyzes the reaction. We consider the case where the microorganism grows by dividing itself. Hence, the growth is proportional to the microorganism (cell) concentration in the reactor $r_p = k\,P$.

For a PFR, a certain quantity of microorganisms needs to be fed to the reactor to have any growth. This is called enting. We define the inlet concentration as P_{in}. The outlet concentration for PFR is then $P_{\text{out}}/P_{\text{in}} = e^{kt_{PFR}}$. Or, if you look along the reactor length ($L = v/t$), then you see an exponential growth of microorganisms reaching its maximum at the reactor outlet. It is assumed, for this simple case, that sufficient nutrients are also fed to the reactor to allow this microbial growth. Beyond a certain concentration of nutrients, the growth kinetics behaves indeed as zero-order kinetics called Monot kinetics.

For the CISTR, however, cells need only to be fed at startup, but as soon as steady state is obtained, the reactor does not need to be fed with microorganisms anymore. The microorganisms obtained by growth in the reactor are backmixed to the inlet, so to say. At steady state, the concentration of microorganisms in the CISTR is the same as everywhere and equal to the outlet concentration. Hence, the reaction rate r_p everywhere in the CISTR is higher than in the PFR. Hence, the CISTR needs a larger volume to reach the same outlet concentration as the PFR. It, furthermore, needs no continuous feed of microorganisms (only a startup amount). As making the enting amount of organisms is expensive, this is an additional advantage of the CISTR.

4.6 Micromixing, earliness of mixing and segregation

The previous section on RTD represents only one of the three key mixing related phenomena in a reactor. These three are: macromixing, the state of aggregation, and the earliness of mixing. RTD only relates directly to macromixing. These are separate factors, but it is not always easy to really separate them! The two factors are part of what is commonly called micromixing.

To introduce this micromixing, let us first distinguish between the three scales of mixing: macro-, meso-, and micromixing:

- Macromixing refers to the mixing on the scale of the whole reactor.
- Mesomixing refers to the mixing phenomena near the feed inlet(s). This takes place via turbulent diffusion and fragmentation of the fresh feedstock elements or “lumps” entering the reactor, where it mixes with the reactor fluid mixture.
- Micromixing refers to the mixing on the scale of the fluid elements. This takes place by deformation, erosion, and fragmentation of these elements or agglomerates, until ultimately, uniformity down to the molecular level via diffusion is achieved .

On micro- and mesomixing

Some confusion is associated with the term micromixing, as it may or may not include the mesomixing. In our experience, it has happened a few times that one engineer said she had a micromixing problem; a colleague then estimated the micromixing time and concluded there was no problem: the timescale for micromixing was around a millisecond and the time scale of reaction was 1 second. It turned out that the millisecond timescale for micromixing was indeed that of the agglomerates at the so-called Kolmogoroff scale of turbulence becoming uniform. The worry of the first engineer was that of the scale of the lumps of the inlet fluid, that is, the mesomixing time. For liquids, this mesomixing time can be of the order of seconds or even minutes. With a timescale for reaction of 1 s, indeed, there would not be a micromixing problem by definition, but a severe problem could occur in case the term micromixing comprises the mesomixing effects.

One other way of distinguishing micromixing effects in a reactor, as a whole, is by looking at it from the "time of mixing" perspective, and on the other hand, from a "spatial" perspective. From the first, the concept of "earliness of mixing" arises. From the second, the concept "degree of segregation" arises.

4.6.1 Earliness of mixing

It is a common misunderstanding that, even for a single-phase system, if one knows the RTD and the reaction kinetics, the conversion is fully determined. But this is only true for first-order reactions, that is, for linear kinetics.

Here is a case for a second-order reaction to proof this point. A CISTR followed in series by a PFR will have the same RTD as that same PFR followed by that same CISTR. However, the overall conversion for a second-order reactor will be higher in the latter case; see Figure 4.12.

Without demonstrating this via mathematics, it can be understood as follows:

For a second-order reaction, the rate of reaction is proportional to the square of the local concentration. The conversion in both CISTR and PFR increases with increasing inlet concentration; see Table 4.1. But always, for a constant residence time, the PFR has a higher conversion than the CISTR. For the sequence CISTR-PFR, the backmixing occurs "early." That implies that the better performing PFR gets a lower inlet concentration compared to the sequence PFR-CISTR. Therefore, the overall conversion will be lower in case the mixing is "early". For the case of a first-order reaction, however, the sequence or earliness of mixing, has no influence because the conversion is each reactor is independent of the two inlet concentrations of the two reactors in series.

Here we used macromixing models to illustrate, in a very simple way, the importance of earliness of mixing for nonlinear kinetics. In a way, this is a micromixing model of the whole system that physically may look very different than a PFR-type tubular reactor followed by a CISTR type mixed vessel (or vice versa). Zwietering [5] was one of the first to construct conceptually simple micromixing models, where,

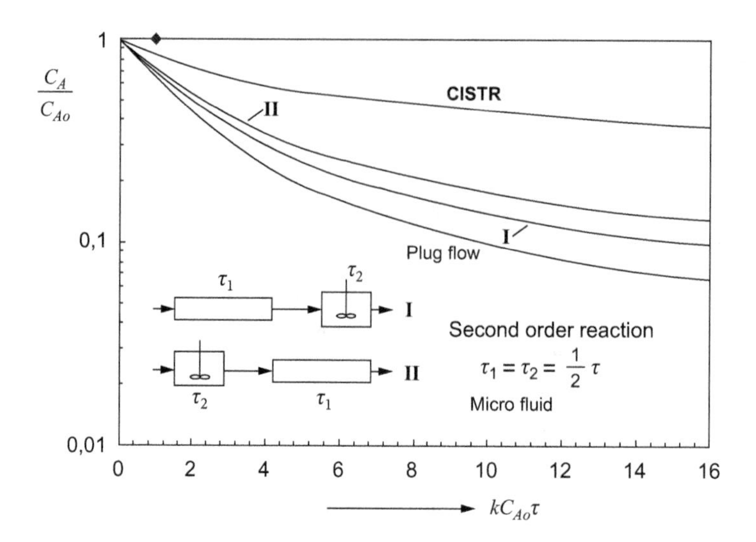

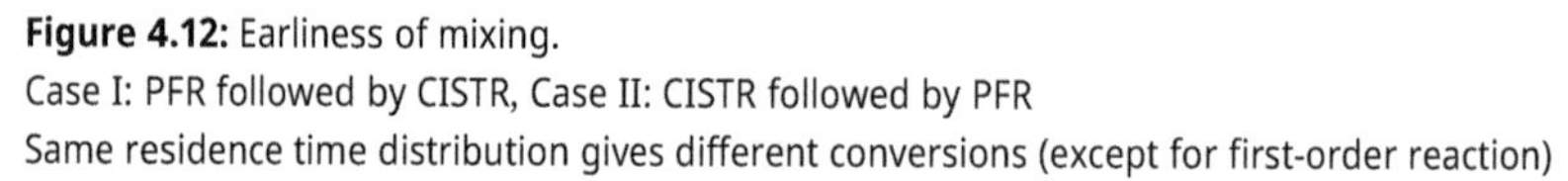

Figure 4.12: Earliness of mixing.
Case I: PFR followed by CISTR, Case II: CISTR followed by PFR
Same residence time distribution gives different conversions (except for first-order reaction)

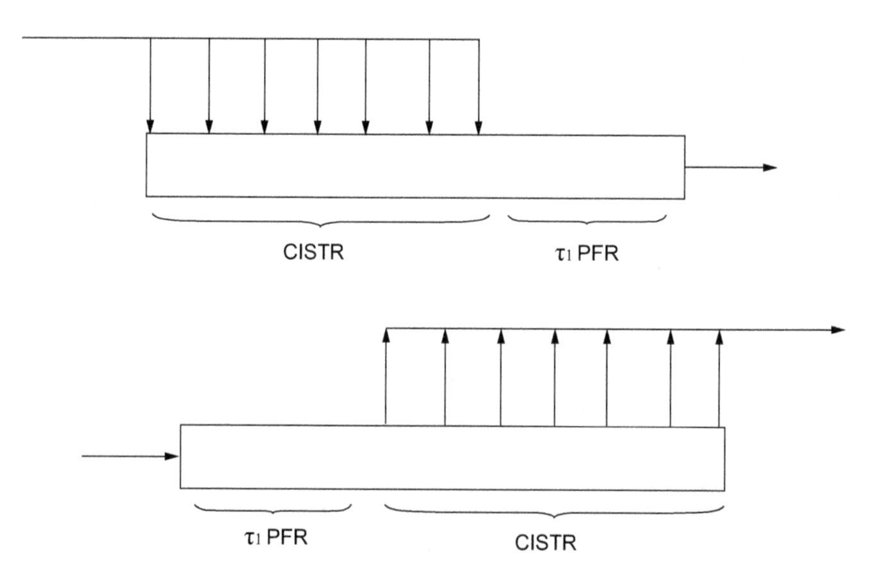

Figure 4.13: Micromixing models: maximum mixedness (top) and minimum mixedness (bottom).

virtually, for all types of RTD's, and thus, for all types of macromixing, different degrees of micromixing "in time" could be modeled. Figure 4.13 illustrates this principle.

In a reactor with maximum mixedness, the mixing of aggregates occurs at the earliest possible moment. Figure 4–12 showed the example CISTR-PFR configuration.

This basic concept can be generalized to match any arbitrary RTD by these Zwietering types of micromixing models, see Figure 4.13.

Both the top and bottom pictures show a PFR with either a large number of lateral inlet points or lateral outlet points. Either way, the RTD can be made exactly the same. By defining the number of these points and flows through all, lateral conversion and selectivity can be calculated. With only lateral inlet points, the mixing occurs "early." This is also called "maximum mixedness." For any reaction with order different from 1, the micromixing model with only lateral outlets (minimum mixedness) will predict a different conversion (and selectivity) than the model with only lateral inlets.

4.6.2 Degree of segregation

The degree of segregation refers to the extent the fluid elements travel as lumps of molecules, together. In a two-phase dispersed system of, for example, a water and oil phase, these two phases remain segregated. In single phase systems too, there may be aggregates. If these stay together, they are called a *macro*-fluid. If there are no aggregates, the fluid behaves like a *micro*-fluid. Another example in multiphase systems is a gas-solid reactor. Figure 4.14 shows various degrees of segregation.

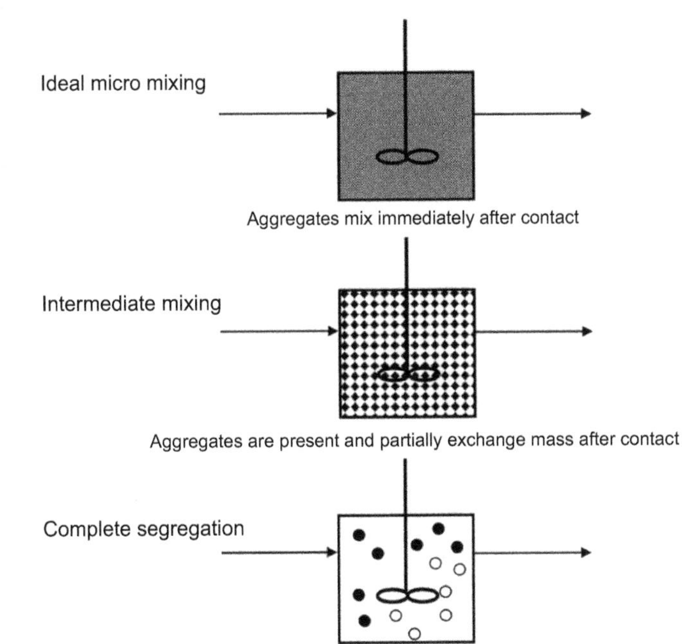

Figure 4.14: Degrees of segregation in a reactor.
Top: no segregation; middle: some segregation; bottom: complete segregation.

From this description, it will be clear that when the reactions occur on or inside particles, the reaction system is always segregated.

A perfectly macromixed reactor can, on one extreme end, be also perfectly micromixed. In that case, everywhere in the CISTR, there will be the same concentration. On the other extreme end, a perfectly macromixed reactor can be fully segregated. In that case, all the macrofluid elements can be regarded as individual batch reactors with a different residence time.

Since, in a CISTR, there is an exponential RTD and all these batch reactors leave the reactor at different times, there is no uniform concentration in the reactor. This, obviously, can affect the overall performance of the reactor – not only the conversion, as we will illustrate, but also selectivity. The latter is usually a more important effect (economically) than the effect on conversion.

We demonstrate the potential huge impact of this micromixing phenomenon by considering a zero-order reaction – for example, a solid material to be converted in a perfectly macromixed reactor. The concentration versus residence time is shown in Figure 4.15.

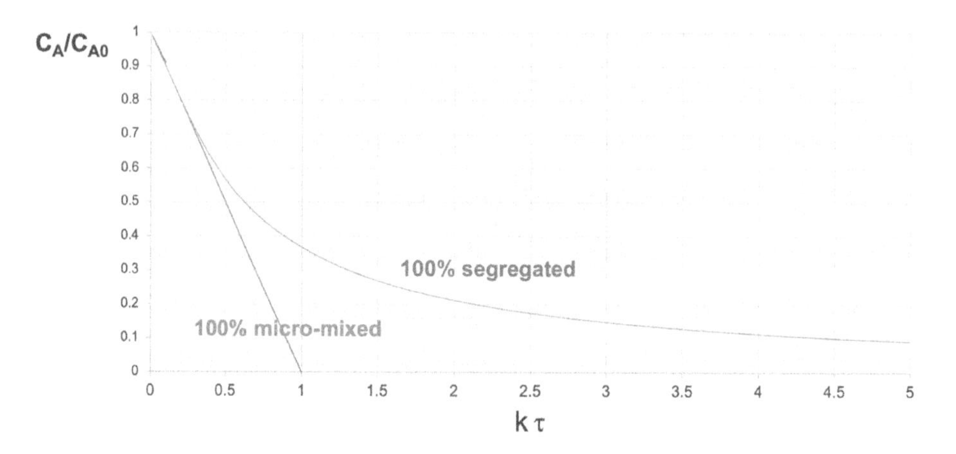

Figure 4.15: Concentration profiles in a macromixed reactor.
For completely segregated and completely micromixed conditions.

Our colleagues from Twente University, Professor Kersten et al., use the elucidating Figure 4.16 to explain the negative or positive effect of segregation on conversion. The black dots in the graph represent the case of complete segregation, that is, first reaction, mixing only after exiting the reactor. Except for $n = 1$, after mixing, the averaged rate (grey dots) differs from the rate obtained in case there is no segregation (white dots). Note how different the zero-order case is.

To conclude this section, Zwietering had suggested the use of a PFR with either many lateral inlet or lateral outlet streams as a micromixing model and has also

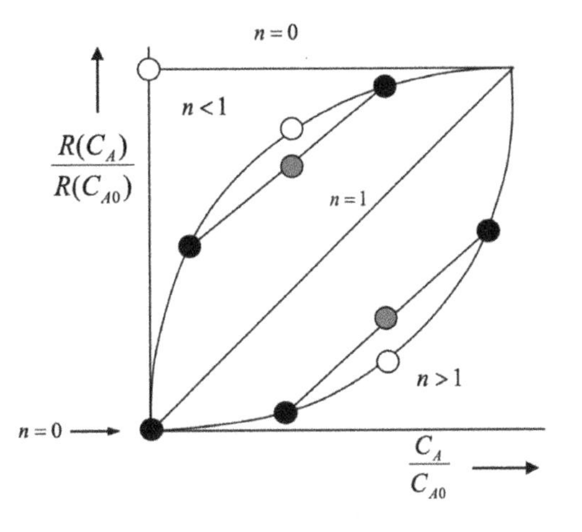

Figure 4.16: Effect of complete segregation in a perfectly backmixed reactor.

introduced a definition for the degree of segregation: the ratio of "variance of the ages of the aggregates" and "variance of the ages of the molecules" [6].

4.6.3 Takeaway messages: macro- and micromixing

- RTD has a big influence on the performance of a reactor (conversion, selectivity); it is one of the main reactor selection considerations.
- Only for first-order reactions, the conversion is calculated from the RTD.
- For orders $n > 1$, early mixing increases conversion; for $n < 1$. it is the opposite.
- The degree of segregation can be very important for zero-order processes.
- For relatively fast reactions, "mesomixing" (feed inlet effects) may become important; only for extremely fast reactions (milliseconds). micromixing is done up to the Kolmogorov scale.

Table 4.3 adapted from the course by Professor Guy Marin of Ghent University nicely summarizes how to deal with the various situations from a modelling perspective. Cases 1–3 reflect macromixing, according to any of the idealized concepts like CISTR or PFR, but with different degrees of micromixing ranging from ideal to zero micromixing (complete segregation). It shows that, except for first-order reactions, the intermediate micromixing case needs additional information, that is, reactor model micromixing parameter(s).

For macromixing or RTD's deviating from that of a CISTR or PFR, in practice, the RTD needs to be determined experimentally and/or modelled using one or more additional mixing parameters. The exception is the complete segregation case, where it suffices to know the RTD.

Table 4.3: Macro- and micromixing cases; adapted from Prof. Guy Marin's course notes.

	Mixing		E(t)	Reactor model mixing parameters
	Macro	**Micro**		
1	Ideal	Ideal	Theory	None
2	Ideal	Nonideal	Theory	Micro*
3	Ideal	Zero	Theory	None
4	Nonideal	Ideal	exp	Macro
5	Nonideal	Nonideal	exp	Micro + macro
6	Nonideal	Zero	exp	*E*(*t*)

*Not for first order.

4.6.4 Application of RTD theory to ideal reactor type selection

RTD plays a major role in reactor type selection, particularly in the concept stage. Chapter 10 treats reactor selection in detail, for all innovation stages, for many aspects, including the role of RTD, also in combination catalyst decay, and therefore, it is not dealt with, here.

4.7 RTD of real reactors

4.7.1 RTD of two- and three-phase fixed bed reactors

In industrial-scale catalytic fixed bed reactors, in most cases, there is axial flow over a bed of relatively small particles. That is, these catalyst particles are small compared to the dimensions of the reactor itself. For adiabatic fixed bed reactors, this holds for both the length of the reactor and the diameter of the reactor. In general, the flow over the catalyst bed will be quite uniform in the radial direction, except for the region near the reactor walls.

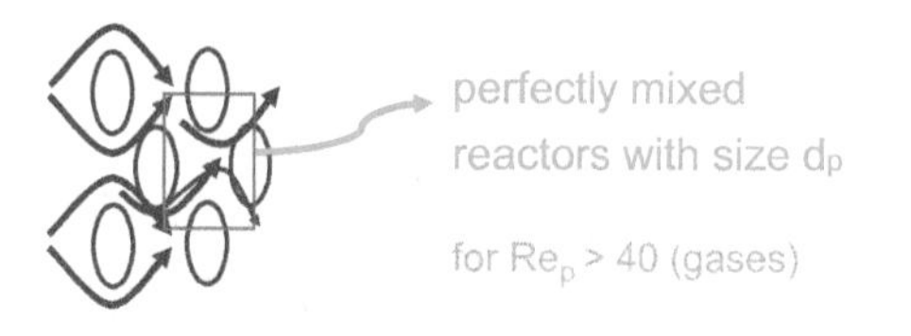

Figure 4.17: fixed bed model of perfectly mixed reactors in series. (for liquids: $Re_p \gg 100$ for near perfect mixing per void).

One simple way of looking at such a fixed bed is that it consists of a huge number of small mixed "tanks-in-series" (see Figure 4.17).

The fluid can only flow through the voids between the particles in the bed. The voids are of the size of the particles themselves and are interconnected. When the fluid is flowing around all these particles, they, in effect, flow from one void to the next via a kind of "mix-split" mechanism. The main flow direction is axial, but there is also a radial component (causing some radial mixing/ dispersion).

For a typical commercial scale catalytic fixed bed reactor catalyst particle size range of 1 to 10 mm and a bed height of 1–10 m, we see that such a reactor would conceptually resemble 1,000–100,000 "voids-in-series". So even if the mixing in each of these voids is limited, intuitively, such a fixed bed reactor will have an RTD that would approach plug flow behavior quite readily. And this is also what numerous studies have shown: because of the high number of particles on a reactor length, for most commercial fixed bed reactors, (close to) plug flow behavior can be assumed. However, for trickle-bed and/or for lab-scale reactors this may not be the case, as we will discuss below.

For low conversions, the plug flow criterion is relatively easily met. Low conversions imply small gradients. Hence, any backmixing or dispersion will have little net impact. For deep conversions, the plug flow criterion will be more stringent. For this, the approach by Gierman is very practical. This approach is particularly useful for laboratory downscaled fixed bed reactors.

For deep conversions, the Péclet number and the Bodenstein correlation with the particle Reynolds number can be used to quickly size the bed length L:

$$\mathrm{Pe} = L/d_p\,\mathrm{Bo}$$

$$\mathrm{Bo} = v_s\,d_p/D_{\mathrm{ax}}$$

For downscaled laboratory and pilot plant scale reactors, plug flow behavior is desired for three reasons. The first reason is that, with plug flow behavior, they have the same behavior as the commercial-scale fixed bed,[1] so they are truly downscaled for the RTD. The second reason is that, by plug flow behavior, the laboratory reactor results can be compared with the kinetic model prediction, so the results can be used to validate the model. The third reason is that the results of lab scale reactors with different catalysts can be used more easily to compare catalyst performance. With plug flow, the differences in conversion between the reactors are reliable, while for reactors deviating from plug flow, the deviations may be different between the reactors, so that a reliable comparison between catalysts is not possible.

1 Of course we here neglect heat effects. At lab scale, the heat removal is nearly always very different. In this section, we assume isothermal behavior.

A method for quickly downsizing the laboratory-scale fixed bed reactor to the minimum scale for plug flow behavior is also provided by Gierman [7].

First, the minimum Péclet (Pe_{min}) is determined from the desired conversion X and the reaction order n:

$$Pe_{min} = 8\,n \ln(1/(1-X))$$

This again shows that to judge whether a reactor can be considered to be operated in plug flow, one has to take the kinetics and the targeted degree of conversion into account.

The actual Péclet number should be higher than this minimum value to allow the reactor to be treated as a PFR. The reactor length-based Péclet number can be estimated from the particle size-based Bodenstein number, Bo:

$$Pe = L/d_p\,Bo$$

$$Bo = v_s\,d_p/D_{ax}$$

$$Bo = 2\,\text{for}\,Re > 10$$

The latter Bo = 2 is not only based on experiments. It has a theoretical basis: the voids being like tanks-in-series, as discussed in Chapter 4 and shown in Figure 4.18. For a bed of 5 m and 5 mm particles, this leads to $Pe_{min} = 2{,}000$.

For trickle-beds and Reynolds number for the liquid typically smaller than 10, Gierman recommends a conservative value of Bo = 0.01 for the liquid phase.

By varying the bed length L (and v_s for constant contact time or space velocity), the required bed length can be determined to obtain $Pe > Pe_{min}$

For a downscaled bed and for deep conversion, this may result in a very long bed. By using bed dilution with fine inert particles, a factor 10 smaller than the catalyst particles, the bed length can be further reduced.

4.7.1.1 Wall flow and its prevention

For fixed bed reactors with a small (<20) ratio of reactor diameter over particle diameter, the wall effects are more pronounced. Small ratios of reactor over particle diameter are often present in multitubular reactors (MTR) that are very common in the petrochemical industry for highly exo- or endothermic reactions.

With relatively few particles on the diameter, there will be significant preferential flow near the wall because, inherently, the void fraction is much higher there.

For short beds, this, indeed, can cause a very severe deviation from ideal plug flow. However, because the packing itself also provides a means for radial mixing in nearly all practical commercial scale applications, the net effect is negligible. Again, it is the huge number of particles on a reactor length (often ≫1,000) that will ensure that plug flow remains a good approximation, in practice.

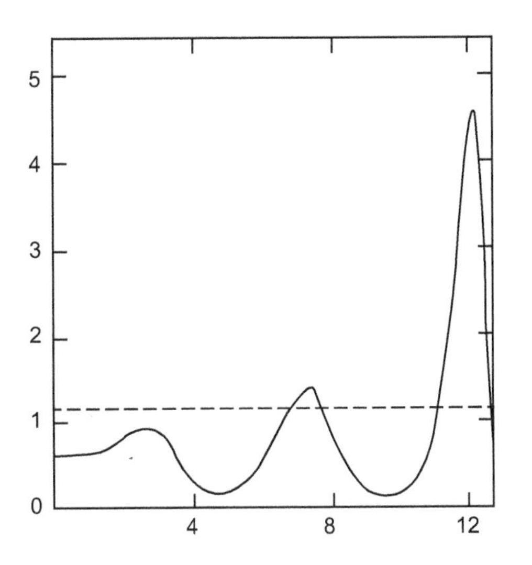

Figure 4.18: Radial velocity profiles: x-axis radial position in mm, y-axis velocity in m/s; dashed line is the average velocity; 4 mm particles in bed of radius 13 mm [3].

For laboratory-scale reactors and, in particular, for trickle-flow, this is not the case. Again, bed dilution with inert small particles (often, nonporous silicon carbon particles of 0.1 mm are chosen) wall flow can be prevented by having a sufficiently high "number of particles on a diameter". For this Gierman [7] suggested $D/d_p > 25$.

Bed dilution is more commonly used for suppressing heat effects. A pitfall is that not only the dilution must be done and must remain quite homogeneously, but too much dilution can also have an adverse effect on conversion (see [8]).

4.7.1.2 Feed distribution lab-scale fixed beds

For downscaled laboratory reactors, an inert layer of small particles should be placed on top of the catalyst bed. The height of this layer should be five times the reactor diameter to ensure even distribution (see [7]).

4.7.1.3 Feed distribution commercial-scale fixed beds

This subject is treated in Section 11.5.3. It is mentioned here as well to stress that feed distribution of gas and or liquid for packed bed reactors (gas–solid, liquid–solid, and gas–liquid–solid) is very important.

4.7.1.4 Origin of trickle-bed catalytic reactor, modeling, and scale-up

As early as in 1957, 't Hoog of Shell presented experimental (cold flow) modelling studies on trickle-flow over particles to study the effect of wall flow and the effect of particle sizes on the flow. In combination with experiments in a hot pilot plant with a height of

15 feet and a diameter of 3 inches for the hydrogenation of Sulphur compounds in oil fractions, a successful scale-up by a factor 1,000 to commercial scale was obtained [9].

4.7.2 Residence time distribution G-L bubble columns

4.7.2.1 Liquid-phase residence time distribution

The RTD of the liquid phase of bubble columns is a not simply to describe as plug flow or backmixed. The liquid flow pattern is complex. Some of the liquid flows upward in the center of the column, and on reaching the top, flows down along the wall. Hence, there is a vertical liquid circulation. Small bubbles also follow this liquid circulation pattern. Author Jan Harmsen has observed this circulation flow in the seventies of the last century in various transparent columns up to a diameter of 0.63 m at Shell research center, Amsterdam.

Professor Rietema of Eindhoven University of Technology (TU/e) explained this phenomenon to Jan, in 1986. The physical phenomenon causing this circulation is that in the center part, the bubble holdup is larger, so that the averaged density of the gas-liquid core section is lower than in the wall part. The density difference then causes the vertical circulation, just as this happens in free convection circulation with density differences caused by temperature differences. The reason for the bubbles moving preferentially to the center part can be found by looking at the overall force balance. The liquid at the bottom "sees" that the center part has a lower pressure, because of a lower value of $\rho_m\, g\, h$ (ρ_m is the two-phase density, g the gravitational constant, and h the height) than the wall section. This pressure difference causes the liquid to flow to the center part. Gas bubbles follow this radially inward liquid flow. In this way, the bubble rich center column stays intact. Wall friction is the only counterforce limiting the liquid circulation rate. This is the reason that, with increasing column diameter, the circulation velocity increases, as the wall surface area per unit volume decreases. For some reason, Rietema has not published this description.

After more than 50 years of discussion, Raimundo [10] reported, in 2019, experimental results of liquid circulation rates for four column diameters: 0.15, 0.4, 1.0, and 3.0 m, using water as liquid and nitrogen as gas. His results show that the axial liquid circulation flow rate increases proportionally to the column diameter to the power 2.5, and that the averaged circulating liquid velocity increases with the diameter to the power 0.4. The liquid velocity in the center could be correlated with the superficial gas velocity V_{sg} and the column diameter D as follows:

$$U_0 = 1.35\ {V_{sg}}^{0.16}\, D^{0.4} (\mathrm{m/s})$$

For the column diameter $D = 3$ m, the averaged central liquid velocity U_0 was 1.5 m/s.

Of course, this correlation only holds for the water/nitrogen system used. It will likely to be okay to use it for liquid phases with density and viscosity similar to water and for gases with a density similar to ambient nitrogen. The key point of the experimental results is that liquid circulation strongly increases with diameter.

This means that for commercial scale reactors, liquid circulation velocities are in the order of m/s, and the circulation time is of the order of seconds. For most cases, the liquid residence time required for conversion will be in the order of hundreds of seconds or even longer, so that as first approximation, the liquid phase can be considered fully backmixed, and so behaves like a CISTR.

4.7.2.2 Horizontal bubble columns

For horizontal bubble columns, the column is placed horizontally. Liquid flows horizontally through the column, and gas is sparged over its length and flows vertically upwards. Hence, it is in crossflow. By placing vertical baffles, the liquid RTD can be staged, representing several backmixed CSTR reactors in series. The gas bubbles flow upwards through the liquid. But special internals are needed to avoid that part of the bubbles flow along the cylinder wall and the part in the center, leaving, in between, a liquid zone that is not sparged. Here, starvation of the gas component can occur. This behavior was determined by Klusener et al. [11] in a commercial scale horizontal reactor D = 4 m, and length of 25 m. The special internals for gas bubble redistribution are found in a patent [12].

A horizontal bubble column should be chosen in case the reaction selectivity and/or conversion is sensitive to the liquid RTD. Commercial scale applications provided by Klusener et al. [11] and Harmsen [12] show that liquid staging is beneficial. The latter reports a case where a conversion of 99.999% was obtained in such a horizontal reactor [13].

Gas phase residence time distribution

The gas phase RTD of large-scale bubble columns also cannot be simply described as plug flow or fully backmixed. The RTD will, in general, depend on the column diameter and on the flow regime. For the small-bubbles regime, gas flows upward through a central core section, and some bubbles flow back to the bottom with the circulating liquid.

This is, in particular, the case for the heterogeneous regime, where large bubbles move with velocities of 1 m/s and higher through the column, while liquid and small bubbles circulate up and down through the column.

Seher [14] measured the gas phase RTD for two columns with diameter 0.45 and 1.0 m. He correlated the results for the 1-meter column with a dispersion coefficient (D_{ax}) and found $D_{ax} = 2.5\ V_{sg}$ (m^2/s). For the 0.45 m column, the dispersion coefficient dropped by a factor 2. Thus, the dispersion coefficient is strongly dependent on the column diameter. The accuracy of the measurements, the usefulness of using the dispersion model for the RTD, and whether the correlation holds for other column dimensions and superficial gas velocities is, however, questionable.

Bubble column reactors are often chosen for reaction systems where deep conversion of the gas component is not required, such as in oxidation reactions using air; a cheap feed is used, so a surplus of feed can be applied without strong economic consequences, or when pure gas is used, so that the gas concentration over the reactor height is not dependent on the RTD

For commercial-scale reactors, given the large uncertainty of the gas phase RTD, it is best to assume a fully backmixed RTD and use that model to design for the gas phase conversion.

An additional design choice can be to have a small single gas phase conversion and recycle the gas externally and add feed gas. In this way, the gas phase composition is nearly uniform and independent of the RTD.

4.7.3 Bubbling fluid bed residence time distributions

Chapter 2 shows an enormous variety of fluidized bed. Here, we focus on two types of bubbling fluidized bed, vertical fluidized beds, and horizontal fluidized beds, as these are used mainly for chemical reactions. These bubbling fluid beds behave very similar to bubble columns [15, 16]. There is a dense phase (also called emulsion phase) comprising particles and gas flowing in between the particles at a dense phase gas velocity. The dense phase has, in some aspects, the same circulation behavior as a liquid in bubble columns. It has the same tendency to flow upwards in the central fluid bed section and flow downwards in the wall section. Also, the vertical circulation velocity increases with column diameter [16].

The fluid flow properties of the dense phase, however, strongly depend on the averaged particle size (and its distribution) and gas density and viscosity.

4.7.3.1 Solid residence time distribution

Abbasi showed experimentally that gas-solid fluidized beds (Class A) behave similar to gas-liquid bubble columns, with respect to axial velocities [15]. Efhaima showed experimentally that the axial solids velocity also increases with the fluid bed column diameter [16]. He only did measurements on the diameter scales 0.14 and 0.40, so a power law could not be determined. It is likely that the correlation of the averaged circulation velocity of the dense phase with the column diameter is also to the power 0.5 and that the order of magnitude for reactor diameters larger than 1 m is 1 m/s.

For average solid residence times of minutes or more, the solids phase can then be considered fully backmixed, so it behaves like a CISTR.

In the case of a catalytic fluid bed, where there is catalyst deactivation, an often overlooked consequence of the very good backmixing solid phase arises. An obvious way to counteract deactivation is to continuously replace the catalyst, that is, there is a continuous, or semicontinuous, flow of fresh catalyst into and "aged" catalyst out of the reactor. As we have seen in Chapter 4, ideal backmixing of the solids implies an exponential RTD. In other words, after steady state has been reached due to this RTD, there also will be a "catalyst age distribution" in the reactor. Thus, some catalyst particles will have an activity close to the activity of a fresh catalyst, some others will have an activity corresponding to a highly deactivated catalyst, and most particles will have an activity in between, that is, corresponding to the "age distribution."

In general, this implies that the performance of the reactor will not be equal to that of a reactor with a catalyst that has the activity corresponding to the average "age" of the catalyst in the reactor. An example of methanol to olefins (MTO), where this is addressed, is given by Bos et al. [17]. At the end of this chapter, we present an exercise based on a recent example of new catalytic bioprocess in a three-phase slurry reactor, where this was identified as a key commercial scale reactor selection, design, and operation challenge.

4.7.3.2 Gas phase residence time distribution

The gas phase RTD in a fluid bed is very wide. Some of the feed gas flows quickly through the reactor as large "bubbles." Some of the feed gas moves inside the dense phase and circles around endlessly. The exchange of the bubble gas with the dense phase is limited.

The best way to deal with the large uncertainty of the gas phase RTD is to avoid dependency of the RTD on the design performance. Here are a few ways to do so.

For commercial-scale reactors, given the large uncertainty of the gas phase RTD, it is best to assume a fully backmixed RTD and use that model to design the gas phase conversion. To avoid shortcut flow via bubbles, the most conservative mass exchange model for bubbles to dense phase should be used, so that the number of transport stages is two or higher.

An additional design choice can be to have a small single gas phase conversion and recycle the gas externally and add feed gas. In this way, the gas phase composition is nearly uniform and independent of the RTD.

Another way is to use a pure gas feed, so that the gas phase concentration does not decrease but only the amount of gas flow in the reactor.

4.7.4 Residence time distribution G-L-S bubble columns and fluid beds

RTDs of three-phase bubble columns and fluid beds are similar to two-phase reactors, if the particles are so fine (typically a hundred micron or smaller) that the liquid-solid slurry behaves like a liquid. Thus, what is stated in section 4.7 about gas phase RTD and liquid phase RTD for two-phase reactors also holds here for three-phase reactors. The RTD of the solids phase follows the liquid phase. Also, horizontal bubble columns and fluid beds with baffles for liquid flow staging and solids flow staging apply here.

Ebullated bed reactors use much larger particles, similar to the size used in fixed beds. These particles have a tendency to settle in quiet areas and hence, need special design configurations. These reactors are not discussed here.

4.8 Exercises

4.8.1 Industrial exercise 1: RTD of a new reactor for a new process

Given: Shell has developed a new and improved process for making a bulk chemical *P*. In this process, the product is made by hydration of an intermediate *I*.

The stoichiometry equation for main reaction is: I + H2O = P

The reaction can be considered to be first order in intermediate I: $r_I = -k[I]$.

The feed concentration of I is about 50% by mass.

The conversion of *I* in the reactor system should be very high: the specification is no more than 1 ppm of *I* in the reactor outlet. This is because traces of *I* cannot be easily separated from the product, and anything more than a few ppm in the final product is unacceptable, for product quality reasons.

Luckily, with the homogeneous catalyst system, the reaction is very fast. In a pilot plant, which consisted of six well-mixed reactors in series, each having a residence time of 10 min, the desired conversion was met. For scale-up to the commercial plant, a horizontal reactor vessel was chosen, which was internally divided into compartments divided by baffles. Originally, the number of compartments was six because a lab scale six mixed reactors in series had proven to work.

An Aspen model was built, and the horizontal reactor was modeled as six CISTR's in series. With the known kinetics of the *I* hydration, it was calculated that the desired conversion was met, that is, the *I* concentration of 1 ppm at the outlet when *I* at the inlet was 50%.

The reaction engineers on the team put forward that the true RTD of the vessel or of each section of the vessel will probably differ from the idealized CISTR concept.

A computational fluid dynamic model was made of the vessel (one section); see Figure 4.19. The colored arrows in Figure 4.19 show the flow direction and velocity (red = high velocity, blue = low velocity). This is the flow field in the center of the reactor. The vertical lines are the baffle plates. For symmetry reasons, the CFD modelling was done on two half "compartments."

To allow for any uncertainties and as confirmed by the CFD model, the design engineers proposed to increase the number of baffles in the reactor vessel to 12, based on the reasoning that the ASPEN model showed that for 12 CISTR's in series, the *I* outlet concentration will be far below the 1 ppm spec, which would already have been met for 6 CISTRs in series.

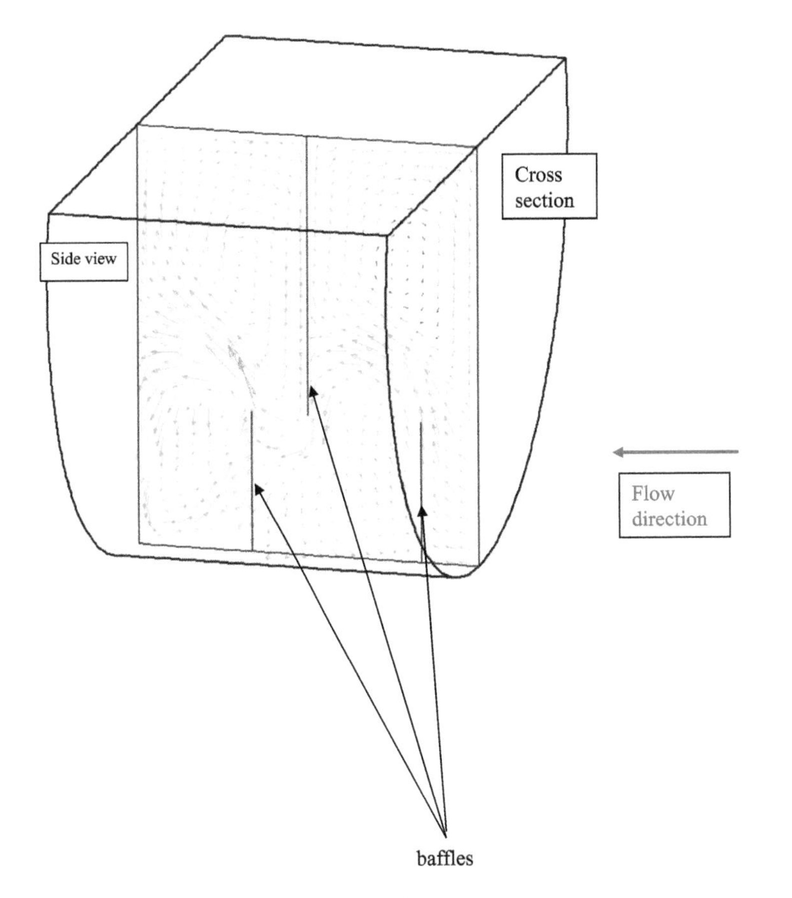

Figure 4.19: CFD modeling of a section of the new reactor.

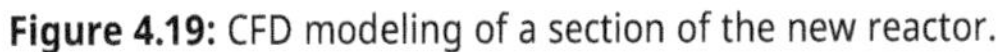

Questions

Q1: Do you think the flow pattern and/or the RTD of the horizontal large scale reactor vessel gives rise to concern? Why or why not?

Q2: What do you think of the mitigation measures proposed by the design engineers (i.e., the doubling of the baffles)?

Q3: Can you think of alternative risk mitigation measures and list the pros and cons?

4.8.2 Industrial exercise 2: fresh coconut drying in a fluid bed

A coconut roasting company has a horizontal gas-solids fluid bed. The bed width is 3 meters, and the length is 10 m. The bed height is 2 m. Hot gas flows through the bed, causing drying. The drying should not be within a narrow range for optimum downstream processing to coconut butter, which is optimally in between 5.5% and 6.5 % moisture content. The fresh material has 60 % moisture content.

The product quality (and, thereby, also product yield on raw coconut in downstream processing) has to be improved, and the process technologist has to come up with a proposal to achieve the higher product quality and higher yield.

In the past, he has tried various gas flows and temperatures. He found the best combination to reach the averaged dryness, however, measuring lots of small samples, the wide distribution on dryness around the average value remains the same. Some samples still had 40 % of moisture and some samples had 1 % moisture.

He has read about reaction engineering and decides that the moisture removal can be seen as conversion. He then wants to know the RTD of the beans. He measures the RTD in the present fluid bed using coconut beans of different color by throwing a bucket of these in the feed to the fluid bed and observing the outlet stream and counting the number of colored particles, in time intervals of 1 min. He plots the number of colored beans versus time. From the curve, he determines that the average residence time is 20 min. He then replots the number of beans versus the dimensionless time ($t/t_{average}$) as shown in Figure 4.20.

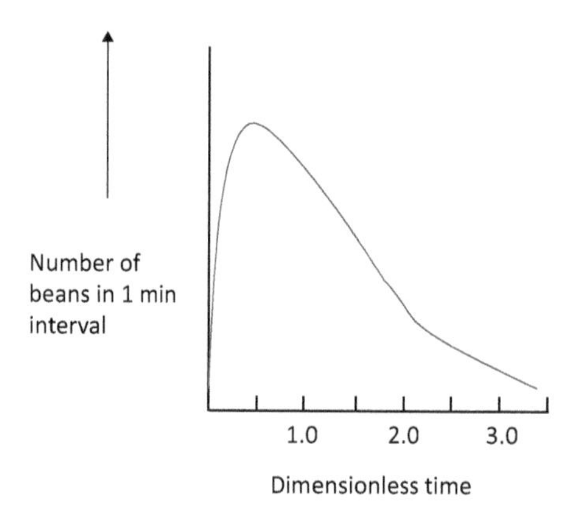

Figure 4.20: Residence time distribution in a fluid bed dryer.

From the reaction engineering theory, he considers the idea of staging the solids flow by placing walls perpendicular to the solids flow in the fluid bed, with small holes for the nuts to flow through, so that a staged flow of solids is obtained, so a lower spread of drying conversion and a higher yield of product in downstream equipment.

Questions

Q1: What is your best estimate of the number of CISTR in series equivalent to this RTD?

Q2: How many walls should be placed (or how many chambers created) to achieve a far better conversion and quality (reduced spread of RTD)?

Q3: How many walls should be placed to obtain the same narrow spread in dryness as the laboratory results by [18]?

4.8.3 Industrial exercise 3: catalyst deactivation in a three-phase slurry reactor

In this process, a feedstock *A* is dissolved in a solvent containing a homogeneous catalyst that is able to convert *A* into an intermediate, *I*. This intermediate, *I* can be oxidized to product, *P* by means of a heterogeneous catalyst, *Catalyst 2*.

Rather than doing these two steps separately, it was found that by carefully choosing the reaction conditions, it can be done in a single reactor. In a three-phase slurry reactor, oxygen is bubbled through the liquid mixture of *A* and solvent, *S*. Dispersed in this mixture is the heterogeneous *Catalyst 2* with a particle size of typically a few hundred microns.

The homogeneous catalyst, *Catalyst 1*, needed to form the intermediate *I*, is very stable. However, experiments at 1-liter scale using an intensely stirred reactor (a CSTR, close to being a CISTR), showed that, initially, there is a very good oxidation activity with high selectivity to *P* and low selectivity to the undesired by-product X_2. However, over time, the heterogeneous *Catalyst 1* deactivates. In these experiments, there is continuous flow in and out of the liquid phase (reactant, solvent, homogeneous catalyst, and in the outlet in addition the main product as well as the many byproducts). The heterogeneous cat_2 is just loaded once and does not leave the reactor.

The real reaction network is quite complex with dozens of reactions, but the Figure 4.21 gives an adequate simplification:

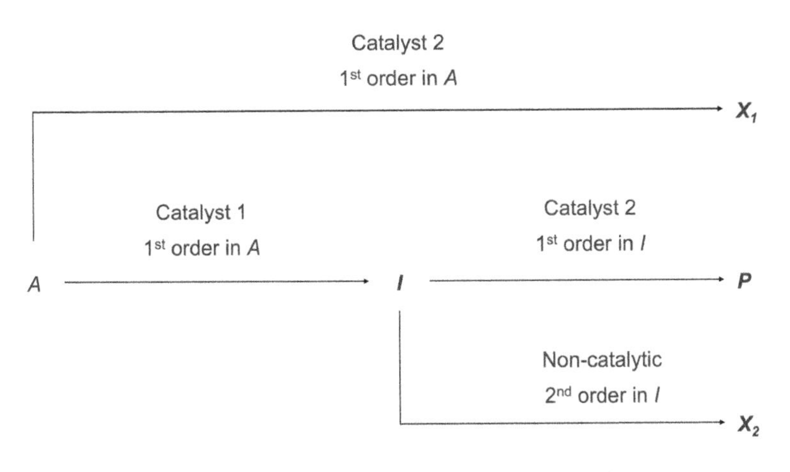

Figure 4.21: Catalyzed reaction network exercise 3.

As can be easily deduced from this network and (simplified) kinetics, there needs to be a fine balance between the activity of the homogenously catalyzed reaction and that of the oxidation catalyst. If the oxidation activity is too high, reactant *A* undergoes too much oxidation to X_1 in favor of the desired reaction to intermediate, *I*. However, if the oxidation activity is too low, too much of the intermediate *I* will react with second-order kinetics into the undesired by-product X_2. In other words, with "age" of *Catalyst 2*, not only the activity goes down, but also the differential selectivity of I to P versus I to X_2.

This reaction network is consistent with the experimental observation in the stirred tank reactor that initially the conversion of A is very high, but the selectivity to P is not fantastic, for example, 70%. Over the course of a few hundred hours, the selectivity to P, however, increases slowly to 90%. After 500 h, however, the selectivity to P starts to decline again, ultimately, even back to well below the initial selectivity of 70%.

As an average selectivity of around 75% is sufficient for the process to become economically viable, a commercial-scale reactor might just be a scaled-up version of the 1-liter reactor. (Of course, one does need to address all potential issues with differences in micro- and macromixing when scaling up both the reactor volume and size and r.p.m. of the stirrer). The operation strategy then being similar to that of the experimental unit: load heterogeneous catalyst cat_2 and then, run the reactor as a stirred tank without continuous replacement of that cat_2. Thus, even at commercial scale, the selectivity is then expected rise from 70% to 90%, and then drop relatively rapidly. An economic optimization will then occur at which time the reactor is taken off-line for complete replacement of the catalyst.

The chemists in the team had what they considered an obvious improvement: after the cat_2 has lost some activity and hence selectivity has increased to 90%, simply remove part of the aged catalyst and replenish it with fresh catalyst. And do this regularly. In other words: top-up and bleed. That way, the total oxidation activity is "controlled" at or near the "optimal" level.

The reaction engineer on the team liked the idea of (semi-)continuous catalyst replacement from an operational point of view. However, this reaction engineer had properly understood section 5.7 of our book on micromixing and segregation. And hence, he expressed some serious concerns and strongly advised to first test that "top-up and bleed" concept at the 1 L scale.

Questions

Q1: If the dosing of cat_2 would be continuous, the deactivation would be linear in time and the macromixing would be perfect: how would the RTD and the activity distribution in the steady state look like?

Q2: From a micromixing theory point of view, the homogeneous catalyst is perfectly micromixed. Can that also hold for the heterogeneous catalyst?

Q3: What would be the main concern of the reaction engineer?

Q4: If the "optimal" average solids residence time would be 500 h: would the concern go away if the dosing was not continuous by which the "top up and bleed" would be done only once per day, replacing each time some 24/500 = 4.8% of catalyst? Or once every 100 h, replacing 20% each time?

Q5: Explain that in case cat_2 was a homogeneous catalyst, the reaction engineer would have embraced the chemists' suggestion and that, in that case, 90% average selectivity could be feasible at commercial scale via continuous dosing of *Catalyst 2*.

4.9 Takeaway learning points

Learning point 1

RTD has a big influence on the performance of a reactor (conversion, selectivity); it is one of the main reactor selection considerations.

Learning point 2

Only for a first-order reaction, the conversion be calculated from the RTD. For orders $n > 1$, early mixing increases conversion; for $n < 1$, it is the opposite.

Learning point 3

A perfectly backmixed reactor ("CSTR") may not necessarily by perfectly mixed on the meso- or microscale. The degree of segregation can be very important for zero-order processes (or other kinetics strongly deviating from first order).

Learning point 4

For relatively fast reactions "mesomixing" (feed inlet effects) may become important; only for extremely fast reactions (milliseconds), micromixing is done up to the Kolmogorov scale.

Learning point 5

In a system with (semi)continuous dosing/removal of a heterogeneous catalyst to counteract deactivation, the system should be considered completely segregated, potentially leading to undesired and very broad catalyst age distribution.

References

[1] U.S.A. Census Bureau, Current population survey. Annual Social and Economic Supplement, 2015.

[2] Danckwerts PV, Continuous flow systems: Distribution of residence times. Chemical Engineering Science, 1953, 2(1), 1–13, https://www.census.gov/library/visualizations/2015/demo/distribution-of-household-income--2014.html

[3] Froment GF, Bischoff KB, Chemical Reactor Analysis and Design, 2nd ed. Singapore, John Wiley & Sons, 1979.

[4] Westerterp KR, van Swaaij WPM, Beenackers AACM, Chemical Reactor Design and Operation. New York, NY, Academic Press, 1984.

[5] Zwietering TN, A backmixing model describing micromixing in single-phase continuous-flow systems. Chemical Engineering Science, 1984, 39(12), 1765–1778.

[6] Zwietering TN, The degree of mixing in continuous flow systems. Chemical Engineering Science, 1959, 11(1), 1–15.

[7] Gierman H, Design of laboratory hydrotreating reactors: Scaling down of trickle-flow reactors. Applied Catalysis, 1988 Jan 1, 43(2), 277–286.

[8] Berger RJ, Pérez-Ramírez J, Kapteijn F, Moulijn JA, Catalyst performance testing: bed dilution revisited. Chemical Engineering Science, 2002, 57(22–23), 4921–4932.

[9] Hoog H, Discussion of papers presented at the fourth session of joint symposium the scaling-up of chemical plants & processes, Proceedings, London, UK, Inst. Chem. Eng., 1957, S131–136.

[10] Raimundo PM, Cloupet A, Cartellier A, Beneventi D, Augier F, Hydrodynamics and scale-up of bubble columns in the heterogeneous regime: Comparison of bubble size, gas holdup and liquid velocity measured in 4 bubble columns from 0.15 m to 3 m in diameter. Chemical Engineering Science, 2019 Apr 28, 198, 52–61.

[11] Klusener PAA, Jonkers G, During F, Hollander ED, Schellekens CJ, Ploemen IHJ, Othman A, Bos ANR, Horizontal crossflow bubble column reactors: CFD and validation by plant scale tracer experiments. Chemical Engineering Science, 2007, 62(18–20), 5495–5502.

[12] Harmsen GJ, Rots AWT, Process for the preparation of alkylene glycol. Patent EC C07C29/12, 2009.

[13] Harmsen GJ, Verkerk M, Process Intensification – Breakthrough in Design, Industrial Innovation Practices, and Education. Berlin, de Gruyter, 2020.

[14] Seher A, Schumacher V, Determination of residence times of liquid and gas phases in large bubble columns with the aid of radioactive tracers. German Chemical Engineering, 1979, 2(2), 117–122.

[15] Abbasi M, et al., Numerical comparison of gas-liquid bubble columns and gas-solid fluidized beds. The Canadian Journal of Chemical Engineering, 2015 Oct, 93(10), 1838–1848.

[16] Efhaima A, Al-Dahhan MH, Bed diameter effect on the hydrodynamics of gas-solid fluidized beds via radioactive particle tracking (RPT) technique. The Canadian Journal of Chemical Engineering, 2017 Apr, 95(4), 744–756.

[17] Bos ANR, Tromp PJJ, Akse HN, The conversion of methanol to lower olefins – Kinetic modelling, reactor simulation and selection. Industrial & Engineering Chemistry Research, 34, 1995, 3808–3816.

[18] Fernando JA, Amarasinghe AD, Drying kinetics and mathematical modeling of hot air drying of coconut coir pith. Springer Plus, 2016 Dec, 5(1), 1–2. The complete experimental work is available as PhD thesis by P.M. Raimundo, Accessed August 12, 2022, by, https://tel.archives-ouvertes.fr/tel-01267349/document

5 Inter- and intraphase mass and heat transfer

The purpose of this chapter is to explain in some detail mass transfer and heat transfer theory. It also describes mass transfer and heat transfer on the scale of bubbles, droplets, and particles.

5.1 Introduction to mass transfer

This chapter is written from a designers' point of view. From a designers' point of view, mass transfer from the feed phase (gas or liquid) to the reaction phase (often a porous catalyst) should keep up with the reaction rate so that the reactor volume required stays minimal. If the mass transfer is limiting the reaction rate, then to reach the required production capacity, the reactor volume needs to be larger than needed if this limitation is absent. The second reason why a design wants to avoid, if feasible, mass transfer limitations is that mass transfer performance of reactors is in general sensitive to scale-up effects and moreover the scale-up effects are not accurately known. So the designer in general wants to stay far away from mass transfer limitations so that the design is robust to uncertainties in the effect of scale on the mass transfer performance.

The designer therefore wants to know how to increase the mass transfer performance of his reactor. This chapter on mass transfer theory is therefore written from this viewpoint. To get an impression of what mass transfer limitations mean, a generic picture of three-phase reactors is drawn in Figure 5.1. It is about a case where a reaction component (say oxygen) is fed as a gas phase (say air) to the reactor. This component has then to be transferred to the liquid phase, and then to the porous catalyst particle, where the reaction takes place.

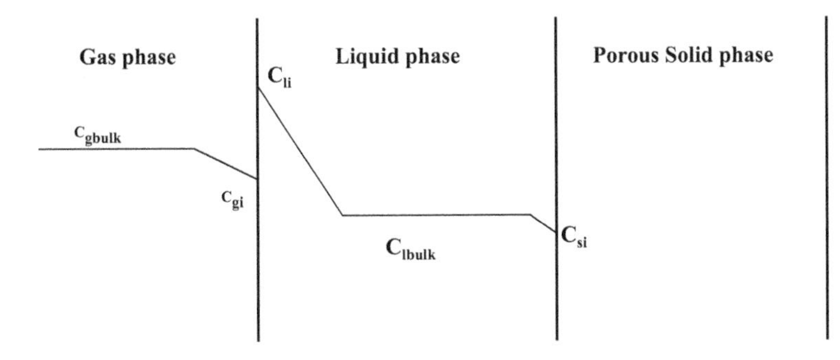

Figure 5.1: Interface concentration profiles in a three-phase reactor.

https://doi.org/10.1515/9783110713770-005

5.1.1 Mass transfer from gas phase to liquid phase to porous solid phase

The transport from gas to liquid and from liquid to solid phase is described as a flux of molecules.

The flux depends on the concentration differences between the bulk and the interface for each phase and on the mass transfer coefficient. It is similar to an electrical current that is proportional to the voltage and the conductivity (reverse of the resistance).

In steady state this flux will be equal to the reaction rate. This means that the concentration differences will adjust themselves to this steady state. Now if the rate of mass transfer is high –compared to the rate of reaction – then the concentration differences will become small, and the concentration of the reacting component from the gas phase will be at its maximum value at the porous catalyst particle surface. This means that mass transfer resistances in the gas and liquid phase are absent.

This chapter treats phase transfer and heat transfer at *microelement level*. A microelement can be a solid particle, or a droplet, or a bubble. It is defined not only by what is happening inside but also by its interface with the macro system surrounding it. The advantage of treating transport inside the microelement is that it facilitates understanding what is going on at this level, without the complex phenomena such as residence time distribution (RTD) outside it. Also a microelement optimization can be obtained by choosing its geometry and local conditions such as temperature. The purpose of this chapter is then to provide theory to understand mass and heat transfer phenomena at microelement level.

5.2 Concept of transfer coefficients

A classic concept in for describing heat and mass transfer phenomena in flow systems is the so-called film model. If, for example, we consider turbulent flow in a simple pipe, the total velocity gradient can be assumed to be fully located in a relatively thin boundary layer of thickness δ.

It is then assumed that all mass or heat transfer resistances are only present over that thin stagnant film. In this film the mass transfer is assumed to take place via diffusion only, for example, Fickian diffusion. For heat transfer this is conduction. Note that the actual transfer mechanisms are far more complication and pure diffusion or conduction may hardly play a role. Nevertheless, the concept has proven to be very valuable.

Figure 5.1 already introduced the concept: the straight lines from the bulk of the gas phase to the gas–liquid (G-L) interface as well as the from the G-L interface to the bulk of the liquid phase and from the bulk of the liquid phase to the surface of the solid catalyst are depicted represent the conceptual film. There are only gradients in these three films.

The mass flux (under nonreactive conditions!) is assumed to be proportional to the driving force, often simply taken as the difference in concentration between the bulk of fluid and the interface. At the interface equilibrium with the other phase, the mass flux J of a component i per square meter interfacial area in mol_i/m^2 s is

$$J_{\text{phys}} = k_m \ (C_{\text{i}} - C_{\text{b}}) \tag{5.1}$$

The subscript "phys" highlights that the relation is only valid in case there is no chemical reaction *in the film*. The mass transfer coefficient k_m coefficient hence follows from

$$k_m = D/\delta$$

where D is the diffusion coefficient and δ is the film thickness. If the film is in the gas phase, then it is usually referred to as k_{G}. For the liquid phase it is referred to as k_{L}. As always, great care should be taken with respect to the units. The flux J is per square meter of interfacial area and the concentrations are per cubic meter of liquid of gas phase. Therefore, the often seen "m/s" as unit for k_{L} can lead to errors. Inspection of eq. (5.1) reveals that the unit actually is cubic meters of liquid film per square meters of interfacial area per second.

One could ask oneself what we have gained by introducing this conceptual film thickness as we only have shifted the problem from estimating a mass transfer coefficient to estimating the film thickness. However, in many cases this film thickness can be derived from hydrodynamic boundary layer theories or at least can be estimated with relevant dimensionless parameters that are suitable for correlating the mass transfer coefficient in dimensionless form. It also "enables" the analogy between mass and heat transfer by *assuming* that the film thickness for heat transfer equals that for mass transfer. Typically for mass transfer we obtain Sherwood–Reynolds–Schmidt relations and for heat transfer equivalent Nusselt–Reynolds–Prandtl relations.

In Figure 5.1 the resistances to mass transfer are *in series* with chemical reaction. No reaction is assumed to take place in the gas and liquid phase; hence also in the three films there is no chemical reaction. A fundamentally different situation arises when there are reactions in one or more of these phases. We then speak of mass transfer and chemical reaction *in parallel*. The characteristics of such a system are significantly different. For example, for a G-L system the presence of a chemical reaction "enhances" the mass transfer. Equation (5.1) no longer holds!

Both mass and heat transfer can occur in parallel with the chemical reaction(s) or in series. Just like for electrical circuits of resistances, putting things in series or in parallel is a major difference. Because this also applies to other common situations in multiphase reactors we will treat this here upfront and will apply it in later sections dealing with the specific situations.

5.3 Multiphase mass and heat transfer: inter- and intraphase effects

Figure 5.2 illustrates the difference between parallel and series resistances for mass transfer with chemical reaction for the case of a heterogeneous catalyst system, but it also applies to other multiphase reactors: simply read parts left to the red interface line as non-reactive phase and the parts on the right-hand side of this interface line as reactive phase.

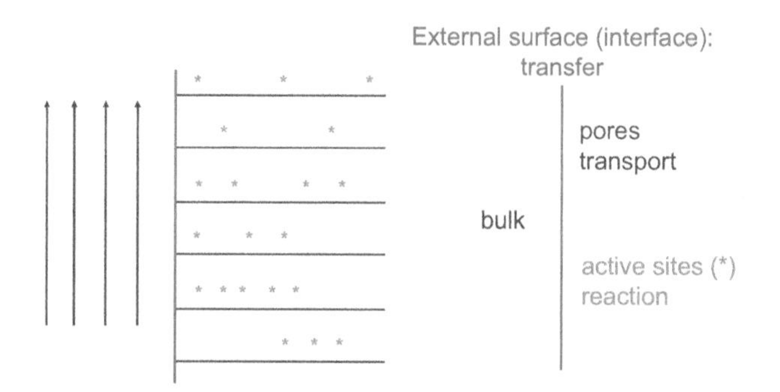

Figure 5.2: Difference between mass transfer in series and in parallel.

The external transport from "bulk" to the interface is an example of in series: in the bulk no reaction is taking place because there is no catalyst. The internal transport from the interface to the active sites in an example of in parallel: in the pores the reactants are diffusing into the catalyst whilst also reacting.

Figure 5.2 is just showing two phases here, but the same concept holds for three or more phases: then we will have two or more interfaces and for each phase one should determine whether one is dealing with mass (heat) transfer in series or in parallel. In multiphase reactors we hence can have for example two times mass transfer in series with reaction combined with one time mass transfer in parallel as would be the case for a classic trickle-bed reactor: in series from gas bulk to liquid film, in series through the liquid film followed by in parallel inside the catalyst pores.

5.3.1 Exercise: mass transfer in series and/or in parallel

In Table 5.1, we have listed 10 different multiphase reactors. Determine for each of these whether there is mass transfer with reaction in series, in parallel, or both.

Table 5.1: Multiphase reactors: ten different multiphase reactors with mass transfer in series, in parallel, or both.

S. no.	System	Series or parallel or both
1	Heterogeneous catalysis by nonporous catalyst (G/S)	
2	Heterogeneous catalysis by porous catalyst (G/S)	
3	Gas **ad**sorber (G/S)	
4	Gas **ab**sorber (G/L, e.g., amine treater)	
5	Membrane reactor (G/S)	
6	Catalytic membrane reactor (G/S)	
7	Chemical vapor deposition	
8	Trickle-flow reactor (G/L/S, e.g., Fischer–Tropsch)	
9	Slurry reactor (G/L/S, e.g., Fischer–Tropsch)	
10	Oxidative coupling of methane in a fixed bed reactor (1) Activation of methane on catalyst surface followed by formation of radicals; (2) Coupling of CH_3 radicals in the gas phase	

5.4 Mass transfer with reaction in gas–liquid reactors

5.4.1 Introduction

Mass transfer in G-L reactors with a reaction in the liquid phase is treated here. The generic reaction stoichiometry equation for which the description holds is: A + bB = P.

The generic reaction rate kinetic expression is given by

$$-r_A = k\, C_A{}^m\, C_B{}^n \tag{5.2}$$

Component A is in the gas phase and is transferred to the liquid phase via the boundary at the liquid side of the G-L interface. Component B is in the liquid phase. Product P stays in the liquid phase.

5.4.2 Chemical enhancement and the Hatta number

Figure 5.3a–c shows concentration profiles in the liquid film layer near the gas phase for three mass transfer limited conditions. For completeness, Figure 5.3d shows concentration profiles of A and B for no mass transfer limitations at all, that is, the reaction is so slow that there are no concentration gradients whatsoever. The bulk concentration of A is in full equilibrium the interface concentration. This is the so-called kinetic regime.

Figure 5.3a shows concentration profiles of A and B for no effect of the reaction on the mass transfer in the film itself, that is, the reaction is slow compared to mass transfer in the film. Hence the gradient of A in the film is linear. This mass transfer limited regime is called the slow reaction regime.

Figure 5.3b shows concentration profiles for the case where the reaction is fast compared to mass transfer: there is already reaction taking place in the film. Therefore, the gradient of A is no longer linear. This is traditionally called the fast reaction regime.

Figure 5.3c shows the maximum effect of the reaction on the transfer rate of A to the liquid phase. Here the transport of the other reactant B becomes limiting. This regime is called the instantaneous reaction regime.

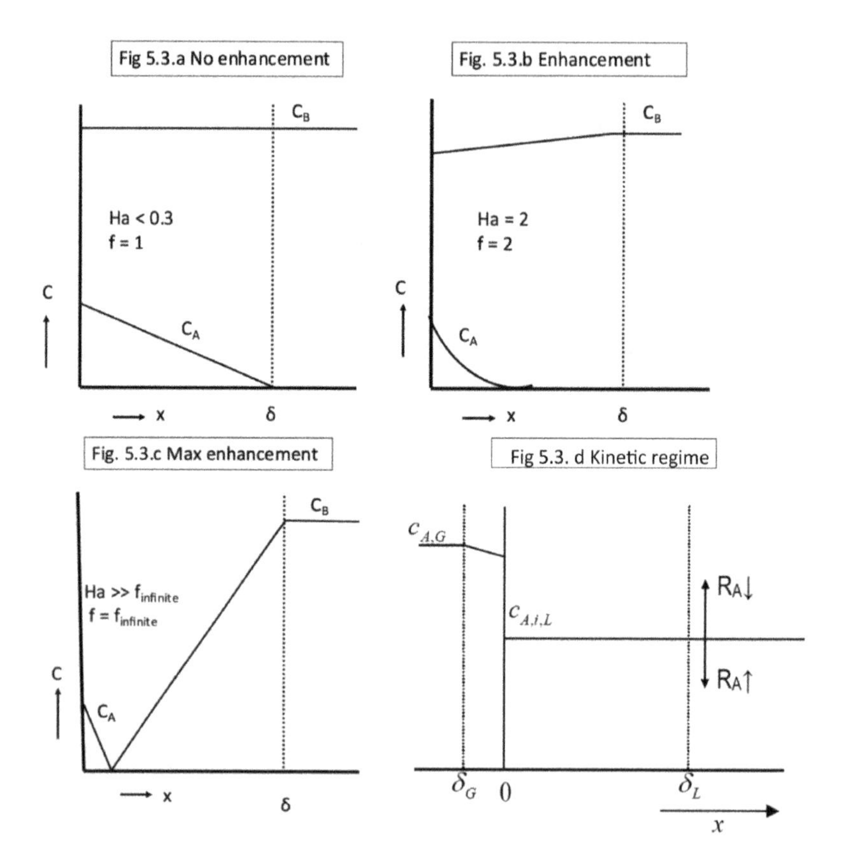

Figure 5.3: The four different regimes for G-L reactions.
(f is the enhancement factor, also known as E_A or F_A).

But first a few aspects of this Figure 5.3 need to be described. The film layer is considered to be stagnant with a thickness δ. So this is the film model for mass transfer described in Section 5.1.

Transport of component A through this film layer only can take place by diffusion. The mass transfer coefficient for component A, k_L is then given by

$$k_L = D_A/\delta$$

Reaction also takes place in the film layer. If the reaction is slow compared to the mass transfer rate hardly any A will react away in the film layer, so the concentration profile of component A will still be a straight downwards line and the mass transfer flux of component J_A will be given by

$$J_A = k_L(C_{Ai} - C_{Abulk})$$

where C_{Ai} is the concentration of component A at the G-L interface and C_{Abulk} is the concentration of A in the bulk of the liquid. In Figure 5.3a this concentration is zero. So the reaction system is already mass transfer limited.

If the reaction rate is fast compared to the mass transfer, then component A reacts away while diffusing through the film layer so that concentration profile will be curved. The slope of the curve at the interface ($x = 0$) will then be higher than for case a. This causes the mass transfer rate at that interface to be higher. This is called enhanced mass transfer. The transport of component A from the gas into the bulk of the liquid phase is higher or "enhanced" in the presence of reaction in the film (at the same driving force). For this an enhancement factor is introduced. This enhancement factor is usually denoted as E_A or F_A or sometimes just f. The mass transfer flux is then given by

$$J_A = E_A k_L(C_{Ai} - C_{Abulk})$$

This enhancement factor depends on the dimensionless number Hatta which is the ratio between reaction rate and mass transfer rate. For the kinetic rate expression (5.2) the Hatta number (Ha) is defined as

$$\mathrm{Ha}^2 = (2/(m+1)\ kC_{Ai}{}^{m-1}\ C^n{}_{Bbulk}\ D_A)/k_L{}^2$$

For a reaction that is simply first order in both A and B this becomes

$$\mathrm{Ha}^2 = k\ C_{Bbulk}\ D_A/k_L{}^2$$

For a reaction that is only first order in A, that is, $r = k\ C_A$:

$$\mathrm{Ha}^2 = k\ D_A/k_L{}^2$$

For Ha < 0.3, the reaction rate is slow compared to the mass transfer coefficient k_L. Then there is no enhancement of the mass transfer so $E_A = 1$.

If the Hatta number is much larger, the reaction rate is significant compared to the mass transfer and the enhancement factor E_A will increase. There are general expressions for E_A as a function of Hatta, but for Ha > 2 the expression becomes simply

$$E_A = \mathrm{Ha}$$

Component B from the bulk of the liquid also has to diffuse through the stagnant film layer. If the reaction also becomes fast relative to the mass transfer of B then the concentration of B in the film layer decreases. If reaction is also very fast compared to mass transfer by diffusion of B then the concentration of B also drops to zero in the film layer, and a reaction "plane" is obtained as shown in Figure 5.3d.

Then the maximum enhancement $E_{A,inf}$ of A is reached. This is called the instantaneous regime and the criterion for this is

$$\mathrm{Ha} > 10\,(E_{A,\,inf} - 1)$$

$$E_{A,\,inf} = 1 + D_B C_{Bbulk}/(bD_A C_{Ai})$$

The enhancement factor then becomes

$$E_A = E_{A,\,inf}$$

So then only the ratio of the concentrations of A and B at the respective boundaries of the film layer and the ratio of the diffusion coefficients determines the enhancement factor. Further increasing the reaction rate will have no effect on the overall reactor performance.

The Hatta number is the ratio of the reaction rate over the mass transfer rate in the liquid layer near the interface. It is similar to the Thiele modulus number which is about the reaction rate over the mass transfer by diffusion in a porous catalyst. For the case that the reaction only involves component A the Hatta number is essentially the same as the Thiele modulus, as then conceptually the process of diffusion plus reaction in a stagnant layer is the same as diffusion plus reaction in a porous medium. There is however one major difference that we will explain in the next section.

5.5 Mass transfer in heterogeneous catalysis

5.5.1 Introduction

Many catalytic reactions for which both reactants and products that are in the gas phase are nonetheless heterogeneous reactions. If a reaction is heterogeneously catalyzed, then the catalyst is indeed in a different aggregation state than the reactants and, in order for the reaction to proceed, gas to solid and solid to gas mass transfer needs to take place. Figure 5.4 shows the different physical and chemical steps, which, originating from a reactant A, lead to a reaction product P via a heterogeneously catalyzed reaction path.

Seven steps can be distinguished even for this simple case of a catalytic reaction A = P:

1. Transfer of A from the fluid phase surrounding the catalyst particle the so-called bulk phase or the mainstream, to the external surface of the particle.
2. Transport of A via the catalyst pores from the external particle surface to the active sites on the internal catalyst surface
3. Chemisorption of A on an active site to A*
4. Surface reaction of A* with formation of P*
5. Desorption of P*
6. Transport of P via the catalyst pores to the external catalyst surface
7. Transfer of P from the external catalyst surface to the bulk phase or the mainstream.

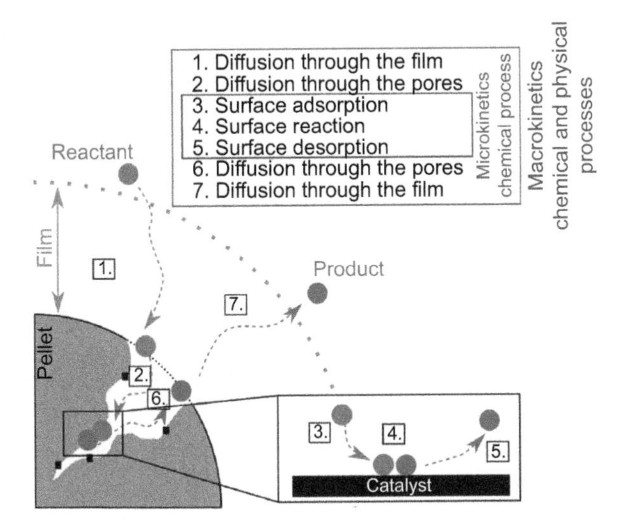

Figure 5.4: The steps in a heterogeneously catalyzed reaction in a porous catalyst (adapted from the course notes by Prof. Guy Marin).

Steps 3–5 are considered exclusively chemical and in series. The corresponding kinetics are called *intrinsic* kinetics. One of the simplest examples of such an A = P catalytic reaction leads to so-called Hougen–Watson or Langmuir–Hinshelwood rate equations, which describe the rate of this chemical phenomenon. The form of the equations will depend on which of these three chemical steps are rate limiting. Other classic examples of such intrinsic kinetics are Mars–van Krevelen and Michaelis–Menten. Regardless which of these chemical steps are rate limiting, the kinetics are referred to as intrinsic, even in case, for example, the desorption step 5 resembles more a physical desorption than a chemical transformation.

In contrast, steps 1 and 2, and 6 and 7 are physical in nature. If any of these necessary steps are affecting the overall production rates, we have mass transfer limitations. The first and last steps involve *external* mass transfer. If any of those two steps are affecting the overall rates, we refer to this effect as external mass transfer limitation. Steps 2 and 6 involve *internal* mass transfer, more commonly known as pore diffusion. If these steps are affecting the overall rates, we refer to this effect as external mass transfer limitation or *pore diffusion limitation*.

This is not just a logical naming: conceptually there are major differences between external and internal limitations. For a fully heterogeneously catalyzed reaction it is the difference between reaction and mass transfer *in series* and reaction and mass transfer *in parallel*. And just like for electrical circuits of resistances, putting things in series or in parallel is a major difference. Because this also applies to other common situations in multiphase reactors we treated this separately in Section 5.1. For the heterogeneous catalysis part we will focus on pore diffusion as this is generally, but not always, the more important one at industrial scale.

5.5.2 Diffusion in porous catalysts

Multicomponent mass transfer (pore diffusion) in porous catalyst particles is actually a very complicated process. Even in cases where the pore structure is assumed to compose of straight channels with a well-defined diameter. Here we will keep it as simple as possible and (a) focus on gas mixtures and (b) only consider ordinary continuum diffusion (or molecular diffusion) and Knudsen diffusion, the driving force being a concentration gradient.

Continuum diffusion in a porous medium occurs when the different species in the gas mixture move relative to each other with predominantly collisions between molecules rather than with the pore walls. By contrast, Knudsen or "free molecule" diffusion occurs when the molecules predominately collide with the pore walls.

For liquid-phase systems, usually continuum diffusion dominates. For gas phase diffusion this often is not the case and collisions with the walls of the pores are significant or even dominant. The corresponding Knudsen diffusion coefficient is much lower than the molecular diffusion coefficient. It is proportional to the pore size and to the square root of the temperature and the inverse of the square root of the molar mass of the diffusing species. Note that unlike the continuum diffusion coefficient the Knudsen coefficient is independent of the pressure.

There are several methods to estimate the diffusion coefficient when diffusion takes place in the intermediate regime. One of the simplest ones is the so-called Bosanquet formula, which is simply treating the two modes of diffusion as resistances in parallel, see, for example, eq. (3.5.2-10) in the textbook by Froment and Bischoff [1].

5.5.2.1 Effective diffusion

Two key aspects of diffusion in porous catalysts that need to be accounted for explicitly are:

Diffusion can only take place in the pores of the catalyst particle and not through the solid walls of the pores (or through the solid microparticles inside the particle), that is, the particle porosity needs to be accounted for.

The pores in such a catalyst are not simple straight channels and hence the molecules travelling inside the particle from point A to point B actually have to follow a tortuous path. So the true distance L_e over which the molecules have to diffuse is longer than the straight-line distance L_s by a factor called the tortuosity factor τ_p.

Figure 5.5 illustrates these two effects.

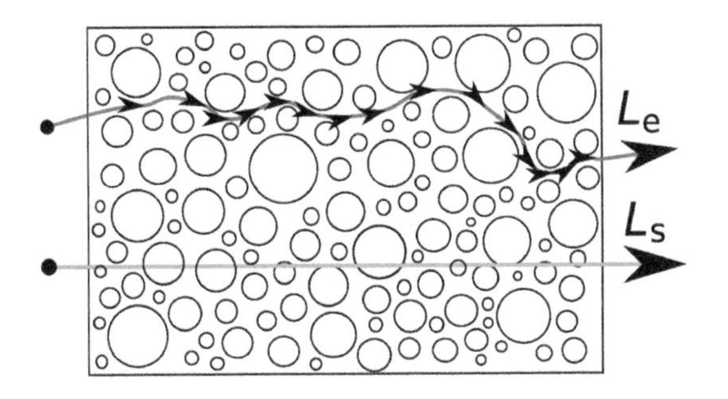

Figure 5.5: Effective diffusion through a porous catalyst particle.
Diffusion path is tortuous. L_e is true diffusion path length; L_s is straight length.

The leads to the simplifying concept of *effective* diffusion coefficient: the diffusion coefficient D of a normal gas mixture needs to be "corrected" with the porosity/tortuosity factor:

$$D_{eff} = \frac{\varepsilon_p}{\tau_p} D$$

Typical values for this porosity/tortuosity factor are 0.1–0.3, that is, typically the effective diffusion coefficient is a factor 3 to a factor 10 lower than the diffusion coefficient in the gas phase only. Ideally, the effective diffusion coefficient should be measured experimentally with the real catalyst particles by one of the various alternative techniques proposed in literature. However, this is far from straightforward, and each technique is prone to pitfalls. So in most cases we believe that it is sufficient to use an estimation based on a measured particle porosity and an estimate value of the tortuosity based on any knowledge of the pore structure.

In summary, for diffusion in porous catalysts, the effective diffusion coefficient is typically at least a factor 3 lower than the molecular diffusion due to the porosity/tortuosity effects. For gas phase systems it is usually one to two orders of magnitude lower, except for small molecules and relatively wide pores.

5.5.3 Consequences for catalyst performance

To explain the potential effects of pore diffusion limitation on the performance of heterogenous catalysts, most textbooks will start with setting up the mass balances over a single particle like a sphere or even a 1-D geometry like a slab. Then via mathematics expressions are derived quantifying the effects of finite rates of diffusion on the so-called *observed* rates. Subsequently more complexity is added, for example, more complex kinetics, other geometries and heat effects. Though there are good reasons to do so, here we will do the opposite and will start with summarizing and explaining the potential qualitative effects without referring to any mathematics.

5.5.4 Effect on catalyst activity: Thiele modulus and the concept of effectiveness factor

In all cases without external transfer limitations, the concentration of reactant A at the external surface of the catalyst is equal to that of A in the bulk of the surrounding fluid. However, since A is reacting away on the surface inside the pores, its concentration inside the catalyst particle must be lower *at least to some extent*. Only in case of *infinitely* fast pore diffusion of A the concentration of A is exactly equal to the bulk concentration.

However, if the diffusion of all reactants as well as all products is very fast compared to the rates of reaction, everywhere in the catalyst particle the concentrations will be almost equal to those outside the particle. Hence, everywhere in the particle the reactions proceed at these "bulk" concentrations. This situation is called absence of pore diffusion limitations or more accurately: negligible pore diffusion limitation. The observed rates of reaction will be essentially equal to the true or so-called intrinsic kinetics.

Figure 5.6 shows typical concentration profiles of reactant A for different degrees of diffusion limitation indicated via the so-called Thiele modulus usually denoted with the symbol ϕ.[1] We will explain this important modulus in detail further.

In the case of significant pore diffusion limitation of reactant A, the concentration of A at the external surface of the catalyst is still equal to that of A in the bulk of the surrounding fluid. However, since A is reacting away on the surface inside the pores and the diffusion cannot fully "keep up" and the concentration inside the catalyst particle is now significantly lower. Hence, instead of reacting everywhere inside the pellet at the bulk concentration of A, it reacts at a correspondingly different rate. Instead of producing product P at the intrinsic rate, the catalyst particle will produce P at the average of all local rates inside the particle. This is then called the apparent reaction

1 Other symbols in use for the Thiele modulus include h_T (Wikipedia), M_T (Levenspiel), ϕ_T (Wijngaarden and Westerterp) and ϕ_n (Fogler).

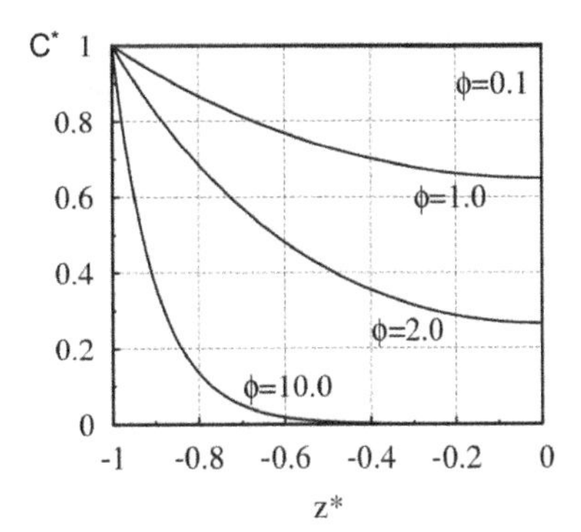

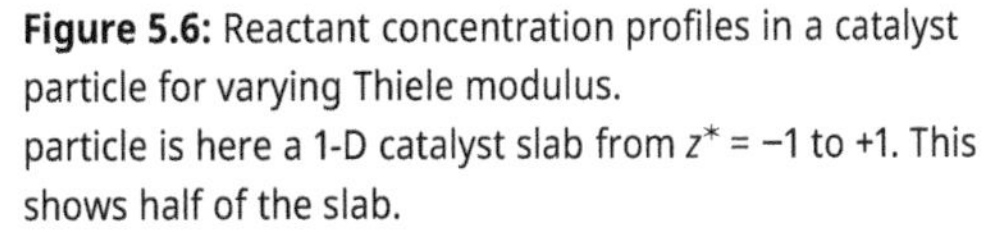

Figure 5.6: Reactant concentration profiles in a catalyst particle for varying Thiele modulus. particle is here a 1-D catalyst slab from $z^* = -1$ to +1. This shows half of the slab.

rate, or rather the rate of consumption of the key component, or most commonly the observed rate R_{obs}.

For simple first-order reactions with intrinsic rate $R_{\text{intrinsic}} = k\,C_A$ obviously the lower concentration of A inside the particle will lead to a lower rate of reaction inside the particle and hence the apparent or observed rate will be lower than the intrinsic rate.

How strong the concentration gradients inside the particle are depends on the relative rates, or rather time constants, of the two competing processes: reaction and diffusion. If the reaction is "fast" compared to diffusion there will be pore diffusion limitation effects.

Note that qualitatively the concentration profiles of the products diffusing out are the inverse or mirror image of the profiles for the reactants, unless their diffusion coefficients are very different. That implies that for kinetics that also depend on the product concentration(s) as may occur for Langmuir–Hinshelwood kinetics with "product inhibition effects" the impact may be even more significant.

The relevant time constant for reaction for a first-order reaction is determined by the rate constant: $t_{\text{reaction}} = \frac{1}{k}$ (with k in 1/s). A high value of the time constant means a slow reaction. More general one could define it at the reaction rate in moles per second per cubic meter catalyst at the bulk concentration of A divided by the bulk concentration of A (in moles/cubic meter in the gas phase).

On the first-order rate constant for a heterogeneous catalytic reaction

Strictly speaking, for a heterogenous reaction, the unit of a first-order rate constant is actually not 1/s. With $R = kC$ being a volumetric rate per unit volume catalyst *particle* and C a gas- or liquid-phase concentration mol/m_f^3 in the pores, the unit of k would actually be $\frac{m_f^3}{m_p^3}\frac{1}{s}$.

Therefore it is usually more convenient and less error prone to express the rate of reaction per mass of catalyst (or per unit surface area) and then for the Thiele modulus convert it to volumetric rates per unit volume particle.

The relevant time constant for diffusion is determined by the relevant diffusion constant of the component(s) and the diffusion length. Pore diffusion can be fast even for relatively low diffusion coefficients if the molecules only have to be transferred to the active sites over a very short distance. It can be easily derived that the time constant for diffusion for simple geometries like slab, spheres or cylinders is proportional to the square of particle size d_p divided by the effective diffusion coefficient.

So for an nth-order reaction we have

$$t_{reaction} = \frac{C_A}{kC_A^n} \quad (s)$$

$$t_{diff} \sim \frac{d_p^2}{D_{eff}} \quad (s)$$

We use "proportional" here instead of "equal" because the time constant for diffusion also depends to a limited extent on the exact geometry, even in case the particle size used in the equation is redefined as the equivalent sphere diameter 6 V_p/A_p. Professor Aris [2] was the first to notice that it was key to define a proper diffusion length, for which he proposed $L = V_p/A_p$, so for a spherical particle $L = d_p/6$. But a small effect of particle geometry, that is, "shape" remains.

A dimensionless number indicative for the relevance of pore diffusion limitation is the ratio of time constant of reaction over pore diffusion. In general, a ratio of a mass transfer time and reaction time is referred to as a Damköhler number. For the specific case of reaction and transfer in porous catalyst however another number is being used. As early as in the 1930s the highly prolific scientist Professor Ernest Thiele from Standard Oil already introduced this concept [3] and hence this dimensionless number is named after him. This Thiele modulus ϕ is the square root of the ratio of the two time constants: that of diffusion over reaction. In other words, it is the square of the second Damköhler number, see Section 1.4.5.

For first-order kinetics we have

$$\phi^2 = \frac{t_{diff}}{t_{reaction}} = L^2 \frac{k}{D_{eff}}$$

If the Thiele modulus value has the value of 1, the time constants of the diffusion and reaction are the same. That means that both are important and there are significant pore diffusion limitation effects, or rather there are significant concentration profiles inside the catalyst particle.

A very small Thiele modulus of $\phi \ll 1$ implies that the reaction is relatively slow compared to the diffusion. Significant concentration gradients can be assumed to be absent in case the Thiele modulus is typically smaller than 0.3. Note that there cannot be a general cut-off point since the (economic) impact of "small" concentration gradients on especially selectivity can be very different for different (commercial) processes.

A high Thiele modulus of $\phi \gg 1$ implies severe diffusion limitation of at least the reactant(s). In the core part of the particles the reactant will be depleted so little or no reaction takes place there. So this part of the catalyst – and reactor volume – is not really utilized for the intended chemical reaction. Especially if the catalyst consists of expensive noble metals, this very poor "utilization" of the catalyst can easily render an otherwise attractive process economically unviable.

Inspecting the effect of the three parameters we see that the Thiele modulus is low in case we have:

- A low value of the rate of reaction
- A high value of the (effective) diffusion coefficient
- A small particle size; this effect is the strongest of the three!

Not only is the particle size the one parameter with the biggest impact because it appears as d_p^2, it is also the one that is the "easiest" reactor/catalyst design parameter for the reaction engineer. Classically, this engineer is stuck with "given" catalyst (intrinsic) activity and the diffusion coefficient is also almost given unless the catalyst pore structure is modified.

On why it is not the "Thiele number"

People have wondered why ϕ is called *Thiele Modulus* and not *Thiele number* Convention in those days was, as we still use in our book, that the reaction rate of reactant *A* to produce *P* was negative. For example, for a first-order reaction one writes $r = -kC$. Hence to make it a positive number the absolute value or *modulus* of the rate was used to avoid having a square root of a negative number.

The strong particle size effect on the Thiele modulus also illustrates another "classic" pitfall in catalyst development. Certainly in earlier years, this was primarily done in small lab-scale reactors with catalysts tested in the form of powders. This means that the particle size is typically small to very small (0.1–0.3 mm). For fixed beds at commercial scale this would almost always give prohibitively high pressure drop. Hence, after the successful selection of a lab-scale catalyst, the catalyst would be "scaled up" for commercial use. Typical particle sizes for fixed bed reactor are in the range of 1–10 mm, so from 0.2 to 4 mm. This factor 20 in particle diameter also implies a factor 20 in the Thiele modulus. So by this scale-up we would move from no significant pore diffusion limitation to severe limitations.

This leads to an intuitively natural introduction of the concept of effectiveness factor (also known as utilization factor) and most commonly denoted as η

$$\text{Effectiveness factor } \eta = \frac{\text{rate with pore diffusion limitations}}{\text{rate without pore diffusion limitation}} = \frac{\text{overage rate over whole particle volume}}{\text{rate at the external surface of the particle}}$$

With this definition the observed or apparent rate is lower than the intrinsic rate by this effectiveness factor: $R_{obs} = \eta\ R_{intrinsic}$.

On Ernest Thiele

Thiele first published his paper on pore diffusion limitation in 1939 entitled: "Relation between Catalytic Activity and Size of Particle" [3].

It is less known that the very same Ernest Thiele is the Thiele of the classic McCabe–Thiele diagrams for designing distillation columns. And he has done lot more, see a presentation by Jonathan Worstell that can be found on the internet: "Ernest W. Thiele – his impact on Chemical Engineering": https://www.researchgate.net/publication/266484272_Ernest_W_Thiele_His_Impact_on_Chemical_Engineering

Figure 5.7 is taken directly from the original thesis by Thiele himself, showing how the effectiveness factor depends on his modulus as well as on the exact kinetics and particle geometry. In this original work the modulus was based on the particle radius of a sphere and half the plate thickness, respectively. Many years after Thiele, famous scientists like Aris [2] and Bischoff [4] have introduced *generalized* Moduli where they tried to bring the plots for different geometries – as well as for different kinetics – closer together.

The most advanced and very elegant generalization we know of was developed by Dr. Ruud Wijngaarden and Professor Roel Westerterp, published in full detail in the book *Industrial Catalysis* [5]. Instead of one Thiele modulus, they introduced two Aris numbers, indeed named after one of the great founding fathers of CRE, who was one of the first to try to bring together, for example, the three lines in the original graph from Thiele.

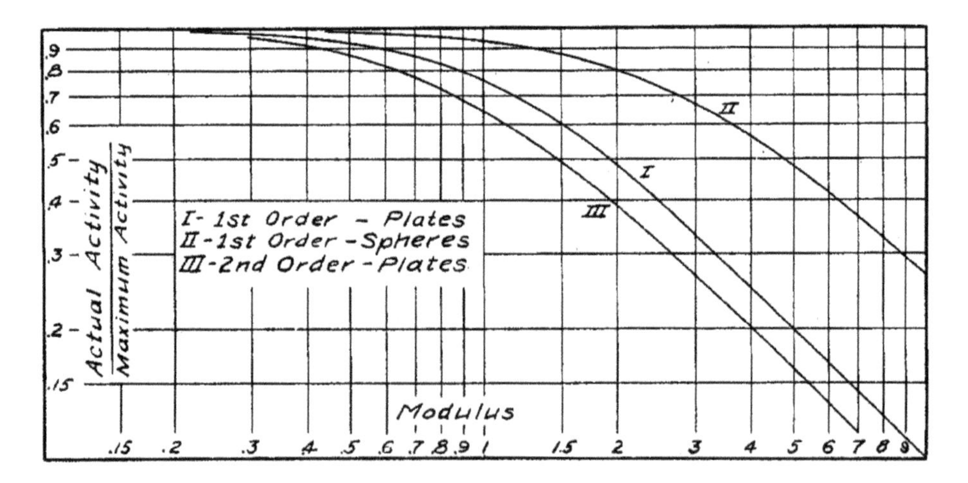

Figure 5.7: Effectiveness factor as a function of the earliest definition of the Thiele modulus. (graph from the original thesis from 1939 of Ernest Thiele [3]).

For simple kinetics and simple geometries one can obtain analytical expressions for the effectiveness factor as function of any well-defined Thiele modulus. These are formulas involving tanh and coth of the Thiele modulus. Only these simple situations allow for easy calculation of η as function of the proper Thiele modulus, see, for example, Levenspiel.

Here we suffice to provide the simplest relation that arises for the case typically for a Thiele modulus greater than 2:

$$\phi > 2 \rightarrow \eta \approx \frac{1}{\phi}$$

In fact, Figure 5.7 of Thiele already indicated that. This is a log–log plot and above a Thiele modulus of 2–3, the lines are linearly decreasing.

Thus for this case of pore diffusion being limiting, that is, the reaction being fast compared to the pore diffusion we obtain for the observed rate of reaction:

$$R_{obs} = \eta\, R_{intrinsic} \approx \frac{1}{\phi}\, R_{intrinsic}$$

For a first-order reaction in a sphere, substituting the relevant Thiele modulus, this then simply becomes

$$\phi > 2 \rightarrow R_{obs} \approx \frac{6}{d_p} \sqrt{kD_{eff}}\, C_A$$

So we see that in the region of pore diffusion region the (apparent!) first-order rate constant changes from k to:

$$k_{app} = \frac{6}{d_p} \sqrt{kD_{eff}}$$

On the (almost) equivalence of Thiele and Hatta numbers

Conceptually the Thiele number and the Hatta number of the previous section are virtually the same; the square of both of these are so-called Damköhler numbers; Da_{II} to be precise, the ratio of reaction and diffusion time scales, see also Section 1.4.5. If the reaction–diffusion equations for a first-order reaction are written in dimensionless form, replacing Ha^2 and Φ^2 with Da_{II} yields exactly the same second-order differential equation, with one major difference!

For a slab-like catalyst, having the same geometry as the liquid film in G-L systems, on both sides of the slab the concentrations are equal to the "bulk" concentration and the concentration gradient in the middle of the slab must be zero, for symmetry reasons. For a reaction in a G-L film on the one side of the slab the concentration is set to be in equilibrium with the gas phase and on the other side equal to the liquid bulk concentration. Therefore, also the solutions of the Hatta and Thiele model are different.

Or simply put: in case of no reaction, for the catalyst slab there can never be any *steady-state* concentration gradient, whereas for the G-L this simply means that the mass transfer is not chemically enhanced, and the concentration profile over the film is linear.

The Thiele modulus approach has been adapted for a variety of different situations for mass transfer and reaction in catalysts with "large pores" where convection inside these pores may also play a role, and even for adsorption processes, see the work by Driessen et al. [6]. In all these cases a modified Thiele modulus has been defined.

5.5.5 Effect on apparent reaction orders

For simple positive-order reaction kinetics, it is obvious that any decrease in local reactant concentration inside the particle yields a reduction in production rate and hence a lowering of the effectiveness factor.

However, for the generic case of an nth-order reaction n can also be negative. Many catalytic reactions, even a simple reaction A + B = P, may proceed via a Langmuir–Hinshelwood mechanism. These can have kinetics that have effective orders ranging from minus 1 to plus 1 or even from minus 2 to plus 1. The effective order depends on the concentration, or partial pressure, of the reactant. At low concentrations the surface concentration of the adsorbed form of the reactant A is low, whereas B may be abundantly available. Doubling the concentration of A could then simply cause a doubling in surface concentration of A and hence simple first-order behavior. However, at higher concentrations of A, the surface may be almost full with one reactant, that is, be inhibiting adsorption of B that is also needed for the reaction. In that case the apparent order in A may become, for example, –1.

For an intrinsic negative order, the rate of reaction actually increases with decreasing local concentration in the particle. So the apparent activity of the catalyst increases if there are significant pore diffusion limitations. Or in terms of effectiveness factor: the effectiveness factor as function of the Thiele modulus starts with and then typically above 0.3; the effectiveness factor η becomes higher than 1.

On negative-order reaction kinetics

If there would be a negative order throughout this can easily lead to physically nonsensible results because it would predict an infinite rate of reaction if the concentration approaches zero. In the authors' experience, even for limited diffusion limitation, using simple negative orders in reactor models can severely affect the (numerical) robustness. Our advice is to always avoid using simple negative orders when modeling chemical reactors and instead either use more sound semiempirical relations not predicting very high rates if one of the concentrations approaches zero or use numerical tricks to force the effective kinetics to become of positive order for low concentrations.

For a zero-order reaction, by definition, the rate of reaction is independent of the concentration. Therefore even if there are significant concentration gradients due to pore diffusion limitation, the net effect can still be negligible. Only in case the diffusion limitation is so strong that in the core of the particle the concentration goes to zero, will there be an observable effect in the apparent activity of a catalyst particle. However, this does not mean that for medium pore diffusion limitation there is no effect at all.

This is because the presence of pore diffusion limitation actually alters the apparent kinetics: for *n*th-order reactions and significant diffusion limitation the effect of changing the bulk concentration no longer follows the expected order based on intrinsic kinetics. It can be derived, see, for example, Levenspiel, that for a Thiele modulus >2, the intrinsic order n converges to an apparent order:

$$n_{\text{app}} = \frac{n+1}{2}$$

This simple relation holds for $n > -1$.

So only for the "special" case of a first-order reaction the apparent order remains equal to the intrinsic order! But for all other orders, the presence of pore diffusion limitations changes the observed order. This can easily lead to incorrect conclusions with respect to the true kinetics because there are often easy chemical explanations for intrinsic orders of ½ or 0.

A simple example here would be a catalytic hydrogenation reaction with dissociative adsorption following a simple Langmuir–Hinshelwood mechanism with a general kinetic equation as follows:

$$R = \frac{k\sqrt{K\,C_{H_2}}}{1+\sqrt{K\,C_{H_2}}}$$

At relatively low surface coverages of hydrogen, the *intrinsic* kinetics will effectively be of order ½, whereas at high surface coverages the *intrinsic* order effectively is zero. For another type of Langmuir–Hinshelwood mechanism/kinetics, with a different rate limiting step, one can effectively get:

$$R = \frac{k^*\,K\,C_{H_2}}{\left(1+K\,C_{H_2}\right)^2}$$

The $n_{\text{app}} = (n+1)/2$ relation shows that a zero-order reaction becomes order ½. An order of minus 1 becomes zero order. So an observed order of ½ can be stemming the intrinsic ½ order kinetics of the first expression at low hydrogen coverages or from the zero-order intrinsic kinetics at high hydrogen coverages under diffusion limited conditions. A zero-order observation can be consistent with the high hydrogen concentration cases of both the second expression with order minus 1 under diffusion limited conditions or the intrinsic kinetics zero-order kinetics of the first expression.

5.5.6 Effect on apparent activation energy

In the previous section we explained that in general the presence of pore diffusion limitations not only can change the apparent activity but also the apparent reaction order. Only for first-order reactions the order does not change. What does change for all orders is the apparent or observed activation energy.

In a nutshell: under conditions with significant diffusion limitation the apparent activation energy will become half of the true (intrinsic) activation energy.

Qualitatively, this apparent lowering of the activation energy, or rather the temperature dependency of the rate of reaction, is easy to understand. In case there is some significant pore diffusion limitation to begin with, increasing the temperature will obviously increase the intrinsic rate of reaction. For a first-order reaction, the rate constant k will increase following classing Arrhenius type of relation:

$$k = k_0 e^{\frac{-E_a}{RT}}$$

with E_a the intrinsic activation energy. This is an exponential increase in intrinsic rate. The rate of diffusion however will only increase slightly because diffusion coefficients are only slightly temperature-dependent. Hence, with increasing temperature the relative importance of pore diffusion increases. In terms of the Thiele modulus it increases and hence the effectiveness factor decreases (for positive order kinetics). So in the absence of diffusion limitations a temperature increase of, for example, 10°C may give a doubling of the reaction rate. In the presence of diffusion limitations, the increased intrinsic reaction rate causes even more diffusion limitation. The same 10° temperature increase may then only yield, for example, a 60% increase in conversion rate instead of a doubling.

We started this section with the nutshell summary on the effect of pore diffusion limitation on the apparent (observable) activation energy: it essentially becomes half the intrinsic activation energy. This can be now by easily reconciled by revisiting the simple expression for the apparent first order rate constant $k_{app} \sim \sqrt{kD_{eff}}$. If we neglect the temperature dependency of D_{eff}, we recall from high school that taking the square root of an exponential corresponds to dividing the exponent by 2. In other words, the exponent E_a of the classic Arrhenius intrinsic rate constant becomes $E_a/2$.

Figure 5.8 shows how pore diffusion limitation would show up in a standard Arrhenius plot:

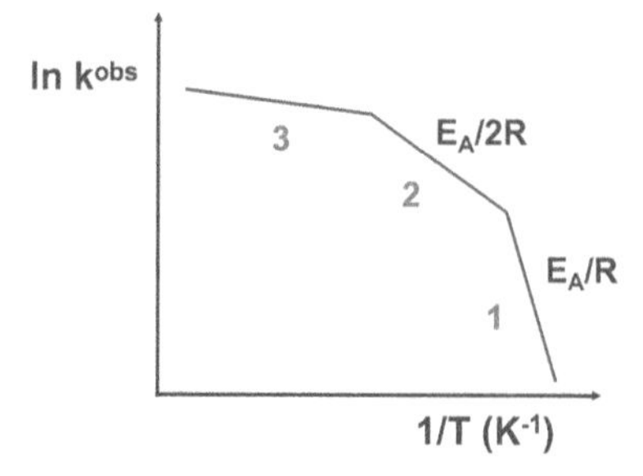

Figure 5.8: Effect of diffusion limitation on an Arrhenius plot.
Region 1: intrinsic kinetics; 2: pore diffusion limitation; 3: external mass transfer limitation.

In region 3, the reaction is so fast that besides pore diffusion limitation also the external resistances start to become important. Since in general there will be no reaction in the gas phase this is a resistance in series. In the limit the observed rate becomes proportional to the maximum driving force, that is, the difference in concentration or partial pressure of the reactant(s) in the bulk and at the interface, that is, the *external* surface of the particle. That latter will approach zero. Hence the rate will become proportional to the external mass transfer k_G (or k_L for liquid phase systems): $R_{obs} = k_G \Delta P_i = k_G P_i$. The temperature dependency of this rate is essentially the temperature dependency of the diffusion. For gases it increases with $T^{3/2}$. This typically translates to an "activation energy" that is very much lower than that of the reaction, for example, only 5–15 kJ/mol, depending on the operating temperature.

5.5.7 Effect of particle size and fluid velocity

The Thiele modulus already nicely illustrates how important the catalyst particle size is. For Thiele modulus >2 it directly follows that the observed rate becomes inversely proportional to the particle size. As long as there are no external mass transfer limitations, the observed reaction rate will not be, as expected by any chemist, affected by the velocity of the gas or liquid flowing around particles. The diffusion inside the pores will not be influenced by that fluid velocity. However, this can change if also external mass transfer becomes significant because in general the mass transfer coefficient will depend on the particle Reynolds numbers. At very low flow, the effect of fluid velocity is still very limited. At higher velocities, typically the mass transfer increases according to correlations for fixed and fluid beds: $Sh \sim Re^x$, where x can range from 0.5 to 1.

For a constant mass transfer coefficient ($x = 1$), it is easy to inspect that the observed rate will become proportional with the total available external area. If the particle size increases, obviously the area increases too, but the area per unit volume of particle decreases. Hence, the volumetric rate will again become inversely proportional to the particle diameter: the specific area of for example a sphere is $6/d_p$.

Table 5.2: Summary of observed parameter values for kinetics, diffusion-limited, and external mass transfer-limited cases (adapted from course notes by Prof. Guy Marin).

	Observed reaction order	**Observed activation energy**	**Observed effect particle size d_p**	**Observed effect fluid velocity v**
Intrinsic kinetics	n	E_a	0	0
Diffusion limited	$\frac{n+1}{2}$	$\approx \frac{E_a}{2}$	$\frac{1}{d_p}$	0
External mass transfer limited	1	0–20 kJ/mol	$\frac{1}{d_p^{2-x}}$	v^x

Table 5.2 (by courtesy of Professor Guy Marin) is a summary of observable effects of transport limitations for an *n*th-order catalytic reaction in a porous catalyst particle. The external mass transfer coefficient is according to $Sh \sim Re^x$, with x ranging from 0 to 1. For fixed beds at $Re_p > 5$ this exponent is typically 0.5, see Westerterp et al. [7].

It should be noted that Figure 5.8 is indicative for catalysts where the active material is more or less evenly distributed over the internal catalyst surface. Then internal (pore diffusion) limitation will kick-in before external resistances can become significant. The simplest rationale for this is that the effective diffusion coefficient is always (much) lower than the molecular diffusion coefficient. For wide pore systems this is due to the effect of porosity and tortuosity. For small pore systems the diffusion may party or completely take place in the Knudsen regime and hence leading to (much) lower diffusion coefficients.

Especially for catalyst with noble metal active material on a support a common solution to prevent inefficient use of the expensive catalyst is to distribute the metal nanoparticles only or predominantly near the external surface of the particles. These are called eggshell-type catalysts. For those systems external limitations may of course kick-in before any internal limitations.

5.5.8 Pore diffusion and catalyst design in terms of size and shapes

Careful selection or "design" of the macroscopic shape and size of catalyst particles for multiphase reactors is an area that requires very good collaboration between catalyst and catalysis experts and reaction engineers. For commercial scale application pore diffusion effects nearly always need to be considered, at least for fixed bed type of multiphase reactor. This is because at least three other key aspects would drive the particle size to larger particles, whereas pore diffusion would drive it to smaller particles. These three other aspects that are important from both a catalyst and reactor design perspective are:

- pressure drop: in general larger particles give lower pressure drop
- heat transfer: in general but not always, larger particles give better heat transfer
- catalyst strength: in general larger particles have a better (side crushing) strength

On catalyst particle strength

Referring to the "strength of a catalyst particle" is rather ambiguous as there are many different types of relevant strengths (e.g., flat plate crushing strength, side crushing strength and bulk crushing strength) and correspondingly different methods to measure it. A particle with a higher flat crushing strength may not necessarily truly stronger in actual applications. For example, such a larger particle with higher crushing strength may actually break more easily when loaded into a reactor, having to withstand the impact of falling several meters to the bottom of the reactor. This is especially relevant when – with increasing particle diameter – the particle length is increased. Commercial catalysts particles, for example, extrudates, often have a relatively high length to diameter ratio. With increasing length, which hardly affects the pore diffusion limitations, typically more breakage will occur despite the higher size crushing strength. In other words, optimizing the strength may be at least as difficult as optimizing for pressure drop or heat transfer.

If only these three were to be considered next to pore diffusion limitation, one would nearly always push toward to onset of where the detrimental effect of diffusion limitations become *economically* more important than the corresponding effects of particle size on pressure drop, heat transfer and strength. The word economically is key here. This may very well lead to a design where on purpose there are significant pore diffusion limitations. Commercial scale Fischer–Tropsch synthesis is just one of many examples here.

Hence, we recommend to always try to determine the effect of pore diffusion limitations either experimentally or using one of the criteria, for example, the Weisz–Prater criterion [1], which essentially uses the Thiele modulus approach as outlined above, but circumventing the problem that of course the intrinsic reaction rate needed for the Thiele modulus may not be equal to the observed reaction rate. Such a (preliminary) assessment serves as input for the optimization balancing the counteracting effects of particle size and shape on pore diffusion limitations, pressure drop, heat transfer, and strength. In specific cases, the particle size may also positively or negatively impact the apparent catalyst stability/lifetime as we will see in Example 5.5.1.

5.5.9 Example: the periodic table of the trilobes

Trilobes are very commonly used in industrial practice because this shape allows for a relatively good compromise between key properties and attributes such as surface to volume ratio, packing density, heat transfer, pressure drop, extrudability, and mechanical strength. In a patent application from 2008 by Shell researchers [8] defined three key geometrical parameters define a range of trilobe shapes. They used these to accurately describe the scope of their invention and to unequivocally define their patent claims to fulfill the so-called "clarity" criterion required for valid patent claims.

Figure 5.9 illustrates how the basic trilobe shape is considered to be an elongated shape particle comprising three protrusions whose cross-section occupies the space encompassed by the outer edges of six circles around a central inner circle. Three of these six circles will be called "lobe circles" and the other three "outer circles" and these alternate as depicted in Figure 5.9. The trilobe particle occupies the space of the inner circle and the three alternating lobe circles equidistant to the inner circle and the six interstitial regions and the particle NOT occupying the space of the three alternating outer circles. The exact size and shape of the trilobe can now be defined by the radii of the central circle, the outer circle and of the lobe circle, respectively. When these radii are defined this in turn define a *nominal size* as depicted in Figure 5.9a.

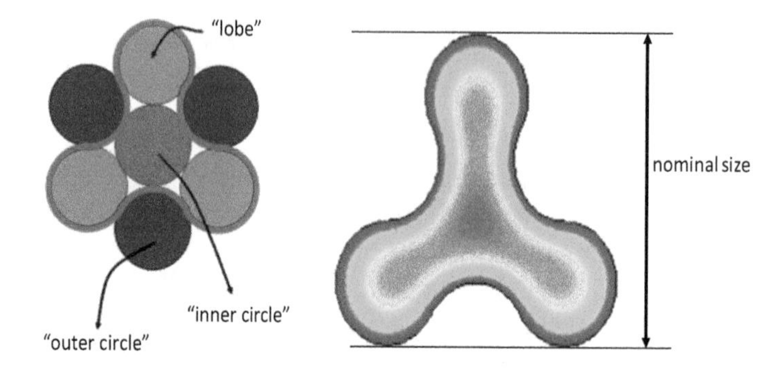

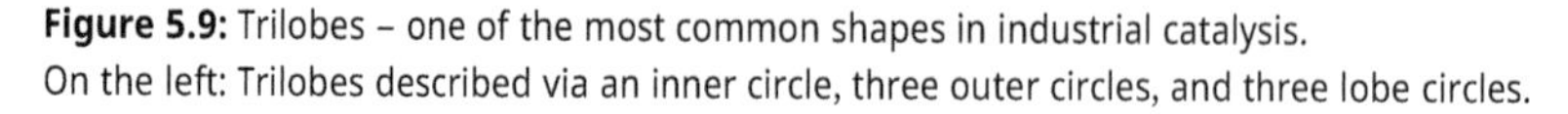
Figure 5.9: Trilobes – one of the most common shapes in industrial catalysis.
On the left: Trilobes described via an inner circle, three outer circles, and three lobe circles.

The colors in Figure 5.9 qualitatively illustrate the profiles of reactant or product concentrations in case of diffusion-limited reactions in such trilobe particles.

The question the researchers posed themselves was: How can these four parameters inner circle, outer circle, lobe circle and nominal size be used to graphically depict the whole "family" of trilobes?

First of all, to describe the whole family we can first separate the particle "size" from its "shape" by having the size determined by the "nominal size" (this is arbitrary but a very convenient and natural choice). Second, the particle *shape* is fully defined by two ratios of the three remaining parameters, for example, the ratio of the radius outer circle/radius lobe and the ratio inner/radius lobe.

For example, if we choose both these ratios to be 1, we obtain the trilobe depicted in Figure 5.9b (arbitrary size). If we choose both these ratios to be 0, that is, no inner and no outer circle, we obtain the more classical trilobe often denoted as TL.

If one would increase the radius of the outer circle, the particle gets a more triangular shape and at very high values of this ratio there are effectively no "lobes" left. At increasing ratio inner circle/radius lobe the shape becomes more and more like three spherical particles held together by a central part.

For both ratios, positive values that make sense are typically in the range of 0–2. Using those ranges for sensible combinations the shapes are depicted in Figure 5.10.

We would like to refer to this as "The periodic table of the trilobes."

Of course, these are not exhaustive ranges, and in fact the ratio inner circle/radius lobe can also be negative and for a value of −1 a simple sphere is obtained; see Figure 5.10b which may be called the *rare elements* of the periodic table, where clearly some of these are not that "rare" at all.

A brief final comment about the colors in these figures: these illustrate the profiles of reactant or product concentrations in case of diffusion limited reactions in such trilobe particles as calculated by solving the reaction-diffusion equations numerically. It should be remembered that in this periodic table we had deliberately eliminated the nominal size, that is, in effect for the concentration profiles the nominal size has been

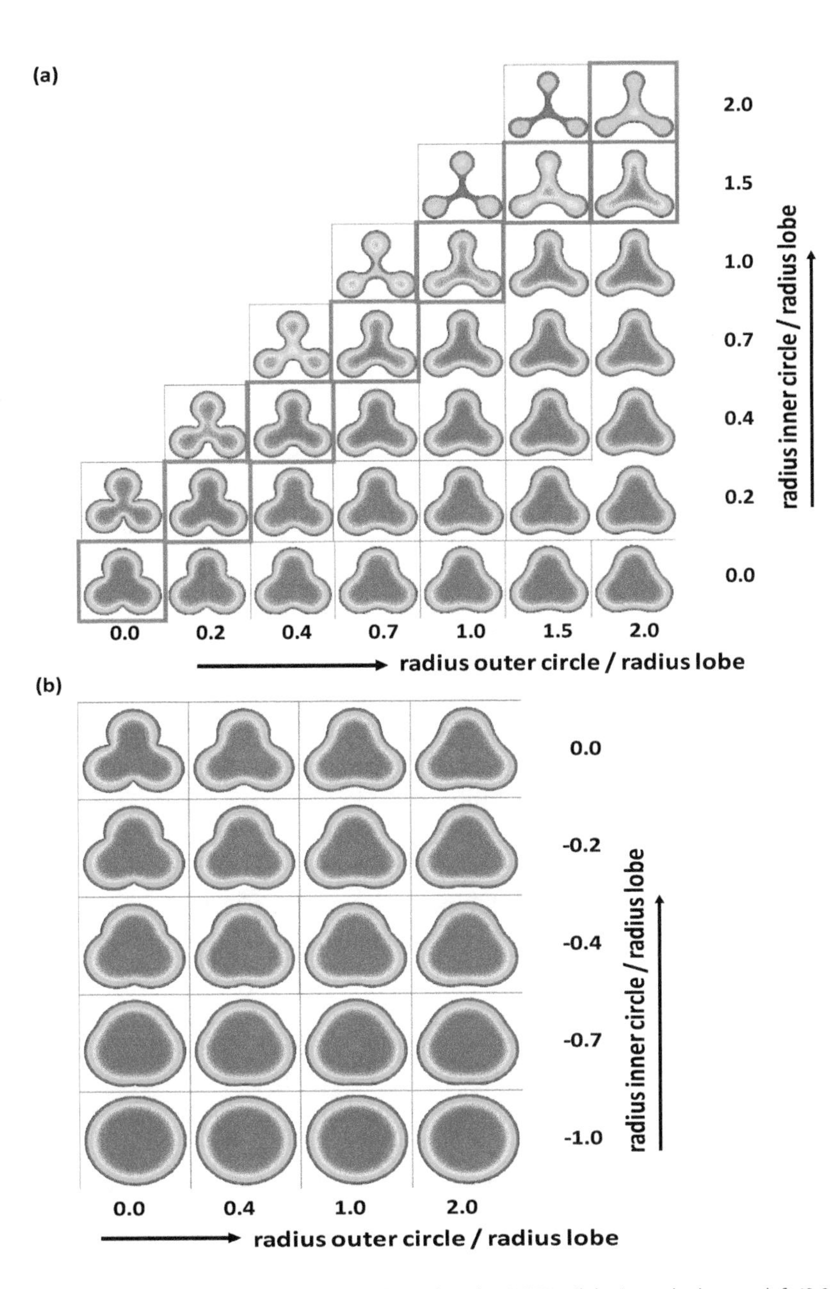

Figure 5.10: (a) The periodic table of trilobes. The "classic" TL trilobe is on the bottom-left (0.0–0.0). (b) The *rare elements* of the periodic table of trilobes.

kept constant. Hence the catalyst volume per particle varies considerably and differences in effectiveness factor deduced from this should not be solely attributed to shape effects. To better assess the shape effect one could repeat the reaction-diffusion calculations where for each shape the nominal size is adjusted to either maintain a constant particle volume (constant mass) or to keep the V_p/A_p constant.

5.6 Exercises

5.6.1 Industrial exercise 1: catalyst particle size and shape for the dehydration of MPC

The gas phase dehydration of methyl phenyl carbinol (MPC, 1-phenylethanol) is an important step in Shell's SM/PO process: coproduction of styrene (SM = styrene monomer) and propylene oxide (PO).

The main reaction of this dehydration step is MPC = SM + H_2O

The reaction is quite endothermic (adiabatic T-rise = −60 °C); therefore a multitubular reactor with "Hot-Oil" of around 250–300 °C for heat input is used. Shell uses an alumina-based catalyst for this and the reaction takes place at/below atmospheric pressure in the gas phase. Shell operates currently at six SMPO plants, and hence a lot of such multitubular reactors, consuming some hundreds of tons of catalyst per year.

The reaction is quite fast and in the course of studying a new version of the catalyst a brief kinetic study was done using cylindrical particles of 5 mm diameter and 5 mm length by measuring under isothermal conditions the conversion as function of the space time by changing the gas hourly space velocity over a wide range both by varying the flow rate and by loading different amounts of catalyst in the lab scale reactor.

A virtually perfect first-order kinetics in MPC was observed.

As it was suspected that the reaction was limited by pore diffusion, we repeated the experiments with crushed particles. Again, the derived kinetics showed a perfect first-order behavior. However, the first-order rate constant found from these experiments was a factor 10 higher.

When even finer crushed particles were used, no further increase in reaction rate was observed.

The chemist then suggested: "we should use crushed catalyst in the plant, this will clearly give you much better activity and thus higher conversion."

The plant engineer said: "well, we cannot do that because particles smaller than 5 × 5 mm will give a higher pressure drop and we cannot cope with that in the plant."

The reaction engineer then said: "perhaps we should have 5 × 5 mm cylinders with a hole in them? (i.e. a hollow cylinder or a ring)."

Questions

Q1: What are the pros and cons of this idea?

Q2: How would you choose the size of the hole?
- What phenomena are to be considered; which ones are dominating?
- Make an estimation/calculation, then select a hole size.

5.6.2 Industrial exercise 2: diffusion and deactivation for bimodal pore size distributions

A far less known and described potential effect of pore diffusion limitation is that on the deactivation of catalysts. This holds in general, and the effect can be positive or negative. In this example we focus on a case where the effect of pore diffusion is changed by changing the pore size distribution.

As described in Exercise 5.6.1 the catalytic dehydration of MPC over an alumina catalyst at industrial conditions can suffer from diffusion limitations. A general option to enhance the (effective) transport within the pores of a catalyst is to design the catalyst pore structure to have a bimodal pore structure, for example, by means of a structure as schematically depicted in Figure 5.11.

The beneficial effects of a bimodal pore size distribution for the MPC dehydration of the previous example have recently been described in a patent application [8]. There the inventors prepared a number of alumina catalysts both with and without a bimodal pore size distribution. Figure 5.11 shows a typical monomodal and a typical bimodal pore size distribution of these catalysts.

Alumina catalysts and/or carriers having a monomodal pore size distribution and a desired specific surface area can be prepared by starting with for example the commercial Alumina SA6x75 and firing that to a temperature in the range of

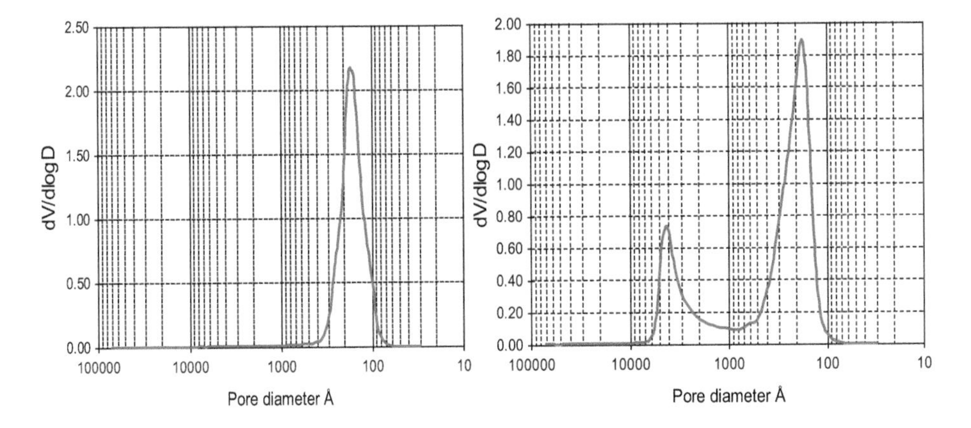

Figure 5.11: Bimodal pore size structure of catalyst.

from 900 to 1,060 °C to produce a specific surface area in the range of 80–140 m^2/g. Alumina catalysts and/or carriers having a *multimodal* pore size distribution and a desired specific surface area can be prepared by starting with a high surface area multimodal alumina carrier such as SA6x76 and calcining that at the appropriate temperature range.

The dehydration of MPC (1-phenylethanol) under industrial conditions is in fact accompanied by the parallel dehydration of 2-phenylethanol (sometimes abbreviated to BPEA or 2-PE) which is also present in the feed stream. In Table 5.3, the tables of the patent application are reproduced [9]. The performance tests were done in an in an isothermal plug flow reactor at conditions similar to those mentioned in the previous example.

Table 5.3: Data for the MPC dehydration catalyst [9].

Catalyst	A	B	C(*)	D(*)
Type of pore size distribution	Multimodal	Multimodal	Monomodal	Monomodal
MPD_V (nm)	19.7	27.6	15.0	18.2
Pore diameter (nm) at highest peak in range of 0–100 nm	14.2	18.2	14.2	17.3
Pore diameter (nm) at highest peak in range of >100 nm	576.0	480.0	No peaks	No peaks
Total pore volume (ml/g)	1.04	0.92	0.70	0.63
% of total pore volume in pores having a diameter of 0–100 nm	71.2	70.4	98.4	96.3
% of total pore volume in pores having a diameter of >100 nm	28.8	29.6	1.6	3.7
Surface area (m^2/g)	132.0	89.3	126.0	98.6
Particle shape	Rod-like; hollow quadrilobal cross section	Solid cylinder	Solid cylinder	Solid cylinder
Particle length (mm)	5.4	5	5	5
Particle diameter (mm)	5.9/5.0 (1)	5	5	5

Table 5.3 (continued)

Catalyst	**A**	**B**	**C(*)**	**D(*)**
Particle bore diameter (mm)	1.5	No bore	No bore	No bore
Side crushing strength (N)	84	55	104	119
Bulk crushing strength (MPa)	1.1	1.2	>1.6	>1.6

(*) Comparative catalyst; MPD_v, median pore diameter calculated by volume; (1), both include the bore; the second does not include the quadrilobal extensions.

Catalyst	**A**		**B**		**C(*)**		**D(*)**	
Conversion of *X* (%)	*X* = 1-PE	*X* = 2-PE	*X* = 1PE	*X* = 2-PE	*X* = 1PE	*X* = 2-PE	*X* = 1-PE	*X* = 2-PE
At *t* = 8 h	99.8	53.1	98.8	45.8	98.7	52.0	98.6	50.5
At *t* = 70 h	99.5	45.7	97.7	36.5	86.1	30.1	85.9	29.4
% decrease	0.3	13.9	1.1	20.3	12.8	42.1	12.9	41.8
Selectivity (%)								
At *t* = 8 h	94.9		95.9		95.3		94.9	
At *t* = 70 h	95.4		96.5		94.9		95.0	

Question

Q1: What are the most striking results?

Below our answer, that is, our briefest summary:

First of all, it can be noted from the data in Table 5.3 that the rate of dehydration of 2-PE is very much lower than that of MPC and hence, in case MPC dehydration is only moderately diffusion limited, the dehydration rate of 2-PE is virtually nondiffusion-limited. As demonstrated in the previous example this is indeed the case and hence under these conditions the dehydration of 2-PE can be regarded as an indicator of the intrinsic activity of the catalyst.

Comparing the results of the monomodal catalyst with those of bimodal catalysts it can be seen that the conversion of 2-PE was virtually the same if also the surface area was approximately the same (catalysts A and C). This supports the assumption that the intrinsic activities of the mono and bimodal catalysts are approximately the same.

The "fresh" catalysts, that is, after just a few hours' time-on-stream, showed under the testing conditions very high conversions of MPC: 98.8% and 98.5% for the bimodal and monomodal catalyst, respectively. However, after 70 h time-on-stream, the MPC conversion of the monomodal catalyst has dropped from 98.5% to 85.9%, which represents an (apparent) activity loss of more than a factor of 2. In contrast, the MPC conversion for the bimodal catalyst was only slightly lower than that of the fresh catalyst (97.7%) which represents a loss of only around 10%.

Careful analyses confirmed that indeed this extension of the practical lifetime of the catalyst, or rather the increase in the time intervals between required in-situ catalyst regeneration, can indeed be attributed to the bi- or multimodal pore structure of said catalyst.

This MPC example is just an illustration of the less widely known effects of bimodal pore structures on catalyst deactivation and hence of the beneficial potential for catalyst "lifetime" or the required frequency of catalyst regeneration. These two are often important economic drivers and can even become more important than the initial catalyst activity itself.

5.7 Takeaway learning points

Learning point 1
Next to RTD, proper understanding of the interactions between mass transfer and chemical reaction is key to any rational chemical reactor analysis, design and operation.

Learning point 2
Apparent kinetics, both in terms of reaction rates, reaction orders and activation energies may be quite different from the true (intrinsic) reaction rates.

Learning point 3
Reaction and mass transfer in series is fundamentally different from reaction and mass transfer in parallel.

Learning point 4
Many reactors operate under conditions that are at least partly limited by mass transfer. And this may in fact be the "optimal" from an economic or sustainability point of view.

References

[1] Froment GF, Bischoff KB, Chemical Reactor Analysis and Design, 2nd ed. Singapore, John Wiley & Sons, 1979.

[2] Aris R, On shape factors for irregular particles – I: The steady state problem. Diffusion and reaction. Chemical Engineering Science, 1957, 6, 262.

[3] Thiele EW, Relation between catalytic activity and size of particle. Industrial & Engineering Chemistry Research, 1939, 31, 916.

[4] Bischoff KB, Effectiveness factors for general reaction rate forms. AIChE Journal, 1965, 11(2), 351–355.

[5] Wijngaarden RJ, Kronberg AE, Westerterp KR, Industrial Catalysis, Optimizing Catalysts and Processes. Weinheim, Wiley-VCH Verlag, 1998.

[6] Driessen RT, Kersten SRA, Brilman DWF, A Thiele modulus approach for nonequilibrium adsorption processes and its application to CO_2 capture. Industrial & Engineering Chemistry Research, 2020, 59 (15), 6874–6885.

[7] Westerterp KR, van Swaaij WPM, Beenackers AACM, Chemical Reactor Design and Operation. Chicester UK, John Wiley & Sons, 1984.

[8] Calis HPA, Verbist GLMM, Catalyst, catalyst precursor, catalyst carrier, preparation and use thereof in Fischer-Tropsch synthesis. Patent application by Shell Internationale Research Maatschappij, WO2008087149, 2008.

[9] Bos ANR, Koradia PB, Process for the preparation of styrene and/or a substituted styrene. WO 2009/ 074461, 2009.

6 Quantification of mass transfer in G-L(-S) reactors

The purpose of this chapter is to provide quantitative mass and heat transfer information for several common gas–liquid (G-L) and gas–liquid–solid (G-L-S) reactor types for designing these reactors in the concept stage. For industrial purposes this means providing information for conservative estimates for selecting and sizing the reactors. Information by which these conservative estimates can be made is usually sufficient.

6.1 Introduction

Mass transfer in multiphase reactors is about transport of components between phases described by mass transfer in series and about transport within a phase where also reaction takes place, called mass transfer and reaction in parallel as described in Chapter 5.

For a three-phase reaction system where one of the reacting components is in the gas phase and actual reaction takes place in the solid phase (a porous catalyst for instance) then the component has to be transported over the gas-side boundary layer, the liquid-side boundary layer of the gas–liquid (G-L) interface, the liquid boundary layer outside the catalyst particle and be transported inside porous catalyst particle. Each of these mass transfers can be described with transfer flux J (mol/m^3 s), by a concentration difference between the bulk and the interface ΔC (mol/m^3), with overall mass transfer coefficients k_m and interfacial area a, see Chapter 5 for more details:

$$J = k_m \, a \, \Delta C$$

All these fluxes in a steady state will have the same value as same to the reaction rate (of course with proper accounting for volume fractions the various coefficients and reaction rates are defined for). It is similar to heat transfer in a heat exchanger where heat is transferred from the bulk of a fluid over a boundary layer to the heat exchanger wall, then is transferred through the heat exchanger tube wall by conduction to the other side, and then via a boundary layer to the bulk of the second fluid. The local temperatures at both sides of the heat exchange surface will adjust themselves until a steady state is established. The same happens with the concentrations at the interfaces.

If the mass transfer coefficients are very high, then concentration differences over each boundary layer will become very small. This situation is called no mass transfer limitations. This means the concentration in the reaction liquid or reaction catalyst will be in equilibrium with the gas phase concentration, and the observed reaction rate will be the same as the kinetic reaction rate.

https://doi.org/10.1515/9783110713770-006

This is the preferred option for the process designer for two reasons. The first reason is that the maximum overall reaction rate is obtained, and the second reason is that accurate information about mass transfer coefficients not needed.

6.2 Mass transfer coefficients and Sherwood numbers

The mass transfer coefficient k depends on the local flow defined by a Reynolds number (Re) and a Schmidt number (Sc) accounting for molecular diffusion over viscosity effects. General expressions for k are based on dimensionless number correlation with Sherwood number ($\mathrm{Sh} = k\, d/D$) as a function of Re and Sc. The general shape is

$$\mathrm{Sh} = \mathrm{Sh}_{\mathrm{min}} + c\,\mathrm{Re}^a \mathrm{Sc}^b$$

Textbooks for a long time stated $\mathrm{Sh}_{\mathrm{min}} = 2$, the theoretical number for mass transfer to a spherical particle by diffusion only. However, for Reynolds numbers below 10 experimental results showed that this may not be the case in reality.

On Sherwood numbers well below the "theoretical" minimum value

It is known for a long time that measurements of mass transfer coefficients in packed and fluidized bed reactor have yielded values of Sherwood orders of magnitude lower than 2. It occurs at low Reynolds numbers, typically when Re drops below 10–50. This was also found for heat transfer and the corresponding Nusselt number. This phenomenon is still not well understood; some explanations suggest a flaw in the various measurement techniques, and other authors pointed out that (very) low Sh or Nu numbers were found by different techniques including local particle-scale measurements. The interested reader is referred to Wijngaarden and Westerterp [1] and to recent work by the group of Professor Lynn Gladden [2].

So, provided the Reynolds number is not too low (say $\mathrm{Re} \gg 10$) the Sherwood number can be conservatively estimated by $\mathrm{Sh} = 2$. For many concept designs in the early concept stage this is sufficient. However, the relevant dimension of the object with which the transfer takes place has to be known: this is the key variable d in the Sherwood number. For solid particles that is relatively easy, but for a gas bubble, or a liquid droplet it is not. For those cases the size may depend on the fluid turbulence and on the interface tension. Moreover, there may be a wide size distribution.

Lab-scale experiments are advocated to observe bubble or droplet sizes and use these as a first estimate of actual sizes in the commercial-scale reactor, keeping in mind that at scale-up this may change significantly.

The second parameter is the specific surface area a. For surface areas of solid particles that is relatively easy. For a sphere $a_p = 6/d_p$, with d being the particle diameter. This is the specific area per volume of particle, not reactor. If the specific surface area is needed per unit reactor, that is, fluid plus particle volume together, then $a = \varepsilon\, 6/d_p$, in which ε is the volumetric holdup fraction of the solids over fluid plus solids.

For a G-L system, conventionally the area a is provided in square meter of interfacial area per cubic meter of reactor volume, excluding any "empty" volume, for example, above the liquid level in a stirred slurry reactor. However, sometimes the interfacial area is actually measured or reported per cubic meter of liquid phase only. Then for the total flux calculation we again need to correct it for the actual liquid-phase holdup, just as shown above for the case of solid particles.

6.3 Quantified mass transfer two- and three-phase bubble columns

Gas–liquid mass transfer coefficients in bubble column reactors are hard to predict accurately. This is due to the complex fluid flow behavior of gas and liquid. For this textbook we concentrate on the concept stage where accurate designs are not needed. A conservative design approach is sufficient for sizing the column reactor.

We start with vertical columns (height/diameter ratio is larger than 1) and for the so-called homogeneous bubble regime. Here bubbles are formed at the gas sparger. They may then rise as individual bubbles and leave the liquid at the top or coalesce to somewhat larger bubbles. G-L interface tension plays a role here, as well as liquid viscosity, the liquid film in between the two coalescing bubbles has to move away. Accurate mass transfer correlations for bubble columns, due to this coalescing behavior, are not available. Available correlations should be used for order of magnitude estimates. These can then be used to determine of mass transfer limitation that are likely to occur for the actual application or not. If the reaction is so slow that mass transfer limitation is unlikely to occur than the design may proceed, and pilot plant tests can be carried out to confirm the absence of mass transfer limitations.

For the heterogeneous regime some bubbles coalesce to flat-shaped large bubbles wiggling upwards through the liquid, while also small bubble flow up and downward in the column. This happens when the superficial gas velocity is above 0.1 m/s (for air–water systems and other low viscosity liquid systems) [4]. It depends on the actual system whether these large flat bubbles exchange mass with the small bubbles quickly or hardly. As a conservative estimate only the large flat bubbles can be taken for G-L mass transfer. Their typical Sauter diameter is 5×10^{-3} m (for air–pure water systems) [3–5]. With Sherwood = 2 as a conservative estimate the gas side and the liquid side mass transfer coefficient can then be estimated.

The specific surface area per unit bubble volume can be directly determined from the Sauter diameter with $a_{\text{bubble}} = 6/d_{\text{bubble}}$.

The specific surface area per unit reactor fluid volume is then given by

$$a_{\text{r,fluid}} = a_{\text{bubble}}\ \varepsilon_{\text{b}}$$

where ε_{b} is the bubble holdup per unit reactor fluid.

The averaged bubble holdup for the heterogeneous bubble flow regime is provided by Raimundo et al. [3, 4]:

$$\varepsilon_b = 0.49\, V_{sg}^{0.49}\, D^{-0.047}$$

Raimundo et al. [3, 4] measured the bubble holdup for air–water systems. So the correlation only holds for similar liquid density and viscosity values. The importance of his correlation is that it shows the effect of column diameter. Raimundo measured this for four different column diameters with the largest diameter of 3 m.

Table 6.1 shows value ranges of mass transfer parameters for conventional bubble columns. If with choosing conservative estimates the reactor is not mass transfer-controlled for the gas-to-liquid phase, then those estimates may be good enough.

Table 6.1: Typical parameter values for commercial-scale bubble column reactors.

Parameter	Dimensions or expression	Value range
k_L	$m_f^3\ m_i^2/s$	5×10^{-5}–5×10^{-4}
a_V	m_i^2/m_L^3	10^2–10^3
a	m_i^2/m_r^3	0.5×10^2–10^3
D_A	$m_f^3/m_L\ s$	10^{-9}–10^{-8}
Sh_m	$k_L/(a_V\ D_A)$	10^1–5×10^3
Ha	$(k_V\ D_A)^{0.5}/k_L$	10^{-2}–10^0
$k_L a$	$m_f^3/m_r^3\ s$	10^{-2}–10^{-1}

For three-phase bubble columns the same reasoning holds but, in some cases, fine solid particles move into the G-L interface and stay there. They can then hamper the G-L mass transfer by forming a barrier for transport. They also can cause foaming. The liquid holdup will be greatly reduced, and the gas liquid separation at the top of the column will not work anymore; hence it will obstruct the reactor performance. Solving this foaming problem is beyond this textbook.

For nonfoaming systems, the Sherwood = 2 approximation in combination with observed bubble size in a "hot" pilot plant with the real fluids and particles present will be a workable approach. So experimental pilot plant experiments for three-phase reactors are even more essential than for G-L bubble columns.

6.3.1 Gas–liquid mass transfer in horizontal bubble columns

For horizontal bubble columns, so columns that are placed horizontally and with a length/diameter >1 the mass transfer is even more complicated. Part of the bubbles flow upwards along the cylindrical wall and part of the bubbles flow through the central part. The liquid section in between these two bubble streams is void of bubbles and can be starved of the component supplied via the gas phase.

Klusener et al. [6] have performed RTD measurements for the gas phase and for the liquid phase of a commercial-scale horizontal bubble column reactor with diameter of about 5 m and a length of about 20 m. The reactor was furthermore equipped with vertical baffles and heating/cooling coils. They also made a CFD model for the gas and liquid flow. The CFD model reveals that bubble leaving the sparger near the cylinder wall moved along the wall, while the bubbles leaving the center part moved vertically upward. The RTD measurements of the actual reactor confirmed this bubble flow behavior [6]. By changing the baffles and gas spargers as described in a patent [7], the starvation area could be largely prevented. An elaborate description of this case is found in Section 13.3.

6.3.2 Liquid–solid mass transfer in three-phase bubble columns

In three-phase bubble columns fine particles are applied. The particles can be porous catalysts. The mass transfer in the boundary layer surrounding the particle can be conservatively estimated with Sherwood number is 2.

For reactions inside the porous catalyst with no significant internal diffusion limitation (Thiele modulus <0.3) the outside mass transfer will also not be limiting because the diffusion coefficient in the liquid outside the particle will at least be a factor 3 larger than the diffusion coefficient inside the porous catalyst, see Chapter 5. For cases with internal diffusion limitations also outside mass transfer limitation can occur. As the fine particles in general move with the liquid turbulent eddies with a small slip stream the Sherwood number will only be a little bit higher than 2 and will depend on the liquid flow pattern and turbulence intensity and liquid viscosity. The reader is referred to literature on the subject for more information.

6.3.3 Shear rate distribution commercial scale on bubbles and droplet size distribution

The shear rate distribution of fluids in commercial-scale reactors is in general not uniformly distributed. This is particularly the case for mechanically stirred reactors, where the stirrer blades' shear rates can be very high causing very small bubbles or droplets. Far away from the stirrer that shear rate is often much lower. Bubbles or droplets can coalesce there, forming larger bubbles or larger droplets. This ratio of high shear rate and low shear rate is higher for the commercial scale than the pilot plant. This can cause a lower interface surface area in the commercial scale and thereby a lower mass transfer coefficient, which in turn can cause a lower production capacity than designed for.

With modern fluid flow modeling the shear rate distribution of large-scale reactors can be reasonably well predicted. However, droplet breakup and droplet coalescence can in general not be well predicted by models. Trace components in the commercial-scale reactor can have a preference for the G-L interface thereby reducing the interface tension by which bubbles coalesce faster or thereby hampering coalescence causing fine bubbles causing foaming or frothing inside the reactor, hampering normal processing. Measuring bubble coalescence behavior of the real reactor fluids at laboratory scale in combination with fluid flow modeling can help to estimate these potential problems.

6.3.4 Particle (catalyst) breakage and attrition

Particle breakage in commercial-scale multiphase reactors is often causing problems in the reactor and in downstream processing. This problem is often caused because it was not highlighted in the development program and implicitly it was assumed not to be a scale-up issue. However, in large-scale fluid bed, bubble columns and mechanically stirred reactors, the particles containing phase move inside the reactor with far higher circulation velocities, often several meters per second, by which the impact of walls and internals on the particles is far higher. This can cause particle attrition resulting in very fine material, slipping through separators and ending up in process sections where it should not be. If the material is catalytically active, it can cause an unwanted reaction at locations where this is a disaster for product quality and process integrity.

Hence fluid flow shear rates of the commercial-scale reactor on particles should be estimated in the development stage and counteracting measures should be taken to avoid particle attrition and breakage.

In commercial-scale fixed bed reactors the forces on particles in the bottom section can exceed the bulk crushing strength. The particles are then crushed into finer particles, which can cause increased pressure drop and fluid flow obstruction. In the development program the bulk crushing strength of particles should be measured and also the pressure on the particles, caused by the catalyst bed weight and the fluid flow pressure drop should be determined. From that information counteractive measures should be taken.

6.4 G-L-S mass transfer in trickle-bed reactors

We only treat here commercial-scale catalytic trickle-bed reactors. In these trickle-bed reactors liquid trickles in thin films over the porous particles. Gas moves in the remaining voids co-currently down. If the diffusion limitation of reacting components inside the catalyst particles is absent (Thiele modulus <0.3) then the mass transfer on the outside the particles in the flowing liquid and gas is also absent. This is easily understood by knowing that the diffusion coefficients outside the porous particles are higher than inside the particles and the penetration thickness in the liquid film and in the gas film is similar to the particle diameter. So if pore diffusion limitation is negligible, then also mass transfer limitations outside the particles must be negligible.

Even for cases where some mass transfer limitations inside the porous catalyst particles exist (Thiele modulus in between 0.3 and 1) mass transfer limitations in the flowing gas and liquid will be negligible as the penetration thickness through the boundary layers of gas and liquid will be much thinner than the particle size. Due to the flow at sufficiently high Reynolds number the Sherwood number will be much higher than 2 and even the diffusion coefficients of the components are at least a factor 4 higher than the effective diffusion coefficient inside the porous particle (see Section 6.7).

The only concern for mass transfer limitations in trickle-bed reactors is that the flowing liquid does not reach all particles due to a poor liquid distributor with insufficient number of feed points per square area. The number of feed points should at least be 1600/m^2 (every 2.5 cm a feed point). We have seen a commercial trickle-bed reactor provided by a world-famous engineering company with far too few feed points resulting in very poor reactor performance. The engineering company could also not find the course of the reactor performance problem. Only when a trickle-bed expert was consulted the problem was solved. Due to the poor performance 50 M$ was lost.

Another potential problem is incomplete irrigation of the catalyst particles by the liquid. For hydrocarbon liquids and conventional alumina or silica porous particles and superficial liquid velocities $>10^{-3}$ m/s the irrigation (wetting) according to Gierman wetting number [8 Gierman] will be sufficient. For other liquids laboratory experiments are recommended to determine the critical liquid velocity for complete irrigation.

There is also considerable amount of literature on flow regimes in trickle beds. For water–air systems, a two-phase flow regime map has been made (see Figure 6.1). For real systems using high pressure gas and hydrocarbon liquids of higher viscosity this map can only be used as a first indication.

However, the flow regime is not very important for commercial-scale trickle-flow reactor performances. For commercial-scale reactors with liquid velocities >1 kg/m^2 s and $L/d_p \gg 100$ mass transfer limitations will in general be absent and residence time distributions for the gas and liquid flows will also be nearly plug-flow, regardless of the flow regime. As all phenomena happen on the scale of particles, any local vertical maldistribution will be removed by mixing and mass transfer on the scale of particles.

The real concern of commercial-scale trickle-bed reactors is radial maldistribution. If the initial gas and liquid distribution is poor with not sufficient feed points per cross-sectional square meter, then the reactor will perform poorly. Radial mixing beyond the scale of particles is virtually absent. So, radial differences in flow will persist

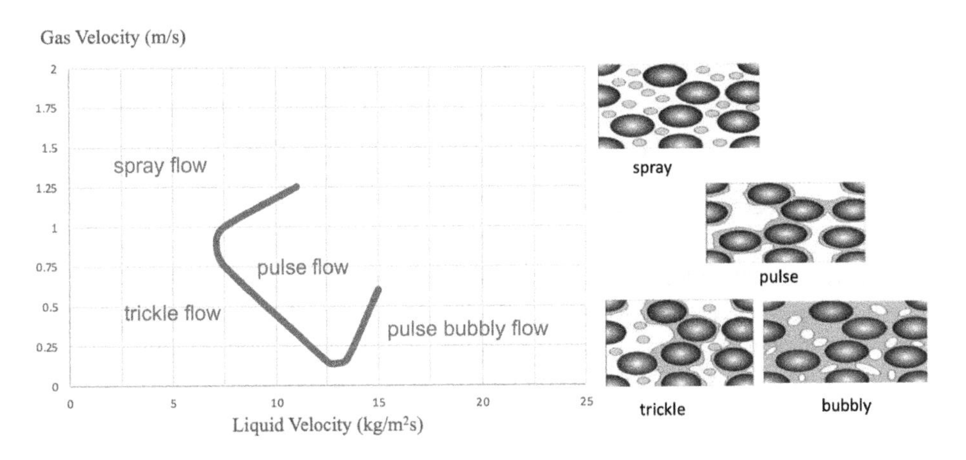

Figure 6.1: Two-phase flow regimes in trickle bed reactors. Adapted from Sie and Krishna [9].

throughout the bed. Poor conversion and selectivity and even local hot spots can occur in such a reactor.

6.5 Process intensification methods for interface transfer

Process intensification provides new concepts for high heat and mass transfer [10]. Here we highlight process intensification methods to increase interface mass transfer. As indicated above interface mass transfer has two parameters: the mass transfer coefficient k (1/s) and the specific interfacial area a (m^2/m^3).

6.5.1 Rotating reactors

The mass transfer coefficient is increased by thinning the boundary layer and by creating a velocity gradient in this boundary layer. The main way to increase interface mass transfer by orders of magnitude over conventional reactors is to increase the slip velocity of the phases involved and to decrease the thickness of the fluid layers. Both are obtained by increasing the gravitation acceleration parameter g by rotating the liquid phase(s). A rotating packed bed and a spinning disk, rotor–stator type. The latter reactor creates very high values of k and a by having thin fluid layers flowing counter-currently, due to the high gravity field and the shearing action of the spinning disk.

The rotating packed bed is used for adiabatic G-L applications. Details for the rotating packed bed are provided by Stankiewicz et al. [10] in their textbook on process intensification.

The rotating plate is applied for G-L, liquid–liquid, and also for three-phase applications and in particular for cases where also high heat transfer is required [10]. As the disk and the stator have chambers for heat exchange medium to flow through.

6.5.2 Other process intensified reactors

If mass transfer limitations in conventional reactors occur than process intensified reactors can be considered. Process intensified reactors for three-phase systems such as high-gravity field rotor–stator reactors orders of magnitude higher values of mass transfer coefficients and specific surface areas can be achieved [11, 12].

Process intensified reactor is a rapidly developing field with many two- and three-phase reactor types. It is beyond the scope of this book to describe these. The reader is referred to the textbook by Stankiewicz et al. [10] to find the description of most intensified reactors.

6.6 Exercises

Exercise 1: A three-phase trickle-bed reactor porous catalyst

Given: reactions 1: A + H = P 2: P + H = X

- Reactions 1 and 2 are catalyzed by a porous catalyst.
- Reaction 1 kinetics: $r_A = -\ k_1 C_H C_A; k_1 C_H = 2\times10^{-4}/s$
- Reaction 2 kinetics $r_P = -\ k_2 C_H C_P; k_2\ C_H = 5\times10^{-6}/s$
- Reaction rates are both defined on porous catalyst volume.
- The porous catalyst particles are spheres with a diameter of 2×10^{-3} m.
- The effective diffusion coefficients of A, H, and P inside the porous catalyst are: $D_{eff} = 10^{-10}$ m^2/s
- Component H is transferred from the gas phase to the liquid phase. Components A and P are in the liquid phase.
- The reaction is carried out in a trickle-bed reactor. Component H is supplied in large surplus to the reactor.

Questions

Q1: Do you expect significant mass transfer limitations of *A* and *P* inside the porous catalyst?

Q2: Do you expect significant mass transfer limitations of *A* and *P* in the liquid outside the catalyst particle?

The chemist finds a 10 times more active catalyst (with the same pore structure) for reactions 1 and 2.

Q3: Do you expect with this new catalyst mass transfer limitations of A and P inside the porous catalyst?

Q4: If the answer is yes, what do you expect to happen to the yield of P on A converted?

Q5: If the answer is yes, what could you do to improve the yield of P on A converted?

6.7 Takeaway learning points

Learning point 1

Mass transfer coefficients can be conservatively estimated with Sherwood = 2 for Reynolds >10.

Learning point 2

Specific G-L surface areas for bubble columns are hard to predict reliably from correlations.

Learning point 3

Specific G-L surface area values should be obtained experimentally.

Learning point 4

Commercial-scale trickle-bed reactors are not external mass transfer limited if no diffusion limitation takes place inside the catalyst particles.

References

[1] Wijngaarden RJ, Westerterp KR, Pellet heat transfer coefficients in packed beds: Global and local values. Chemical Engineering and Technology, 1993, 16, 363–369.

[2] Elgersma SV, Sederman AJ, Mantle MD, Gladden LF, Measuring the liquid-solid mass transfer coefficient in packed beds using T2-T2 relaxation exchange NMR. Chemical Engineering Science, 2022 Feb 2, 248, 117229.

[3] Raimundo PM, Cloupet A, Cartellier A, Beneventi D, Augier F, Hydrodynamics and scale-up of bubble columns in the heterogeneous regime: Comparison of bubble size, gas holdup and liquid velocity measured in 4 bubble columns from 0.15 m to 3 m in diameter. Chemical Engineering Science, 2019, 198, 52–61.

[4] Raimundo PM, Analysis and modelization of local hydrodynamics in bubble columns. Doctoral dissertation, Grenoble, France, Université Grenoble Alpes, 2015.

[5] Augier F, Raimundo PM, Effect of rheology on mass transfer and bubble sizes in a bubble column operated in the heterogeneous regime. The Canadian Journal of Chemical Engineering, 2021 May, 99(5), 1177–1185.

[6] Klusener PAA, Jonkers G, During F, Hollander ED, Schellekens CJ, Ploemen IHJ, Othman A, Bos ANR, Horizontal crossflow bubble column reactors: CFD and validation by plant scale tracer experiments. Chemical Engineering Science, 2007, 62(18–20), 5495–5502.

[7] Hollander ED, Klusener PAA, Ploemen IHJ, Schellekens CJ, Horizontal reactor vessel. Patent WO2006024655, 2006.

[8] Gierman H, Design of laboratory hydrotreating reactors: Scaling down of trickle-flow reactors. Applied Catalysis, 1988, Jan 1, 43(2), 277–286.

[9] Sie ST, Krishna R, Process development and scale up: III. Scale-up and scale-down of trickle-bed processes. Reviews in Chemical Engineering, 1998, 14(3), 203–252.

[10] Stankiewicz A, van Gerven T, Stefanides S, The Fundamentals of Process Intensification.Weinheim, Wiley-VCH, 2019.

[11] Meeuwse M, van der Schaaf J, Schouten JC, Multistage rotor-stator spinning disc reactor. AIChE Journal, 2012, 58(1), 247–255.

[12] Martínez ANM, Assirelli M, van der Schaaf J, Droplet size and liquid-liquid mass transfer with reaction in a rotor-stator Spinning Disk reactor. Chemical Engineering Science, 2021, 242, 116706.

7 Heat management

The purpose of this chapter is to highlight the important aspects related to heat management in a reactor. Due to the interaction between these heat effects and reaction kinetics, complex reactor behavior and design challenges arise.

7.1 Introduction

In this chapter we provide:
- Awareness of the importance of controlling reaction heat for reactor design and operation.
- Understanding of nonisothermal and runaway behavior of reactors.
- Design methods for heat control.
- Experimental guidelines for obtaining runaway behavior information of reaction systems.
- Information about the safe operation of reactors in view of counterintuitive behavior.
- An industrial case to give additional insight into nonisothermal behavior.

7.2 Theory nonisothermal behavior reactors

In Section 3.5 on "reaction runaway" we indicated the difference with one of the topics of this chapter: "reactor runaway." More general, in this chapter we mainly focus on single reactions with simple first-order kinetics in idealized reactors. From the relatively straightforward mass and heat balances for such systems quite complex phenomena can arise. Like in the rest of this book, rather than treating the mathematics from which these complexities can be understood as well, we focus on the concepts themselves and provide qualitative explanations.

On a very high level, the energy balance for any reactor is shown in Figure 7.1.

The main three aspects governing the nonisothermal effects, and thus design and operation, of nonisothermal reactors are:
- Thermodynamics: the reaction enthalpy, preferably expressed as the adiabatic temperature rise ΔT_{ad} in °C or K. This relates to the "Thermal power" box in Figure 7.1.

  This reaction enthalpy is not a constant but depends on the actual temperature T of the reactor. The enthalpy dependance is relatively strong for gas phase reactions. Reliable physical property data models are needed to account for this temperature difference. Experimental confirmation is needed if the reaction heat is strongly heat integrated with the whole process.

https://doi.org/10.1515/9783110713770-007

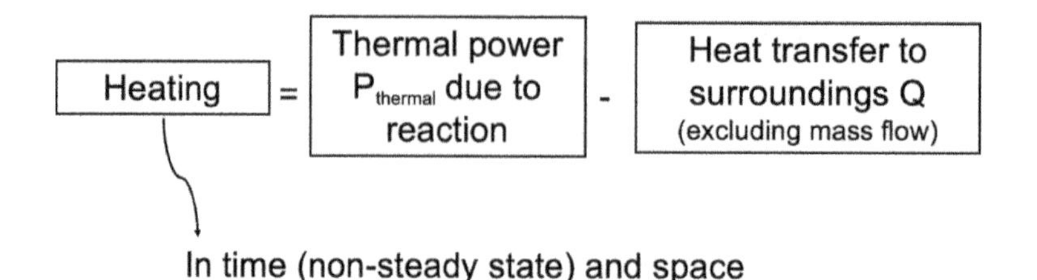

Figure 7.1: Conceptual energy or enthalpy balance for a nonisothermal reactor.

- Kinetics: the activation energy E_a (Arrhenius law): $r \sim e^{-E_a/RT}$. This also relates to the "Thermal power" box in Figure 7.1.

  The rates of reaction generally show a (very) strong nonlinear dependency on T.

- Heat transfer: the net heat transfer from the flow into and out of the reactor $F_m C_p$ $(T - T_0)$; the "heating" box of Figure 7.1 the net heat transfer to the environment (external cooling or heating): UA $(T - T_c)$; the "surroundings" box of Figure 7.1.

  In general, these show a near-linear dependency on T.

On the merit of adiabatic temperature rise

Why look at the "adiabatic temperature rise" even more so than reaction enthalpy?
Rather than a reaction enthalpy of −100 kJ/mol, from your boss's perspective and from a reactor stability point of view, that number does not mean much even both know very well what the definition of "reaction enthalpy" is! But a ΔH_r of −100 kJ/mol can correspond, without any cooling, to both just a 10 °C rise or 500 °C or almost any other positive number. Just mentioning the ΔT_{ad} will both make your boss or your reaction engineer happy (or not).

On diminishing sensitivity of increasing the temperature at higher temperature

An almost high school chemistry type of rule of thumb is that rates of reaction double with every 10 °C temperature rise. This is correct for reactions with a typical activation energy of 80 kJ/mol at 100 °C. However, at 500 °C for the same activation energy the reaction rate increases with less than 20% upon a 10 °C increase and an increase of almost 50 °C is needed for a doubling of the rate. At 1,000 °C a 10 °C increase only yield a 6% increase in rate. Therefore, at higher temperature levels, the strong nonlinearity of the reaction rate and hence of the heat production is diminishing.

Now let us just first look at the so-called phase plane of a simple batch reactor in which an exothermic reaction takes place, and that the reactor is completely adiabatic, so all the heat generated will go into the reaction medium. In such a phase plane the variable time is eliminated, and we just plot the temperature as function of conversion or vice versa. Because the reaction enthalpy is virtually constant, there is a unique and linear relation between the temperature in the batch reactor and the conversion (see Figure 7.2a).

In Figure 7.2b, the phase plane is shown in case the reactor is nonadiabatic. Here too there is a unique relation between the temperature and the conversion, but the curve now deviates from the linear adiabatic line. Part of the heat generated will "leave" the reactor. This phase plane shows a curvature. At low conversion (low temperature) the deviation with the linear adiabatic curve is small. This deviation increases ever more strongly with increasing T simply because with increasing conversion the temperature *difference* with the environment increases and hence the driving force for heat loss.

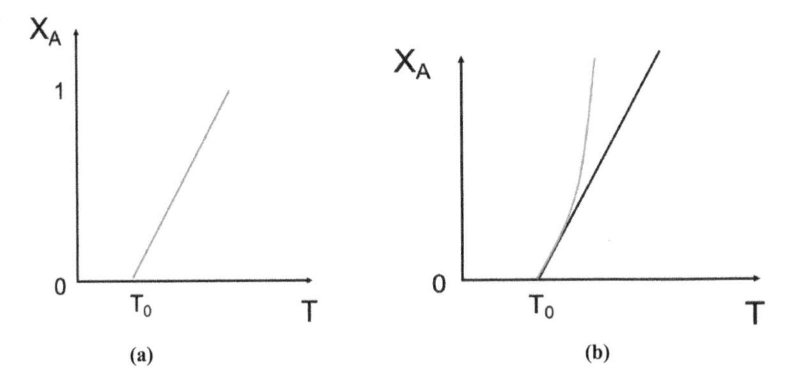

Figure 7.2: Unique relation conversion of A (X_A) versus temperature: (a) adiabatic batch or plug flow reactor; (b) nonadiabatic batch or plug flow reactor, blue line deviating from linear adiabatic line.

As we already learned from Chapter 4, *conceptually* the perfectly mixed *batch* reactor is the same as an ideal plug flow reactor. What time is in a batch reactor is space in a PFR. So also for a PFR there is a unique relation between conversion and temperature, also in case of external cooling or heating.

7.2.1 Nonisothermal backmixed reactor

For a nonisothermal perfectly backmixed reactor (CISTR), the situation can be different and even for simple first-order reactions not always will be a unique relation between conversion and temperature. There is a very easy way to illustrate this graphically by looking at both the heat production rate (HPR) and the heat removal rate (HWR) in the CISTR. The HPR is the product of the volumetric reaction rate, the reactor volume and the reaction enthalpy, expressed in watts. Figure 7.3 shows how this HPR depends on the temperature in the reactor. With increasing temperature the rate of reaction therefore increases exponentially. Of course, with increasing temperature also the conversion in the reactor has increased and therefore the reactant concentration has decreased. That in itself will reduce the exponential increase. At a further increase of the temperature, the conversion will ultimately approach 100% (or the thermodynamic equilibrium); therefore, there is a maximum HPR.

As mentioned at the beginning of the section, the HWR is typically virtually linearly dependent on the temperature: with increasing (yet unknown) reactor temperature both the temperature difference with the coolant and the difference between the outlet and inlet fluid flow increase linearly. So the HWR curve in Figure 7.3 is a straight line. In case the coolant temperature and the inlet temperature are equal and T_0 the HWR is zero at reactor temperature T_0. The slope of the HWR line is determined by the cooling rate, that is, it can be influenced via the heat transfer area and heat transfer coefficient. But do note that also for an adiabatic reactor the HWR curve has a certain slope: the HWR is then simply the heat removed with the hot outlet process stream itself.

In the steady state of the reactor the HPR and HWR balance exactly: the HPR must equal the heat removal rate. The steady-state reactor temperature is the temperature where the HPR and HWR cross.

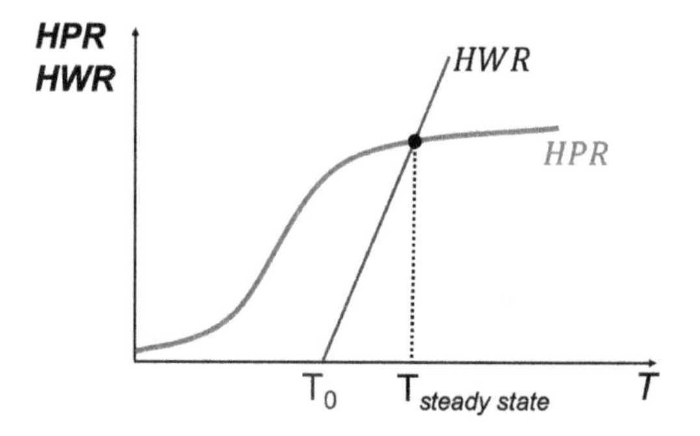

Figure 7.3: Heat production rate (HPR) and heat withdrawal rate (HWR).

Now in the example of Figure 7.3 there is only one point where HPR = HWR. There is a unique steady state. Figure 7.4 gives a simple example where this no longer holds. If compared to Figure 7.3 the inlet/coolant temperature T_0 would be lower, close to T_1 in Figure 7.4, then we see that there are three points where HPR equals HWR. So, all three can be valid steady states. In the field of chemical reaction engineering, this is called "multiplicity of steady states" or simply "multiplicity."

Closely related to this phenomenon of multiplicity of steady states are the phenomena ignition, extinction and hysteresis. These are illustrated in Figure 7.5. Starting up the reactor with a low T_0 the reactor temperature will rise up to the first point where the heat production equals the heat removal. The relatively low temperature T_1 is shown in Figure 7.5. If the inlet/coolant temperature is gradually increased up to T_{0i} the reactor temperature, HPR, HWR and conversion, also increase gradually. However, beyond this critical temperature T_{0i} the graph shows that there no longer a lower operating point. Only the higher operating point exists. This T_{0i} is therefore called the *ignition temperature*.

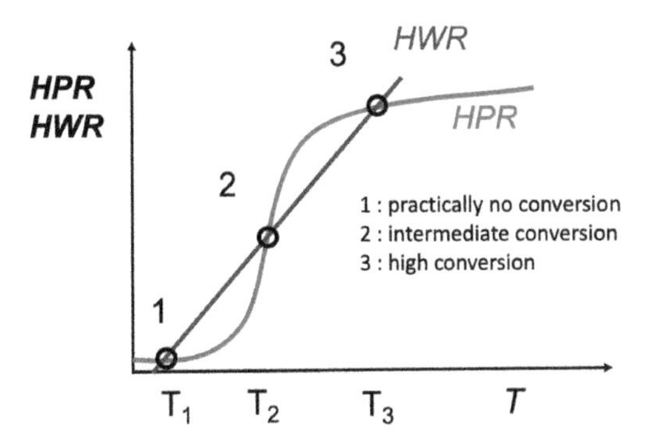

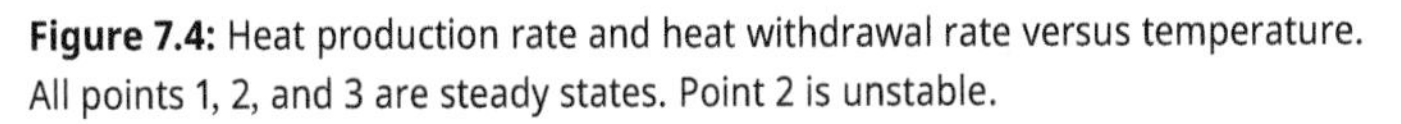
Figure 7.4: Heat production rate and heat withdrawal rate versus temperature.
All points 1, 2, and 3 are steady states. Point 2 is unstable.

On the middle steady-state

The middle steady-state operating point is actually not a stable operation point. It does not fulfill the so-called slope criterion. Any small (random) variation of the temperature will push the reactor towards either the lower or higher operating point. If for a second the temperature is slightly higher than that middle steady-state temperature, the HPR is higher than the heat removal. Therefore, the temperature will rise further and subsequently the HPR becomes even higher than the removal: the temperature will rise until the upper operating point was again HWR = HPR. The upper and lower points are stable, that is, do fulfill the slope criterion but might still not be so-called asymptotically stable and may exhibit oscillations around the operating point. The theory on this however is quite rich and goes way beyond the scope of our book.

As this phenomenon can occur not only at commercial scale but also at lab scale, in our experience, it has happened several times that chemists talk about having observed an ignition temperature or “light off” temperature of a catalytic reaction, which is then interpreted as the temperature where the reaction “starts” or the minimum temperature needed for the reaction to really “light off”. This can certainly be true, but it can also be the consequence of the balance between heat production and heat removal. In that case this “light off” temperature is highly dependent on the testing equipment. An example here is oxidative coupling of methane (OCM) and this explains at least part of the enormous scatter in literature data despite testing the same catalysts under seemingly the same conditions, see Vandewalle et al. [1]. These authors also show how this multiplicity of steady states can be used as an advantage for OCM: once ignited one could feed the reactants at near ambient conditions while the reactor itself operates at temperatures above 800 °C required for the reaction to run at sufficient activity and selectivity.

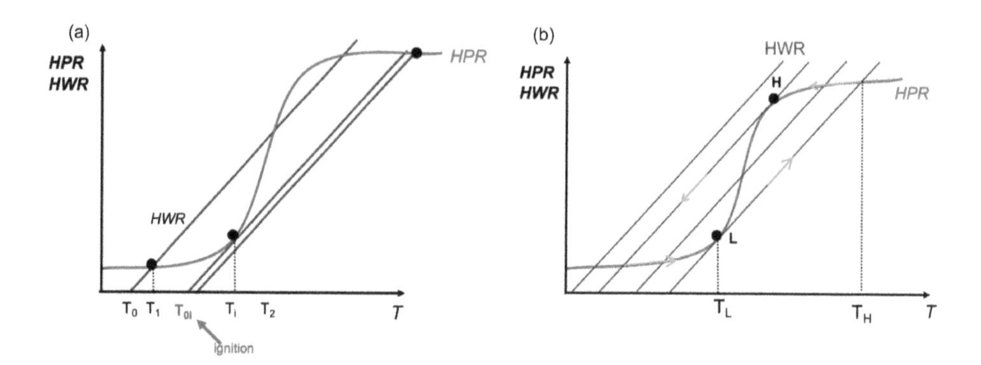

Figure 7.5: Ignition, extinction, and hysteresis: (a) ignition at increasing T_0 beyond T_{0i}: no lower steady-state anymore; (b) hysteresis: ignition temperature at point *L* and extinction point *H*.

7.2.2 Nonisothermal tubular reactor

Many catalytic or noncatalytic reactions are so exothermic or endothermic that active heat management is needed. Because also often plug flow-type behavior is preferred, we see in industrial practice many wall-cooled or wall-heated tubular reactors, or rather multitubular reactors (MTRs), reactor 4 in Figure 2.1b. Typically these consist of hundreds, thousands, or even tens of thousands of tubes, typically 1–4″ in diameter, in a reactor shell filled with a cooling or heating medium. Inherently, each of these tubes is then a nonisothermal reactor, with a temperature hot-spot or cold-spot typically in the first part of the reactor tube: the balance between the local heat production and heat removal.

In the next section on reactor design, we will briefly discuss the phenomena of parametric sensitivity and runaway for nonisothermal reactors. In case of an exothermic reaction, for certain design or operating conditions, any small increase in the coolant temperature can lead to an increasingly more sensitive hot-spot.

It is very important to realize that these are fundamentally different from the ignition/extinction/hysteresis phenomena found for backmixed reactors. For those type of reactors ignition or reactor runaway is caused by exceeding a distinct critical value, above which the (lower) steady-state simply no longer exists. For the nonisothermal reactor there is no such critical value. Unlike the nonisothermal backmixed reactor, the basic nonisothermal tubular reactor does not exhibit the phenomenon of multiple steady-states and ignition and extinction. There is a unique relation between the conversion and temperature simply because there is no backmixing of heat, which is the root cause of the phenomena as described in the previous section.

Note, however, in real nonisothermal reactors there may still be multiplicity of steady states. This can be caused due to some backmixing, like in the axial dispersion model, or in fact can be caused by the complex, nonlinear kinetics.

On multiplicity in tubular reactors

In real nonisothermal reactors there may still be multiplicity of steady states. This, for example, can be due to some degree backmixing, like there is in the axial dispersion model. Or in fact it can be caused by local interaction between reaction and heat/mass transfer, such as multiplicity on the scale of a catalyst particle. It can even be due to the kinetic network itself.

7.2.3 Reactor design to avoid temperature runaway

7.2.3.1 Adiabatic reactors

In adiabatic reactors all heat generated by the reaction goes into heating up the reactor fluids flowing through the reactor. So, the temperature increases with conversion. Beyond a certain conversion the reactor will no longer be stable. This means that the single pass conversion is limited to this temperature.

In fact, it is more complex than just limiting the design conversion. The outlet temperature of the adiabatic reactor will be sensitive to inlet temperature variations. And also to some extend to flow rate variations. A sensitivity study has to be applied to find a safe operating window for the reactor design.

A simple qualitative example will illustrate this high sensitivity by applying high school "rule of thumb" on temperature dependencies of reaction rates: with every 10 °C increase the rate doubles. So if we would try to operate an adiabatic reactor at moderate conversion of 20% corresponding to temperature rise of 60 °C, the rate constant at the outlet of the reactor will be a factor $2^6 = 64$ higher than at the inlet. That implies that most of the 20% conversion actually takes place in the last 10–20% of the reactor. An increase of the inlet temperature of just a few degrees may then increase the conversion from the target 20% to 40%, and hence doubling of the temperature increases over the reactor, 60–120 °C. As a rule of thumb, for adiabatic reactors or beds typically the delta-T is limited to around 30 °C. Of course, this strongly depends on the activation energy, as it is the Arrhenius-type dependency that is responsible for this high sensitivity in the first place.

The sensitivity of the reactor outlet temperature on the inlet temperature also depends on the residence time distribution of the reactor. A fully backmixed reactor can be more sensitive than a perfect plug-flow reactor. Unlike the plug flow reactor, a reactor with complete-or some-backmixing can exhibit the ignition phenomenon described in Section 7.2.1.

The designer can add more inert fluids (a liquid solvent or a gas to allow for a deeper conversion but the conversion will be limited to the maximum temperature). In fact a safety margin has to be taken into account.

It is beyond the scope of this textbook to provide safe design procedures for adiabatic reactors. The reader is referred to an excellent review article by Kummer and Varga [2].

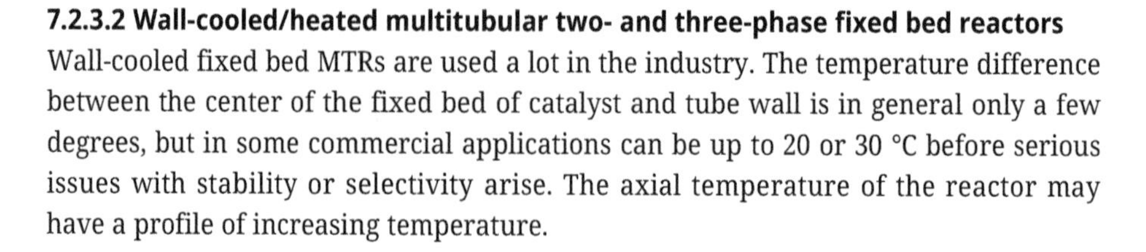

7.2.3.2 Wall-cooled/heated multitubular two- and three-phase fixed bed reactors

Wall-cooled fixed bed MTRs are used a lot in the industry. The temperature difference between the center of the fixed bed of catalyst and tube wall is in general only a few degrees, but in some commercial applications can be up to 20 or 30 °C before serious issues with stability or selectivity arise. The axial temperature of the reactor may have a profile of increasing temperature.

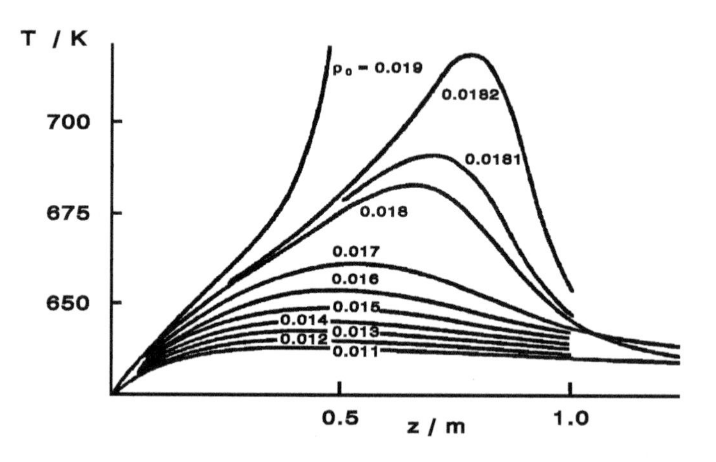

Figure 7.6: Parametric sensitivity in wall-cooled tubular reactors [3].
Effect of increasing inlet concentration p_0 at constant coolant temperature on axial temperature profiles.

Figure 7.6, taken from the classic textbook of Froment and Bischoff [3], illustrates the phenomenon of parametric sensitivity. It shows the axial temperature profiles for different conditions, in this case for increasing inlet concentration of the limiting reactant, expressed as inlet pressure p_0. For low inlet concentration, the temperature initially rises somewhat and then becomes more or less flat. At increasing the concentration to 0.017 in Figure 7.6, a classic "hot-spot" appears, which becomes more pronounced at a further increase. In fact, with any further small increase the hot-spot temperature becomes ever so sensitive toward such an increase. Where an increase from 0.017 to 0.018 "only" leads to around 20 °C higher hot-spot, an increase from 0.018 to 0.0182 yields a more than 50 °C higher hot-spot temperature. At 0.019 the temperature rise will be close to the adiabatic temperature rise. Typically this is referred to as *runaway* or rather *reactor runaway*, see also our discussion in Section 3.5.

Note that the relation between the temperature, or hot-spot temperature, and the inlet temperature is a continuous one. There is no "ignition point" as observed for backmixed reactors. The definition of *in runaway* versus *not in runaway* is arbitrary. It could be defined in a most pragmatic way as the maximum allowable temperature anywhere in the reactor (e.g., to prevent catalyst deactivation or severe selectivity loss) applicable to the specific reaction system at hand.

Barkelew was one of the first to compose a runaway diagram [5], now also known as a Barkelew diagram. Figure 7.7 shows an example for the first-order reaction. Here the adiabatic temperature rise ΔT_{ad} and activation energy E_a have been rendered dimensionless using the coolant temperature T_c and multiplying this gives N_{ad} accounting for the heat production potential. There is another number N_c as measure for the cooling capacity using the variables U (heat transfer coefficient), d_t (tube diameter), ρc_p (fluid phase heat capacity) and the reaction rate constant at the coolant temperature $k(T_c)$. Using such a diagram allows for a quick assessment whether a particular reactor design is "safe" in the sense that it is well above the critical line in Figure 7.7.

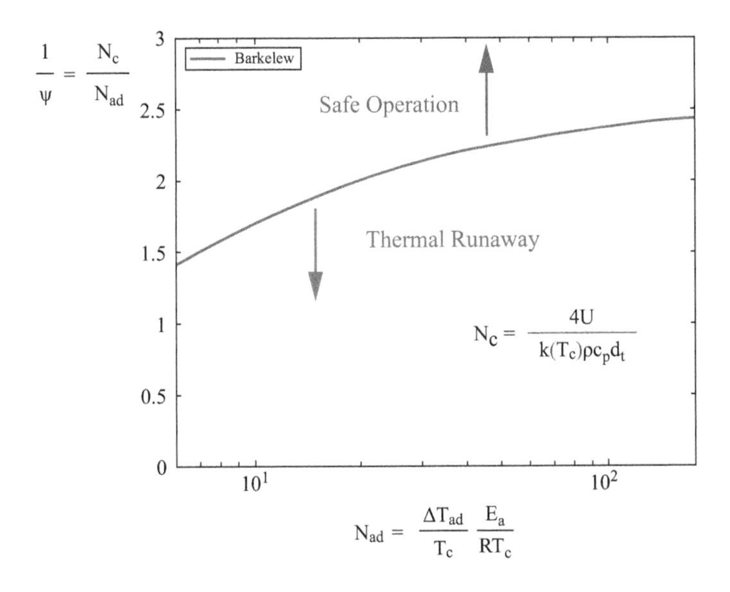

Figure 7.7: A modified Barkelew diagram for a first-order reaction.
The x-axis: dimensionless adiabatic temperature rise times the dimensionless activation energy.
The y-axis: dimensionless ratio heat removal potential and heat production potential (ψ is the Semenov number).

There have also been a variety of more generic definitions, for example, the on-set of an inflection point before the hot-spot: in Figure 7.6 this would be around an inlet concentration of 0.018. For all concentrations lower than 0.018 we see that from $Z = 0$ onwards the temperature rises relatively fast, a steep profile, but then rises less and less up to the hot-spot maximum. However, for inlet concentration of 0.018 we see that at around $Z = 0.4$, before the hot-spot, the steepness of the temperature profile starts to increase again. This is the inflection point. Hence, simply because of this mathematical property, this has become one of the several clear – but still arbitrary – definitions of reactor runaway. Other definitions include the point where the position of the hot-spot changes direction as function of the process parameter, see Froment and Bischoff [3].

It is worthwhile to note that in the various analyses and (modified) Barkelew runaway diagrams, it is always the product of the adiabatic temperature rise and the activation energy, B in Figure 7.7, that is a main parameter. This is independent of the reactor design itself. Moreover, this explains not only the trivial observation that for reaction enthalpy of zero there can never be a runaway but also that for an activation energy of zero there cannot be a reactor runaway in the above definition. There can be a significant hot-spot, but there is no parametric sensitivity.

The heat transfer from the fixed bed to the tube wall is a complex phenomenon. We will discuss this in some detail in Chapter 8. The radial heat transfer through the bed depends on the particle size and shape and packing density as well as the liquid and gas flow velocities. Near the wall the bed porosity is higher than in the central bed section, but whether this has a negative or on fact a positive net effect depends on at least the particle size, the number of particles on a diameter and the liquid and gas velocities.

The "royal route" approach to designing and developing this reactor type is to make a reasonable guess of the heat transfer coefficient using literature information and correlations for the specific catalyst size and shape and to measure the heat transfer in a dedicated test unit. The exact type of material is usually less important because usually the heat transfer under relevant industrial conditions is dominated by the convective heat transfer and solid phase conduction typically hardly plays a role. From the experimentally determined heat transfer parameters as function of the flow rate for the specific packing a better model can be made for the heat transfer coefficients under the conditions in the commercial-scale packed bed.

However, this "royal route" is full of pitfalls, much more time consuming than anticipated, and hence quite expensive. In our experience, dedicated heat transfer measurements only pay out when done with test tubes with a diameter close to the diameter of the envisioned commercial-scale reactor and covering a wide enough range of Reynolds numbers allowing for an extrapolation to the usually much higher Reynolds number in the commercial unit. If there is already a commercial reactor, then a cold-flow test unit may be used for optimizing catalyst particle size and shape. Obviously, in case there already is a single-tube "hot" pilot tube, such cold-flow tests are generally considered superfluous despite the fact that these heat transfer measurements could still help in decoupling the reaction kinetics and mass transfer effects from the heat transfer effects.

7.2.4 Quantified heat transfer for two- and three-phase slurry and fluid bed reactors

7.2.4.1 Introduction

Reliable heat transfer correlations for fluid bed and three-phase slurry phase reactors are not provided in open literature [4]. The first option to consider then in designing heat exchangers for these reactors is to avoid the problem all together and have an external recycle loop with a heat exchanger. Heat exchange for gas phase or liquid phase or slurries can then be reliably designed for most fluid systems. This design approach is applied for commercial-scale polyolefin reactors; see Chapter 14.

If still the designer wants to have the exchanger inside the reactor to avoid the cost of the external loop with a compressor, or pump, piping and the external heat exchanger and she can overdesign the internal heat exchanger or can accept a considerable uncertainty of the heat exchange performance, then the heat transfer correlations provided in this section can be useful.

This chapter then provides convective heat transfer correlations which can be used for conceptual reactor design of fluid bed and slurry phase reactors. So the solid phase is part of the fluid. For reliable detailed engineering design however we advocate to use heat transfer performance parameters experimentally validated for the actual process fluids and solids, as heat transfer of multiphase reactors depends on many individual parameters which interact in a complex way.

Heat transfer between the fluid and heat exchange tubes for multiphase reactors (nonfixed bed) can be described with the same theoretical base. That base is a dimensionless number correlation for convective heat transfer; the Nusselt number as a function of the Reynolds and the Prandtl number. Radiation effects start to play a role for temperatures above around 600 °C [5], so for most chemical reactors design for convective heat transfer is sufficient.

The reactor fluid is outside the tube. The heating or cooling medium flows inside the tube. We focus on the heat transfer coefficient at the outside of the tube. For this the Nusselt correlation for fluid heat exchange with cylindrical tubes is applicable [6]:

$$\mathrm{Nu} = 0.57\,\mathrm{Re}^{0.5}\,\mathrm{Pr}^{0.33} \tag{7.1}$$

The dimensionless numbers are defined as

$$\mathrm{Nu} = hd/\lambda$$

$$\mathrm{Re} = \rho v d/\eta$$

$$\mathrm{Pr} = C_p \eta/\lambda$$

where λ is the thermal conductivity, ρ the density, η the dynamic viscosity, C_p the heat capacity, h the heat transfer coefficient, d the relevant diameter (length scale), and v the fluid velocity.

For the physical parameters of the fluid λ, ρ, η, and C_p, the effective fluid mixture properties have to be used. From the correlation it is clear that the heat conductivity is an important parameter as the heat transfer coefficient h depends on λ to the power 0.67. Viscosity, however, is less important as h depends on it to the power −0.17. So for the latter any reasoned expression for the fluid mixture viscosity will usually suffice.

The remaining important parameter is then the velocity v. Two velocity options are available: a fluid circulation velocity and an averaged eddy velocity v_e. The fluid circulation velocity is available for bubble columns and fluid beds as a function of column diameter; see Chapter 4. The eddy velocity is obtainable from Kolmogorov isotropic turbulence theory. Both velocities are available for two- and three-phase reactor types, bubble columns, fluid beds and mechanically stirred tanks. So both can be used. The lower of the two may then be taken for a conservative estimate.

The Kolmogorov isotropic turbulence theory to obtain an averaged eddy velocity v_e relevant to the scale of the heat exchange tube diameter d depends on the energy dissipation per unit fluid mass E and the local tube diameter d according to Kolmogorov expression (7.2):

$$v_e = (E\ d)^{0.33} \tag{7.2}$$

With these two relations heat transfer coefficients can now obtained for the various reactor types. For each reactor type the energy dissipation is provided below and also the effective physical properties of fluid mixture.

The Kolmogorov approach has been experimentally proven for gas–solid fluid beds by Prins [5]. It correctly predicted the effect of the heat exchanger diameter and also the effects of particle size and gas velocities. The same approach has also been used by Beenackers and van Swaaij [7] for three-phase slurry reactors.

7.2.4.2 Quantified heat transfer two- and three-phase bubble columns

The heat transfer coefficient values for two- and three-phase bubble columns can be obtained using the liquid (or slurry) circulation velocity inserted in expression (7.1). This velocity is provided in Section 4.7. This circulation velocity is a strong function of the bubble column diameter so the heat transfer coefficient is then also a strong function of the column diameter. Jhawar and Prakash [8] reported a positive effect of column diameter on heat transfer of slurry bubble columns. This indicates that this approach is fruitful.

Because of this the eddy velocity approach is likely to be less relevant. For a comparison it may be used. For the eddy velocity first the energy dissipation per unit mass is needed. This is given by $E = g\ v_{sg}$, where v_{sg} is the superficial gas velocity. The averaged eddy velocity v_e is now obtainable from expression (7.2).

7.2.4.3 Bubble column two- and three-phase heat transfer

For two-phase bubble columns, the liquid fluid properties can be directly used in expression (7.1).

For three-phase slurry bubble columns effective slurry fluid viscosity, heat conductivity and density have to be obtained. Several complex expressions for these parameters are published, but all have their limitations. Providing reliable expressions for all fluid mixtures is outside the scope of this book. The reader is referred to finding specific property expressions and data for their fluid mixtures for the early concept design stage.

7.2.4.4 Gas–solid fluid bed heat transfer

For gas–solid fluid beds the heat transfer approach can be analogues to the bubble column approach. For the velocity in the Reynolds number the large circulation velocity as function of column diameter can be inserted. Alternatively, the Kolmogorov eddy velocity can be inserted.

For the latter the energy dissipation per unit mass E causing the turbulence in the dense phase originating from large gas bubbles has to be known. For this gas bubble flow, the superficial bubble velocity is the total superficial velocity U minus the dense phase gas flow velocity U_d.

The energy dissipation E is then given by [5]

$$E = g\ (U - U_d) \tag{7.3}$$

Reliable expressions for dense phase velocity U_d with a large particle size distribution are not available and for the most important Geldart powder class A the dense phase velocity strongly depends on the fraction of fine particles. It is advised to measure the dense phase velocity rather than use a correlation. To still use the energy dissipation approach U_d can be assumed to be much smaller than U so that $U - U_d$ is approximated to be equal to U in expression (7.3).

With E now known the Kolmogorov velocity at the scale of the heat exchange tubes can be calculated using expression (7.2).

The effective viscosity of the dense phase needed for the Reynolds and Prandtl number is provided by expression (7.4) obtained from Prins [5].

$$\log \eta_e = 6.25\,(1 - \varepsilon_d) - 3.88 \tag{7.4}$$

The dense phase gas holdup ε_d in this expression, however, depends strongly on the particle size distribution and most strongly depends on the fine particle fraction of class A powers. So it is recommended to measure ε_d. As a conservative value, $\varepsilon_d = 0.4$ can be used.

The effective heat conductivity of the dense phase is also a complex function of gas and liquid heat conductivity. Westerterp et al. [9] provide an expression IX.51 on page 628 for fixed beds, which Prins also used for fluid beds. But this cannot be a reliable expression because the gas velocity in the dense phase plays an important role and depends on the gas viscosity and the latter parameter is not in the expression. So, it is far better to measure the heat transfer coefficient of the actually used dense phase. This can be done for instance with the method that Prins used [5]. The method is based on emerging a heated silver sphere into the fluid bed and measuring the temperature drop as a function of time.

7.2.5 Mechanically stirred reactor heat transfer

Heat transfer in mechanically stirred reactors depends on a large number of design options. The heat exchange area can be a part of the reactor wall or be a coil inside the reactor. Heat transfer can also be obtained by evaporating part of the liquid. The vapor is then condensed by a heat exchanger, and the condensed liquid is sent back to the reactor. The latter option is often applied in fine chemicals, and pharmaceutical reactors as heat exchanger fouling are greatly reduced and also the risk of local overheating at the heat exchange surface.

For a concept design, the following indicative values for the overall heat transfer coefficients are provided by Carpenter [10]: for the coil side 400 and for the wall side 100 W/m^2 K [9]. Overall it means that the heat transfer through the heat exchanger wall and the heat exchange at the heat exchange medium are also taken into account.

For three-phase slurry reactors it is recommended to determine the heat transfer coefficient at the fluid side experimentally. The Kolmogoroff eddy velocity as function of the power dissipation of expression (7.2) in combination with Nusselt relation (7.1) can be used to compare the experimental results with the theoretical expression.

7.3 Reactor operation and dynamic behavior

During the exothermic gas phase activation (e.g., reduction) or passivation (oxidation) of a catalyst in a commercial-scale fixed bed, the interaction of the heat wave propagation phenomenon and the (desired) exothermic gas–solid reaction(s) may give rise to local temperature excursions way beyond the adiabatic temperature rise of the reactions, even in case the only reaction occurring is the desired one. This is particularly significant for gas phase activation due to the large ratio of heat capacities of the stagnant solid phase and flowing gas phase. Obviously, too high catalyst temperatures can cause significant deterioration of catalysts, undesired/unexpected side reactions and operational problems such as difficult spent catalyst unloading and damage to the reactor walls.

The root cause of the several counterintuitive phenomena that can occur during such as gas phase activation is the high ratio of heat capacity of the flowing gas phase and stagnant solid phase. This gives rise to two distinct waves of fronts travelling through the reactor: the reaction front and the heat front, each with their own velocity. This gives rise to an accumulation zone where heat is "trapped". This in turn, after a relatively short start-up period, creates a pseudo-steady-state "plateau" in the bed, with a constant plateau temperature between locations of the two moving fronts. This plateau temperature can be (much) higher than the adiabatic temperature rise, see Figure 7.8.

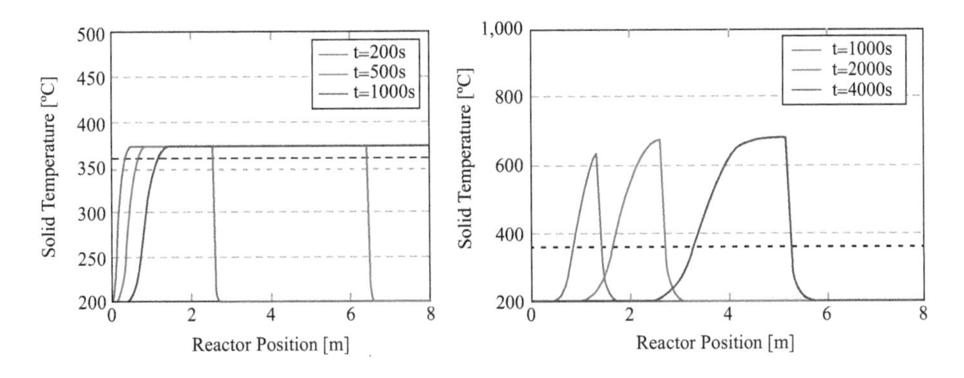

Figure 7.8: Transient axial profiles during CuO reduction in a fixed bed reactor.
Left: 100% hydrogen; right: 10% hydrogen in N_2.
Bold dotted horizontal line indicates the adiabatic temperature rise.

If the two front velocities are getting closer to each other, the maximum temperature in the bed increases further and eventually no plateau is reached anymore and theoretically the temperature would rise to infinity if the two fronts travel at exactly the same velocity.

Another (counterintuitive) effect is that for a certain range of conditions a decrease in reactant concentration can lead to a (large) increase in this plateau temperature. So one may think that the reduction is performed "more carefully" and "more safe" when choosing 3% of hydrogen in nitrogen instead of 10%, whereas in reality the 10% H_2 reduction case would yield an exotherm of 50 °C and in the 3% H_2 case an exotherm of 200 °C (see [11]). Figure 7.8 shows this effect for 100% H_2 versus 10% H_2, the latter giving much higher local temperature rises.

This phenomenon is related to the more well-known counterintuitive effect in catalyst fixed bed reactors during normal operation, that is, not during a gas–solid reaction but a catalytic gas phase reaction. In such a reactor, a sudden decrease in inlet temperature leads to a temporary increase in local catalyst bed temperatures. This is known as *ignition temperature*; see Westerterp et al. [9] and Bos et al. [12] who showed this experimentally or the excellent series of theoretical papers from the group of Professor Dan Luss (see [13] and references therein). The underlying root cause is the same as described for the catalyst activation example: the high ratio of heat capacity of the stagnant catalyst phase and that of the flowing gas phase.

7.4 Exercises

Exercise 1

Figure 7.3 shows the first-order irreversible reaction, the HPR curve, in a fully back-mixed reactor. For the same reactor make a qualitative drawing of the HPR for a reversible first-order reaction.

Exercise 2

Starting from Figure 7.5, draw the hysteresis phenomenon in a plot with outlet temperature versus inlet temperature.

7.5 Takeaway learning points

Learning point 1

Heat management is one of the key aspects for reactor selection, design and operation.

Learning point 2

The Arrhenius-type dependency of reaction rates on temperature is a root cause for relatively complex behavior and challenges for reactor design even for very simple reactions networks and simple reactor concepts.

Learning point 3

The interaction between heat transfer and chemical reaction can lead to counterintuitive effects, especially during transients.

References

[1] Vandewalle L, Lengyel I, West D, Marin G, Catalyst ignition and extinction: a microkinetics-based bifurcation study of adiabatic reactors for oxidative coupling of methane. Chemical Engineering Science, 2019, Apr 28, 198, 268–289.

[2] Kummer A, Varga T, What do we know already about reactor runaway? – A review. Process Safety & Environmental Protection, 147, 2021, 460–476.

[3] Froment GF, Bischoff KB, Chemical Reactor Analysis and Design, 2nd ed. John Wiley & Sons, Singapore, 1979.

[4] Hulet C, Clement P, Tochon P, Schweich D, Dromard N, Anfray J, Literature review on heat transfer in two-and three-phase bubble columns. International Journal of Chemical Reactor Engineering, 2009, 7, 1.

[5] Prins W, Maziarka P, De SJ, Ronsse F, Harmsen J, Van Swaaij WPM, Heat transfer from an immersed fixed silver sphere to a gas fluidised bed of very small particles. Thermal Science, 2019, 23(Suppl. 5), 1425–1433.

[6] Van den Akker H, Mudde RF, Transport Phenomena – The Art of Balancing. Delft, The Netherlands, Academic Press, 2014.

[7] Beenackers AACM, Van Swaaij WPM, Slurry Reactors, Fundamentals and Applications. In: de Lasa HI, ed. Chemical Reactors Design and Technology. Dordrecht, The Netherlands, Martinus Nijhoff Publishers, 1986, pp. 463–538.
[8] Jhawar AK, Prakash A, Heat transfer in a slurry bubble column reactor: a critical overview. Industrial & Engineering Chemistry Research, 2012, 51(4), 1464–1473.
[9] Westerterp KR, Van Swaaij WPM, Beenackers AACM, Chemical Reactor Design and Operation. New York, USA, John Wiley & Sons, 1984.
[10] Carpenter KJ, Agitated vessel heat transfer. In: Thermopedia. Begel House Inc., 2011. https://dx.doi.org/10.1615/AtoZ.a.agitated_vessel_heat_transfer
[11] Zhu K, Bos R, Hellgardt K, Activation of catalysts in commercial scale fixed-bed reactors: Dynamic modelling and guidelines for avoiding undesired temperature excursions. Chemical Engineering Journal, 2020, 382(15), 122962.
[12] Bos ANR, van de Beld L, Overkamp JB, Westerterp KR, Behaviour of an adiabatic packed bed reactor, part 1: Experimental study. Chemical Engineering Communications, 1993, 121, 27–53.
[13] IPin A, Luss D, Wrong-way behavior of packed-bed reactors: Influence of undesired consecutive reactions. Industrial & Engineering Chemistry Research, 1993, 32, 247–252

8 Multiphase reactor modeling

The purpose of this chapter is to explain what modeling is, what modeling can mean for reactor development and design and what modeling methods are available for some main reactor types.

8.1 Introduction

One could argue that chemical reaction engineering (CRE) as an academic discipline was born because of "modeling". The concepts developed within CRE would have no concrete value if not "translated" into a mathematical model. Only then the concepts can be used for reactor design or systematic analysis of steady state or dynamic reactor behavior.

All reactor models are for practical use coined in mathematics, typically sets of nonlinear ordinary (ODEs) or partial differential equations (PDEs) often coupled to nonlinear algebraic equations (ADEs). Therefore, it can be appreciated that for us, the authors, this was rather a challenging chapter to write because this book is intentionally void of any math beyond basic high school level. Therefore, in this book we have given a special focus on modeling, in line with the general focus of this book: providing understanding without resorting to the math itself.

Professor Rutherford Aris was one of the great men of CRE. In 1978, he published a monograph entitled "Mathematical modelling techniques." He defines the term model as "any complete and consistent set of mathematical equations which is thought to correspond to some other entity, its prototype. This prototype may be physical, biological, social psychological conceptual entity." Like Aris, we limit ourselves of course here to physicochemical systems of reactors.

First of all, we cannot stress enough how important it is to clearly establish first the purpose of the model. Or as Aris had already noted: "the purpose for which a model is constructed should not be taken for granted but, at any rate initially, needs to be made explicit." In our experience this is more of often than not "forgotten."

Second, the often-seen phrase "all models are wrong, but some are useful" may have become a cliché, but in CRE, especially in multiphase CRE, it is quite appropriate.

Aris illustrated the difference between "a simulation" and "a model" using the example of flow in a packed bed reactor. One could write down the Navier–Stokes equations, continuity, and diffusion equations. Quoting Aris himself: "this model is admirably complete and founded on the fewest and most impeccable assumptions, but, for two reasons, it is not a very useful model." The reasons being essentially that one would have to do the (then) extremely time-consuming calculations for each and every actual bed and the results would be peculiar to just that bed "making it a good simulation but a bad model." The alternative axial dispersion model is, for instance, much more simple and does not

https://doi.org/10.1515/9783110713770-008

hinge on the details of a specific packed bed geometry, at least not in such an extreme way as for the more complete Navier–Stokes type of model.

Of course, when Aris wrote this in 1978 the computational power was incomparable to what is possible today. Computational fluid dynamics (CFD) and computational chemistry have progressed enormously. Nevertheless, the modeling concepts developed in CRE remain very useful from an educational point of view.

A few more words on a high-level view on "the art of reactor modelling"; the picture we have had in mind for a long time was recently nicely represented in Figure 8.1 (taken from Maher and Mayer [1]). These authors come from the subsurface sciences, but their picture applies very well to reactor modeling.

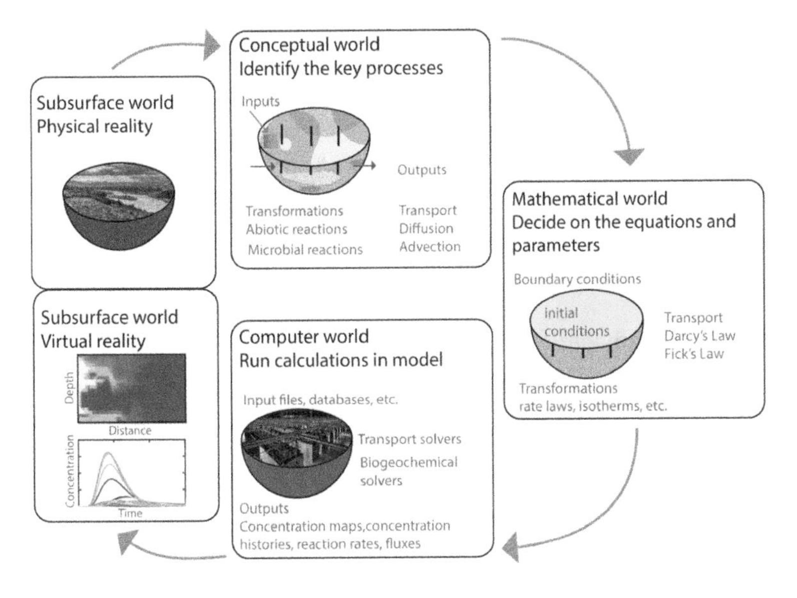

Figure 8.1: The conceptual steps in modeling [1].

For the reactor model equations themselves we refer to the many textbooks that describe these in detail or review papers. We only describe modeling methods which have direct physical, fluid flow, and chemical meaning. Typically, a full reactor model comprises submodels. Examples of such of models are discussed in Chapters 5 and 6 for mixing, residence time distribution, mass transfer and heat transfer.

Finally, in this chapter we will focus on steady-state modeling and will hardly address dynamic models. Our focus is one the type of modeling in support all kinds of design decisions. Steady-state models are most relevant for reactor selection, design, and optimization of operating conditions. These can also be used as a prestep for much more detailed mathematical modeling as the essential ingredients of modeling are identified, but this approach may also reveal which elements are missing or lack quantitative accuracy.

Dynamic models are very useful to explore the controllability of the reactor for input disturbances. However, these models require usually significantly more complex modeling approaches and additional – often poorly known – model parameters. For a new reactor or new process in general this modeling – if at all – is done in the development stage. For reactors such as the reverse flow reactor, dynamic modeling is essential to study the reactor performance. Then the modeling should be done already in the concept stage.

8.2 Models for and two- and three-phase fixed bed reactors

In this section, we will discuss various model approach decisions to be made for gas–solid, liquid–solid, and gas–liquid–solid fixed beds (i.e., trickle-bed reactors). These involve choices such as adiabatic versus nonadiabatic, pseudo-homogeneous versus heterogeneous, plug flow versus plug flow + axial dispersion, 1-D versus 2-D, and traditional CRE versus reactive CFD.

8.2.1 Adiabatic versus nonadiabatic

The way heat is managed in a reactor is very important for each modeling. If the reactor is adiabatic, hence no heat exchange, then reaction heat will cause temperature changes in the reactor, so a heat balance generally has to be part of the modeling for accurate design of reactor performance predictions.

However, not uncommonly adiabatic reactors or reactors with multiple adiabatic beds are in fact intentionally designed and operated such that virtually complete conversion or equilibrium conversion is achieved. Exactly because of the strong nonlinearity of the kinetics the modeling may actually be strongly simplified. The exact temperature profile may be of little concern in practice here. For example, if the adiabatic temperature rise is 50 °C and the activation energy is 100 kJ/mol, the rate constant at an inlet at 100 °C is a factor 50 lower than at the reactor outlet. Therefore, there will either be a low conversion or a nearly maximum conversion. When operated, a simple 1–5 °C increase in inlet temperature can move the reactor to the desired conversion, so there is no real need for a detailed model provided some of the basics like catalyst activity and reference conditions and activation energy are known with some accuracy. We deliberately elaborated a bit here to repeat the first point of our Section 8.1: "to first of all clearly establish the purpose of a model."

The opposite may be true for reactors where the cooling or heating is essential. One of the main purposes of the model would be to properly design the reactor in terms of tube diameter and length. And for that prediction the temperature gradient inside the reactor is key.

If the reactor is cooled or heated, then there are (high level) two ways to model the reactor.

- The heat exchange capacity can be so large that the reactor has very small temperature differences. The reactor can be modeled as an isothermal reactor with a uniform temperature. For exothermic reactions this is generally a conservative approach with respect to conversion, but care should be taken for cases where small losses in selectivity are important (economically or otherwise).
- The heat exchange capacity is limited and temperature gradients usually will have to be part of the modeling because of the strong nonlinearity of the kinetics (Arrhenius).

8.2.2 Pseudo-homogeneous models

In the so-called pseudo-homogeneous reactor modeling approach, the two or three phases present in the real reactor are considered to be just one pseudo-phase. This implies that no temperature or concentration differences between the phases are taken into account explicitly. The heat and mass balances are set up as if there is only one phase flowing. Of course, this does mean that the presence of the solid (catalyst) phase or an additional gas or liquid phase is not taken into account at all. For example, the catalyst bed density or void fraction appears in the mass and heat balances.

Even the effect of mass transfer limitations can be accounted for by the methods described in Chapter 6. In pseudo-homogeneous models these are simply the "correction" terms for the intrinsic reaction rates using the effectiveness factor. Moreover, Professor Vortmeyer [2] has even shown that gas-to-solid heat transfer can remarkably well be modeled under transient conditions by including an additional axial dispersion of heat term. This pseudo-dispersion relates directly to the G-S heat transfer coefficient and not at all to the axial dispersion mechanism as discussed in Chapter 5. A pseudo-homogeneous model without such an "equivalent" axial dispersion term would predict unrealistic block-type responses to step changes.

Pseudo-homogeneous models are very useful in the concept stage as these are still relatively simple models although these are typically still sets of nonlinear differential equations. A process engineer with CRE knowledge at masters' level can implement it in a simple tool such as Excel. For cases where the reactor is (close to) isothermal it may even be quickly put in existing process flow sheet models by which optimizations of the reactor and optimizations of the reactor with the surrounding process unit operations can be carried out. Then also reactors and unit operations can be sized, and investment cost, variable cost, and energy requirements can be determined.

Beyond the concept phase, pseudo-homogeneous models may still be sufficient in many industrially relevant cases. For processes that are still in development stage one should again be considering the main purpose of the model, that is, the added value of more “accurate” models, such as the one described in the next sections. Current practice still is that for catalytic reactions in fixed beds that require strong cooling or heating it is deemed necessary to de-risk the scale-up via demonstration in a (close to) full-size single tube; see Chapter 11. The purpose of a reactor model at that stage is then not to de-risk the process, but more guiding original design and operation of the pilot plant. In the discussion of the more “accurate” models we will provide more considerations with respect to fixed model selection.

8.2.2.1 Plug flow or axial dispersion of mass and heat

Here we will make a rather bold statement: for nearly all commercial-scale two-phase fixed bed reactors axial dispersion of both mass and heat do not need to be taking into account. The simple explanation for this comprises:

- Most commercial-scale reactors operate at particle Reynolds numbers $Re_p \gg 100$
- Most commercial-scale reactors have typically at least 200–1,000 “particles on length”

As explained in Chapter 4, for high enough Re_p each of the voids between the particles is close to a perfectly backmixed “tank”. So for most fixed beds there are at least 200–1,000 of such “Tanks-in-series”. This nearly always justifies the plug-flow assumption. This also holds for a low number of particles on diameter. The very clear wall effects may seem problematic, see Figures 4.4–4.18 in Chapter 4. However, because of the presence of radial mixing in combination with the very high L/d_p counteracts this more than sufficiently.

In contrast, for lab-scale reactors the theory, guidelines, and references in Chapter 4 need to be carefully considered. Both the Reynolds numbers and L/d_p are typically two to three orders of magnitude lower.

8.2.2.2 1-D versus 2-D models

Both 1-D and 2-D fixed bed reactor models are used very commonly. The math for 2-D being less easy to implement for the less-experienced engineer. See Figure 8.2 for a graphic picture on the differences in model setup for 1-D and 2-D, respectively.

A more important consideration here is whether making the model 2-D significantly improves the model predictions. For near-isothermal fixed bed this is hardly ever the case. Radial concentration profiles nearly always can be neglected. Unlike for heat transfer, the reactor wall is of course “closed” for mass transfer. If radial (mal) distribution effects are suspected, other types of modeling would be required anyway (e.g., CFD).

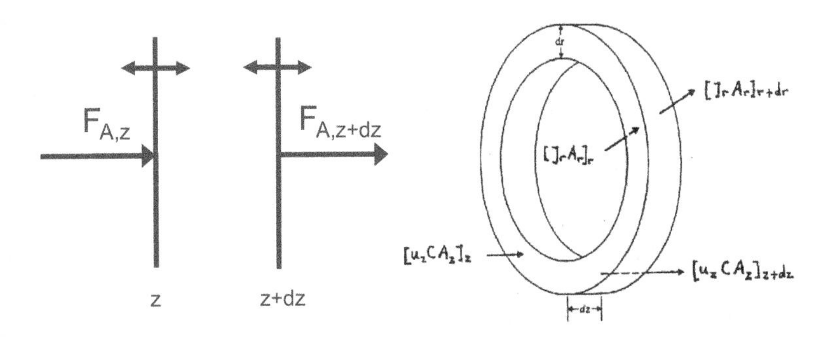

Figure 8.2: Setting up mass balances for 1-D (left) and 2-D models.

For wall-cooled or heat multitubular reactors obviously significant radial temperature profiles will be present. Hence it is tempting to jump to the conclusion that a 1-D model cannot capture this, and a 2-D model is always an improvement. This is a misconception. In a 1-D model radial profiles can be accounted for by using a classing two-parameter heat transfer model. This model makes use of the approximation that roughly speaking there are two regions in the bed:

1. The core of the bed, with average velocities and porosities, where heat transfer can be described with an effective bed conductivity λ_{eff} and
2. A region near the wall, with strongly varying velocities and porosities where the heat transfer can be described with a wall heat transfer coefficient α_{w}. Figure 8.3 depicts this graphically.

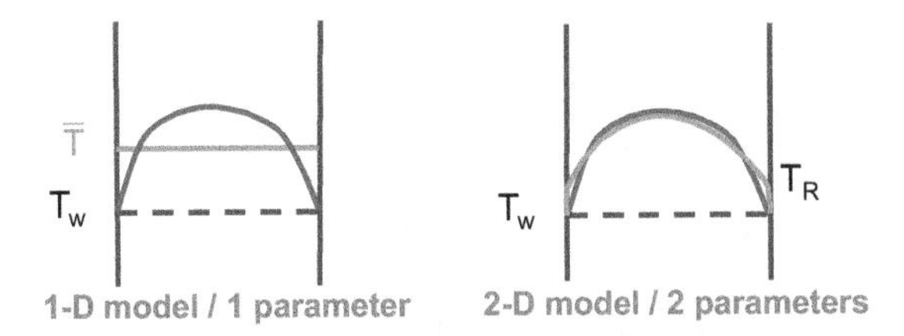

Figure 8.3: Difference between 1-D and 2-D heat transfer models for tubular reactors.
Red lines represent "true" averaged radial profiles, green lines represent profiles as used within the model. T_{w}, wall temperature; T_{R}, reactor temperature at the wall (process side); $\overline{T}$, average T in 1-D model.

The essence in a 1-D model is finding the best possible combination of the two 2-D model parameters into a single overall heat transfer coefficient U_{ov}. This is known as *lumping*. After an extensive study of the differences in model predictions between 1-D and 2-D models under reactive conditions, Westerink et al. [3, 4] recommend the use of the following lump relation for tubes with a diameter d_t:

$$1/U_{\text{ov}} = 1/\alpha_w + 6.12\, d_t/\lambda_{\text{eff}}$$

They also showed that the differences between a 1-D and 2-D model are often very small provided the correct lump equation is used. In that sense, the 1-D model may be more appropriately called a pseudo-2-D model. Moreover, it is relatively straightforward to calculate explicit radial profiles for each axial position from the 1-D model results. Deviations between this pseudo-1-D and full 2-D model become more significant if the so-called Biot number $\text{Bi} = \alpha_W\, d_t/\lambda_{\text{eff}}$ is much larger than 2. However, in most commercially deployed multitubular reactors this Biot number is relatively close to 1.

8.2.3 Heterogeneous models

In heterogeneous models for fixed reactors the interphase mass heat transfer resistances are taken into account explicitly, see the simplified picture of Figure 8.4 depicting possible temperature gradients on the scale of a single particle in the catalyst bed. Then submodels described in Chapters 4 and 5 are used and the methods to obtain estimates for the corresponding model parameters are summarized in Chapter 6.

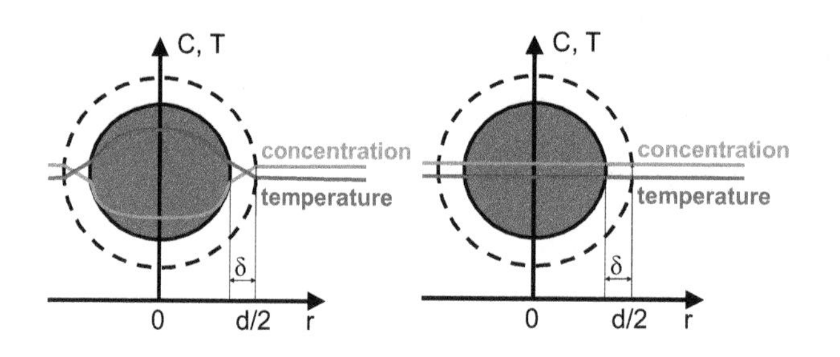

Figure 8.4: Heterogeneous model (left) for mass and heat profiles on a catalyst particle scale.

The main reason to consider a heterogeneous model instead of a pseudo-homogeneous model is the existent significant temperature and/or concentration differences between bulk of the fluid and the catalyst (external surface). There are criteria available to estimate whether or not these effects are significant or not, see Westerterp et al. [5] or Froment et al. [6].

Another more fundamental reason to choose for a heterogeneous model is its capability to describe what is often referred to as "particle runaway" or "multiple steady states" at the scale of a single particle, see Wijngaarden and Westerterp [7] and a recent paper by Vandewalle et al. [8] that includes a real photograph of an ignited particle surrounded by nonignited particles in an experimental fixed bed setup. This scale-dependent phenomenon can be understood in the same way described for an exothermic reaction in a cooled CISTR as explained in Chapter 7 and Figures 7.4 and 7.5. With a pseudo-homogeneous model this cannot be modeled at all. A heterogeneous model can help to explore whether this phenomenon not only may occur or not but also whether this is detrimental or in fact highly beneficial. Examples of the latter include oxidative coupling of methane, see Vandewalle et al. [8] and recent patents by Sabic [9] and Shell [10].

A few words of caution with respect to heterogeneous versus pseudo-homogeneous modeling:

- Modern modeling tools, such as Matlab, Aspen Custom Modeler or gProms, have provided a step-change reduction in the time and effort needed to implement a heterogeneous model. Nevertheless, from a (numerical) mathematical point of view, these models are considerably more complex even in 1-D: instead of sets of nonlinear ordinary differential equations (ODEs) for the equivalent pseudo-homogeneous model, typically one obtains sets of coupled differential and nonlinear algebraic equations (ADEs) that are notorious for yielding numerical problems for chemical/process engineers without sufficient experience. Hence, for development within an industrial setting we recommend that such models are only developed by experts.
- Heterogeneous models may appear more "accurate" because more relevant phenomena are taking into account, especially the gas to catalyst heat transfer. However, not only does this imply more model parameters but also the potential improvement obtained by this may cause that the quantitative description of the bed-scale heat transfer in nonadiabatic reactor actually deteriorates. The vast majority of heat bed-scale heat transfer correlations have been derived using a pseudo-homogeneous model. We doubt that use of those correlations in a heterogeneous model can be justified and likely will further decrease their limited accuracy.

8.2.4 CFD models

In the last decade, progress in CFD modeling for two-phase fixed beds has progressed enormously. One classic barrier in the past seems to have been overcome: the realistic generation of a packing. Recent work now has coupled packing generation, 3-D flow simulation and catalytic reactions. See, for example, the recent Quo Vadis paper by Wehinger et al. [11]. Impressive as it is, it remains to be seen whether from a reactor selection, design, and operation point of view embarking on developing such models in an industrial setting pays off. Again, we refer to what we generally see as "Achilles' heel" of catalytic fixed bed modeling: the reaction kinetics. In our judgment so far it seems that big differences with classic 1-D pseudo-homogeneous modeling only occur for rather extreme reactions and conditions. A real added value of CFD we could envision is that it would allow for screening of the potential of more advance catalyst particle shapes. Experimental determination of especially heat transfer parameters is really time-consuming, not as straightforward as it may seem and expensive. Once properly validated, CFD may be a far more efficient tool for this in the future.

8.3 Models for trickle-bed reactors

Most considerations discussed above for two-phase fixed bed reactors also hold for three-phase fixed bed reactors, that is, trickle-bed reactors. But obviously there are more.

8.3.1 Co-current trickle-bed

Commercial-scale co-current trickle-bed reactors can in most cases be described with a pseudo-homogeneous model. The gas and liquid flow will be in near plug flow. Mass transfer limitations at the gas–liquid interface and liquid–catalyst interface will be absent for most applications. So a simple pseudo-homogeneous plug-flow reactor model can be used for the reactor performance. That plug-flow model can be selected from the library of reactor models in the flow sheet program.

8.3.2 Adiabatic trickle-bed

For the pseudo trickle-bed reactor model of the previous section the adiabatic option should be chosen if the reactor is not a multitubular heat exchange reactor. In the detailed engineering stage precaution measures will be chosen to avoid heat losses via the reactor wall.

8.3.3 Multitubular heat exchange trickle-bed

If the reactor is a multitubular reactor, then for the early concept stage, the reactor may first be modeled as an isothermal reactor. The modeling purpose is then to establish the optimum temperature and residence time for conversion and selectivity.

In the feasibility stage the radial temperature gradient can be modeled using a heat transfer model. Information from Chapter 6 can be used for this modeling.

8.3.4 Countercurrent trickle-bed flow

Countercurrent trickle-bed reactors can be modeled with the gas in plug flow and the liquid in plug flow. Countercurrent operation is chosen because of a specific benefit. Perhaps deep conversion of a component in the gas phase is required while the reaction is an equilibrium reaction. If that is the case, then the modeling should include accurate chemical equilibrium information.

8.4 Models for bubble columns

Bubble column reactors can be modeled with pseudo-homogeneous models if an analysis shows that the reactor system is not mass transfer limited and the reaction system is not sensitive to residence time distribution for gas or liquid phase. In that case a backmixed standard model from a flow sheet program can be chosen. For most projects however that may not be the case. For those projects the remainder of this section may be useful.

8.4.1 Models for G/L bubble columns

For gas–liquid bubble column reactors where mass transfer and or residence time distribution and mixing play a role CFD modeling can be very useful to optimize the reactor configuration. If the main concern is a narrow residence time distribution of the liquid phase for selectivity and or deep conversion and a horizontal baffled bubble column is chosen, then the number of baffles and the configuration can be optimized with a CFD model. An example of this type of modeling is shown in Section 13.3. Model validation of the RTD can be done in a cold flow test facility. Validation of the mass transfer plus reaction can be done in a hot pilot plant reactor.

8.4.2 CFD models for G/L/S (slurry) bubble columns

Pan [12] provides the state of the art of modeling three-phase fluid beds. This is a good entrance into this field.

Li reports a CFD model of a three-phase slurry bubble column [12]. He compared the CFD results with the experimental results. He found that his model better described the bubble coalescence regime that superficial gas velocity had a large effect on the gas holdup of the bed. He also found that the effect of solid volume fraction (0.03–0.30) and particle size (75–270 μm) on the distributions of time-averaged solid holdup and liquid axial velocity was greater than that for particle density (2,500–4,800 kg/m^3), particle size was ≥150 μm and the solid volume fraction was ≥0.09. The larger the values of the solid holdup, particle size, and particle density were, the larger the axial solid concentration gradient was.

So CFD modeling of three-phase slurry bubble columns can now be used for a range of parameter values [12].

8.5 Models for fluid beds

8.5.1 Models for G/S fluid beds

Modeling of gas-solid fluid beds is studied by many research groups for over 50 years. Nearly all models are validated with class B powders. So the models are not valid for class A powders.

Modeling particle attraction forces relevant for class A powders is still a major challenge.

Chapter 14 shows how a commercial-scale design can be made where these attraction forces are relevant and what modeling can still be done.

8.5.2 CFD models for L/S fluid beds

Peng et al. [13] made a CFD model of a tapered liquid–solid fluid bed and validated it with RTD measurements. This means that different reactor configurations can be modeled to optimize the RTD of the reactor.

8.5.3 CFD models for three-phase mechanically stirred fed-batch reactors

Three-phase mechanically stirred fed-batch reactors are used mostly in biotechnology and pharmaceuticals. The phases involved are gas (often air), liquid (often water), and solids (often microorganisms).

The operation mode is often fed-batch. This means that the reactor is filled at the start of the batch with water and nutrients and some microorganisms. Then the reactor is continuously sparged with gas (air) at the bottom via a sparger and continuously fed from the top with an aqueous solution of sugar. The reaction growth rate of the microorganism is first order in the microorganism concentration, so the reaction rate increases with batch time.

By making a CFD model the glucose concentration distribution can be studied to see whether it is uniform or not. In the last stage of the batch the microorganism concentration is the highest and also the glucose reaction rate is then the highest. The model can then show that glucose concentration can then be higher near the glucose feed point and lower in other parts of the reactor. Also the model can show mass transfer limitation of oxygen to the water and to the microorganisms.

The reactor can be optimized for uniform glucose concentration, for instance, by also having an axial pumping propeller on the stirrer axis and having more feed points. The mass transfer can be increased by a higher stirrer speed and or optimize stirrer blades.

8.6 Exercises

8.6.1 Industrial exercise 1: trickle-bed reactor

A company delivers small-scale trickle-bed reactors to oxidize organic components present in rain water so that the water reaches drinking water standards. The reactor design is that water is sprayed on the top of the trickle-bed reactor. The water spray also entrains air so that gas and liquid flow down through the bed. It appears that the organic content of the treated water is a factor 3 lower than the feed water. However a factor 5 is required.

Questions

Q1: What model type do you propose to describe the reactor?

Q2: What is the purpose of that model?

Q3: Which parameters should the model contain?

8.7 Takeaway learning points

Learning point 1

It is essential to articulate the purpose of any (reactor) model. The expediency of the modeling approach and the corresponding level of detail and complexity required strongly depends on the purpose of the model.

Learning point 2

In many cases a pseudo-homogeneous model can be used for two- or three-phase fixed bed and trickle-bed reactors. The net advantages of more complex models should not be overestimated.

Learning point 3

A pseudo-homogeneous model can be used for bubble columns and mechanically stirred reactors only if mass transfer limitations are negligible.

Learning point 4

CFD modeling can be applied successfully for multiphase reactors to study the effect on residence time distribution and mixing in combination with reactions, but the quantitative results must be critically assessed.

References

[1] Maher K, Mayer U, The art of reactive transport model building. Elements, 2019, 15(2), 117–118.
[2] Vortmeyer D, Reaktionstechnische Modellierung von dispersen Zweifasensystem fest/fluid, Dechema-Monographien, 114, Weinheim, VCH Verlagsgesellschaft, 1989.
[3] Westerink EJ, Gerner JW, van der Wal S, Westerterp KR, The lumping of heat transfer parameters in cooled packed beds: Effect of the bed entry. Chemical Engineering and Processing, 1993, 32(2), 83–88.
[4] Westerink EJ, Koster N, Westerterp KR, The choice between cooled tubular reactor models: Analysis of the hot spot. Chemical Engineering Science, 1990, 45(12), 3443–3455.
[5] Westerterp KR, van Swaaij WPM, Beenackers AACM, Chemical Reactor Design and Operation. New York, USA, Academic Press, 1984.
[6] Froment GF, Bischoff KB, De Wilde J, Chemical Reactor Analysis and Design, 3rd ed. Wiley, UK, 2010.
[7] Wijngaarden RJ, Westerterp KR, The role of pellet thermal stability in reactor design for heterogeneously catalysed chemical reactions. Chemical Engineering Science, 1992, 47(6), 1517–1522.
[8] Vandewalle L, Lengyel I, West DH, Van Geem K, Marin GB, Catalyst ignition and extinction: A microkinetics-based bifurcation study of adiabatic reactors for oxidative coupling of methane. Chemical Engineering Science, 2019, 199, 635–651.
[9] Sarsani S, West D, Balakotaiah V, Liang W, Banke J, Oxidative coupling of methane at near ambient feed temperature. WO2018146591A1, 2018.
[10] Bos ANR, Dathe H, Horton AD, Mesters CMAM, Pekalski AA, Schoonebeek RJ, Process for the oxidative coupling of methane. WO2017009449, 2017.
[11] Wehinger GD, et al., Quo vadis multiscale modeling in reaction engineering? – A perspective. Chemical Engineering Research and Design, 184, 2022, 39–58.
[12] Li W, Zhong W, CFD simulation of hydrodynamics of gas–liquid–solid three-phase bubble column. Powder Technology, 2015, 286, 766–788.
[13] Peng J, Sun W, Han H, Xie L, CFD modeling and simulation of the hydrodynamics characteristics of coarse coal particles in a 3D liquid-solid fluidized bed. Minerals, 2021, 11(6), 569.

Part C: **Stage-gate innovation methods**

9 Stage-gate innovation methods

The purpose of this chapter is to introduce the reader into Part C on reactor innovation. It describes the stage-gate innovation model for new reactors to be developed, which is used to structure Chapter 10 on reactor selection and Chapter 11 on reactor development.

9.1 Introduction

Companies have many different ways of developing new reactors from idea to commercial-scale implementation. They also have different words for each step in the innovation trajectory. Bakker et al. [1] provide an elaborate overview of steps (stages) and wordings for various industry branches. But even this overview is far from complete and it is written from an engineering point of view. Vogel of BASF describes only process development, and plant erection including start-up as two formal names for steps from idea to commercial-scale implementation [2].

In the academic field Cooper [3] introduced an elaborate stage-gate innovation method from idea to commercial implementation originally for new product development. Later his method and names were also used for process development. An elaborate stage-gate description for product and process innovation is provided by Harmsen et al. [4].

Nasa developed the Technology Readiness Level (TRL) method ranging from 1 to 9 to determine the maturity of any technology development from idea to and including commercial implementation [5]. Later European institutes also adopted this method and Harmsen translated the terms for use in process innovation and linked the TRL to innovation stages [4].

For this book we have chosen the Cooper method as adapted for process development by Harmsen. It is the most elaborate model and through the last decades more and more companies are using it. We also indicate different names for the same innovation stage so that industrial practitioners quickly recognize which innovation stage is discussed. In the sections of Chapter 11, alternative names are also mentioned.

It should however be stressed that this sequential stage-gate approach is here mainly used to describe in a structured way all scale-up activities. This sequential stage-gate is not the best way to do reactor research and development. Lerou and Ng [6] advocate in his paper to do several items in the early stages of discovery and concept in parallel rather than sequential. Agreda et al. [7] describe in the successful development of the reactive distillation column for methyl acetate production by Eastman Chemical that at the end of the development stage, they went back to the ideation stage to change the design of 11 unit operations into a single column idea and then developed the new

https://doi.org/10.1515/9783110713770-009

reactor with many aspects of modeling and experimentation and pilot plant testing with many activities in parallel.

Chapter 10 discusses what reactor type selection can be taken and also how it can be taken in each innovation stage. Chapter 11 discusses what experimental effort, what design effort and what modeling effort are involved in each innovation stage. In Chapter 13 historic cases of new reactor development are interpreted with this stage-gate method. In Chapter 14 this stage-gate method is used to describe reactor development for polyolefin production from ideation to commercial-scale implementation. Special attention is paid to a downscaled pilot plant for trouble shooting.

9.2 Innovation stages overview

9.2.1 Discovery stage

The discovery stage is also called "ideation stage." Some companies have no formal discovery stage but take it as part of the concept stage. Some companies call this stage early research stage or even research stage. Some call it scouting stage. The discovery stage for a new reactor is about proof of principle experiments and perhaps a reactor family-type preselection.

Chapter 10 treats reactor type selection and Chapter 11 provides information on activities in this stage.

9.2.2 Concept stage

The concept stage is sometimes called research stage. The stage is very important for reactor innovation. Chapter 10 treats various methods of reactor type selection. Chapter 11 covers extensively experimental programs, design, and modeling in this stage.

9.2.3 Feasibility stage

In some companies the feasibility stage is part of the development stage and not treated as a separate stage. However, it is very advantageous to treat it as a separate stage. In this stage a commercial-scale plant, as well as a pilot plant, or a mini-plant is designed. The benefits of the commercial-scale plant are assessed against the cost of the pilot plant and its risk reduction is treated. In Chapter 10 special attention is paid to make a choice between reactor types with large differences in benefits, risks and cost of development.

9.2.4 Development stage

The development stage term is used by most companies. Some call it the pilot plant stage. Some call in the front-end engineering design stage. Chapter 11 provides information on activities in this stage.

9.2.5 Engineering procurement construction stage

This stage is sometimes called execute phase, implementation phase, commitment phase, and detailed engineering procurement construction phase, depending on the industry branch. The majority of manufacturing companies involve an EPC contractor for this stage. Chapter 11 provides details on activities in this stage.

9.2.6 Operation stage

This stage is also called operate phase, operation phase, results phase, start-up and operation phase or deployment stage. It is also called demonstration phase when for the first time the new reactor and process is in some form of commercial-scale operation. Oil and gas petro-chemical companies often use the 4D stage method: discovery, development, demonstration, and deployment. The deployment stage is the subsequent stage after the demonstration stage in which the subsequent processes are designed and constructed years later, containing learning points from the demonstration phase. Chapter 11 describes some elements important to the commercial-scale success.

9.2.7 Abandon stage

This stage is also called abandon and demolition phase. It is not treated in this book. Although in process innovation projects this stage is nowadays included. It involves designing the process for end-of-life demolition in such a way that the ground underneath the process is not contaminated and that construction materials can be reused.

9.3 Takeaway learning points

Learning point 1
The stage-gate innovation model of Cooper is applicable to reactor selection, design, development and implementation.

Learning point 2

For each stage, specific actions for new reactor development can be defined.

References

[1] Bakker HLM, de Klein JP, Management of Engineering Projects – People are Key. Nijkerk, Netherlands, NAP, 2014.

[2] Vogel GH, Process Development – from the Initial Idea to the Chemical Production Plant. Weinheim, Wiley-VCH, 2005.

[3] Cooper RG, Stage-gate systems: A new tool for managing new products. Business Horizons, 1990, 33(3), 44–54.

[4] Harmsen GJ, de Haan AB, Swinkels LJ, Product and Process Design – Driving Innovation. Berlin, De Gruyter, 2018.

[5] Mankins JC, Technology readiness levels: A white paper. NASA, Office of Space Access and Technology, Advanced Concepts Office, 1995.

[6] Lerou JJ, Ng KM, Chemical reaction engineering: a multiscale approach to a multiobjective task. Chemical Engineering Science, 1996 May 1, 51(10), 1595–1614.

[7] Agreda VH, Partin L, Heise W, High-purity methyl acetate via reactive distillation. Chemical Engineering Progress, 1990, 2, 40–46.

10 Multiphase reactor selection

The purpose of this chapter is to provide an introduction and overview of and the authors' view on reactor selection methodologies and practices in an industrial setting.

10.1 Introduction

Reactor type selection is of enormous importance for not only for new reactor design but also for new process design. The selected reactor type determines yield of product on feedstock, investment cost, and also efforts in time and money to develop the reactor and its surrounding process. Reactor selection is often carried out several times during an innovation project when it is about a major innovation. In the early stages little information is in general available, so reactor selection is hardly feasible. From stage to stage, however, the information increases in quality and quantity allowing better reactor choices. We therefore have devoted an entire chapter on this subject so that the subject can be comprehensively treated.

It starts in Section 10.2 with critical reviews academic reactor selection methods so that it is clear which method is suitable for academic education and which for industrial use. Then in Section 10.3 it provides a general picture of reactor selection for industrial innovations. It also gives a bird's eye view of the whole innovation trajectory. In Sections 10.4–10.8 it provides specific methods for each innovation stage. Academic education of reactor selection will mostly benefit from Sections 10.2 to 10.4.

10.2 Critical review some academic methods reactor selection

Here we only review reactor selection methods described scientific literature which we found useful for the purpose of our book.

10.2.1 Reactor family tree selection

The first selection method is that of van Swaaij et al. [1] for gas–solid reactors. Its usefulness is in showing a few pictures most relevant reactor types and in showing when to apply fixed bed reactors or fluid bed reactors only depending on the length of catalyst lifetime in between catalyst regenerations. His method is given in Chapter 2 of this book.

We think that his method can be extended to solids conversions such as in mineral processing and fermentations, where the microorganisms are the product. Also

https://doi.org/10.1515/9783110713770-010

beyond the capacity of the solids to be processed, a fluid bed or slurry reactor type is preferred over a fixed bed.

We also think that his method can be extended to three-phase reactors to select between fixed bed and fluid bed (slurry inclusive) for catalyst lifetime in between regenerations and for solid processing reactors.

The relevance of this method to choose between fixed bed or fluid bed (including slurry) based on the catalyst lifetime is shown in the case heave residue upgrading Section 13.5.

10.2.2 Three-level multiphase reactor selection method

Krishna and Sie [2] provide an easy to understand step-by-step "rational" approach for selecting the solids particle size and the flow regime. Figure 10.1 summarizes the method.

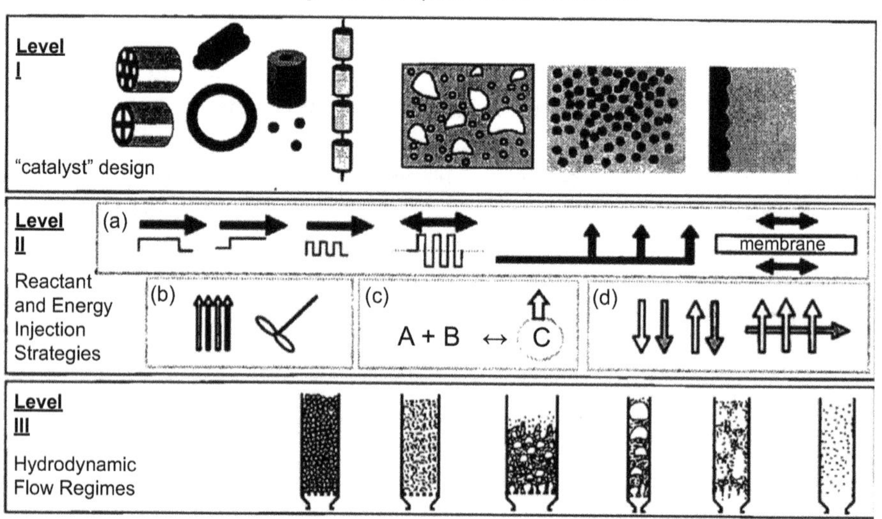

Figure 10.1: The multiphase reactor selection strategies method by Krishna and Sie [2].

One aspect of reactor selection in industrial practice that this method does not explicitly address is catalyst deactivation. In our experience, the strategy on how to deal with changing catalyst performance over time often has a huge impact on the final selection.

The method further has its limitation in that the costs in general and specifically the cost of making the selected particle size are not taken into account. For minerals processing however the cost of grinding and milling to the desired particles size strongly depend on the hardness of the minerals. The cost of milling should therefore be part of the selection method. The method should therefore not only be applied in the concept stage but also be applied in the feasibility stage where information about commercial-scale investment and operational cost are included in the selection.

For soft materials however Krishna's method is very useful.

10.3 Reactor selection method when scale-up risk is low for reactor types considered

The selection method described here is for cases where all critical phenomena are sufficiently known for the large scale for all reactor type options under consideration. Often this is not the case for new reaction systems. Typically, only some basic information is at hand for the reaction system such as the type of catalyst, the phase(s) involved, and the envisaged commercial-scale capacity. During the research and development more information will become available and more critical phenomena can be defined.

The company involved in the reactor and process development can also choose to go for the low-risk scale-up options only, knowing that it may well be not the lowest cost commercial-scale option. The company may choose to limit themselves to these options because they want a short development time and also because they want a low-risk option anyway. This option can be considered in the feasibility stage described in Section 10.5.2.

10.4 Introduction to industrial reactor selection and its practice

10.4.1 Introduction

Reactor selection for industrial applications is a very challenging subject to treat systematically. Reactor selection inside a company depends not only on clear criteria and hard information available but also on the knowledge, and commercial-scale experiences on certain reactor types, of experienced people inside the company that have to make the decision. Arguments of risks and development time needed to reduce those risks or choose a different reactor type with a lower risk are all relevant for the industry but are not treated in the academic methods described in Section 10.1. Moreover, the reactor selection is not made once in the innovation trajectory but made several times during the development and can change radically late in the development.

To illustrate this, we summarize a well-documented history in the reactor selection made inside Shell for the gas-to-liquid (GTL) process. Ton van Helvoort et al. [3] describe in his book of 332 pages the 50 years of development of this Shell process. In the concept stage a slurry-type reactor was chosen [3, p. 149]. And a bench-scale slurry-stirred cascade reactor was operated in the lab. Then a slurry bubble column (a sparged bubble column with suspended catalyst) was designed and experimentally tested in a mock-up facility of the reactor train, containing several vessels in series each with a diameter of 5 m.

Engineering studies had shown that the reactor should be operated at high volumetric catalyst holdup 18–28% and at high superficial gas velocities, outside the usual range of gas-sparged reactors [3, p. 150]. The behavior of the gas, liquid, and catalyst particles extensively tested in the mock-up facility. In particular residence time distributions of the gas and liquid phase were determined as well as G-L mass transfer and bubble sizes (large and small). Also the effect of liquid viscosity on these phenomena was tested.

Then in 1979, due to the oil crisis, a call for synthetic fuels from coal was made by the president of the USA. Due to this Shell decided to speed up the development program. A pilot plant was designed with two sections a gasoline synthesis section with a G-S fluid reactor and for the second stage a three-phase slurry bubble column was designed to produce kerosine and gasoil. The syngas feed should come from coal gasification. The pilot plant has a capacity of 1–2 bbl/day. The investment was 10 million US$ and was mechanically complete by September 1982 [3, p. 155]. It was then mothballed due to the fact that the ManuFacturing (MF) Oil division foresaw that synthetic fuel from coal would not be commercialized before the end of the 1990s. So they decide not to provide a budget for its further development.

Then the gas division of Shell foresaw a future for liquid fuels and notably paraffinic hydrocarbons from natural gas, by partial oxidation of natural gas also syngas could be made and then using the Fisher–Tropsch synthesis catalysis, already developed by Shell, the paraffinic hydrocarbons could be produced. The gas division wanted a rapid development starting with a 100 bbl/d pilot plant. Four reactor types were considered for commercialization [3, p. 181]:

- Two-phase fluidized bed (G-S)
- Three-phase fluidized (ebullated) bed
- Three-phase slurry bubble column
- Multitubular three-phase fixed bed

An analysis with the Thiele modulus showed that the particle size applicable for the fixed bed (and the ebullated bed) had mass transfer limitations and also that the heat to be removed from the fixed reactor using a multitubular design would require a higher investment cost than the G-S fluid bed reactor with smaller catalyst particles and a higher heat transfer coefficient. However, formation of heavy molecules could cause condensation on the particles resulting in agglomeration and defluidization. Therefore G-S fluidization was removed from the options.

The three-phase slurry bubble column had no mass transfer limitations with a particle size of 0.2 mm, but the volumetric catalyst holdup was much lower than in a fixed bed. Also erosion of catalyst and heat transfer tubes inside the reactor could cause problems and also the development time would be longer than the fixed bed. Multitubular fixed beds were already applied for over 30 years in the ethylene oxide processes of Shell, so a lot of commercial-scale experience was already available as well as design knowledge.

So after a lot of analysis a table was made comparing the three remaining reactor types, using 12 criteria with a simple qualitative score. The fixed bed reactor scored 7 out of 12 positive. The three-phase fluid bed scored 4 positive, and the slurry bubble column also scored 4 positive. The overriding argument to choose the multitubular fixed bed reactor appeared to be the shortest development time. However, the last sentence on the paragraph on the reactor choice is:

> "and there are those who suspect that the preference for slurry or fixed bed is based more on belief, or indeed taste, than on reason" [3, p. 185].

In 1985 a multitubular (19 tubes) pilot plant was built. In 1989 a decision was reached to construct a first commercial-scale GTL process in Bintulu, Asia, based on a multitubular fixed bed reactor technology [3, p. 232]. The Bintulu plant with this reactor type, with a capacity of 550 kt/year, started up in 1993. A second commercial plant with the same reactor technology, but with a 10 times larger capacity, was constructed and started up in Qatar in 2011, see Figure 10.2. So in the end the multitubular trickle-flow fixed reactor technology was very successful.

Figure 10.2: Timeline of Shell's gas to liquid technology.

It is clear from this history that reactor choices are not made once but choices made several times during the development, with several options were reconsidered and major changes in reactor type were made.

Sasol Company has also developed and commercialized a process converting syngas to liquid transport fuels. Their Fischer–Tropsch catalyzed reactors are three-phase slurry reactors. So it is also clear that different companies can choose different reactors for the same feed to the same product.

Because of this company dependence on experiences on making the reactor selection and that this selection is carried out several times during the innovation trajectory, we provide guidelines for reactor selection for each innovation stage. In the early innovation stage the reactor family type may be chosen, and the reactor selection may be narrowed in the later innovation stages. The feasibility stage can be the most important point as then also development cost and time are used as elements in the decision process.

We use the term guidelines and not theory as for reactor selection not just one single rational theory can be applied. Reactor selection is dependent on the specific experience and knowledge of the company. It may even involve intuition of experienced reaction engineers. We think however that the guidelines will be useful in the selection process.

Academic methods described in Section 10.3 are mainly applicable in the concept stage. The reason for applying the academic methods mainly in the concept stage is that the methods require a lot of information which is not available in the ideation stage. The academic methods do not treat a risk assessment nor the time and investment needed in the development stage. So only in the concept stage academic reactor selection methods can be considered.

10.4.2 Ideation stage reactor type selection

Reactor selection already at the ideation stage has some advantages. The selection may be just a preselection of the reactor family rather than a precise reactor type selection. The preselection may facilitate a crude estimate of the capital cost of the commercial scale, as the reactor family is clear. The preselection may also help to pass the stage gate to the concept stage. As the information provided to the gate panel may convince them that the researcher has some notion of various reactor options and provide an indication of an attractive reactor family.

A sharp selection in the ideation stage will however not be possible in most cases due to the lack of information. However, selecting between families of reactors using Table 10.1 may be doable. To apply this table very limited information regarding the reaction system is needed.

10.4.2.1 Ideation stage reactor families to choose from

In the ideation stage and also in the concept stage little information will in general be available about the new case. Selecting rationally a specific reactor type may therefore not possible. However, having some idea of what reactor types should be considered for further investigation is enormously helpful in making a crude estimate of the commercial-scale investment cost and the development time and effort needed.

To aid in this selection we made a few families of reactor types. Several of their family features are very different from each other. So a preselection between the families may be possible. The idea of using families for making categories stems from the philosopher Wittgenstein. He states that a family has one or more common features, which distinguishes them from a different family. There will be many other features the families share with each other and with the human race. There is no need for describing all the latter features. There will be an intuitive understanding of the latter. The whole philosophical description is found in [4].

We have a flying start with reactor families using the description of van Swaaij et al. [1] for gas–solid reactors. We take only his classification of reactor families for fixed bed and for fluid bed. We think however that these families can also be used gas–liquid–solid reactors. The slurry reactors then also belong to the fluidization reactors. The three-phase fixed bed reactors belong to the fixed bed reactors.

Table 10.1 shows the two major reactor families which differ on features on at least five parameters. A selection between the two families can now be made in the ideation stage if for a few parameters it is known what is required for the case at hand.

For each parameter some guidelines will be provided to determine the relevance.

Table 10.1: Reactor families and their common features.

Parameter	Fixed bed family	Fluid bed family
Heat transfer	Expensive	Cheap
Residence time distribution	Plug flow	Wide or staged by multibed
Mass transfer G/S	Low	High
Mass transfer G/L	High	Low
Catalyst regeneration rate	Slow	Fast and can be heat integrated
Solids processing	No (some in moving bed)	Yes, high capacity low investment cost
Scale-up development	Low cost and rapid	High development cost and slow

10.4.3 The power of reactor selection in the ideation stage: Shell shale fluid bed case

This case is described in some detail in Section 13.8. It shows the power of having the selection in the ideation stage using the economic criterion in combination with Table 10.1.

Let us walk through this case using Table 10.1 parameters as steps:

Step 1: Heat transfer
The Shell case requires enormous amounts of heat to be transfer to the shale particles for drying and heating up to the reactor retort temperature of 500 °C. So the fluid bed family is preferred for this parameter.

Step 2: Residence time distribution
A deep conversion of kerogen to oil is desired. So plug flow or multistage is required for the solid phase. The gas phase should have short residence time and also be plug flow. Here the fixed bed is preferred.

Step 3: Mass transfer G/S
The mass transfer of oil vapor from the shale particles to the gas phase should be high to avoid oil component cracking. This calls for fine particles.

Step 4: Mass transfer G/L
Gas–liquid mass transfer is not present in the Shell case.

Step 5: Solids processing
The remaining coke on the shale needs to be combusted and the generated heat needs to be used for steam generation and retort feed heating. This calls for fluid beds in which hot particle recycle is feasible. It needs to be at a scale of 160,000 ton/day (to reduce the investment cost per ton of shale processed). This calls for a fluid bed.

Step 6: Scale-up development
Scale-up does not need to be fast, as the application of the process is for the long-term future. So also high-risk options taking long development time can be considered.

All in all the fluid bed option was preselected without any experiment carried out. Economic studies just assuming reaction times showed that the investment cost per tonne of oil produced was a factor 2 lower (40 $/bbl) than the best process technology available.

It also shows the limitations. Grinding cost for instance can be overriding cost for reactor selection for minerals processing. These costs however may be not considered or not known in the ideation stage. They only appear in the feasibility stage when commercial-scale investment cost and operating cost of the whole process are taken into account.

10.5 Reactor type selection in the various innovation stages

10.5.1 Concept phase reactor selection

10.5.1.1 Stepwise selection

In the concept stage information about the reactions will be gathered and a more precise selection of a reactor type can be made. As in many companies no formal distinction between the ideation stage and the concept stage is made, we will treat reactor selection for both stages in this chapter.

For selection following elements are needed:

- Selection criteria
- Options to choose from
- Information of the reaction system
- An easy-to-understand selection method that can be shared with nonchemical engineers

As little information may be available in the ideation stage elaborate criteria needing lots of information cannot applied in this stage. We will therefore sketch the selection procedure resulting a preliminary selection of the reactor family and optionally a subselection of some specific reactor type options.

The first step in the selection procedure is stating the external criteria to be applied for the reactor selection. The criteria items are found in Section 9.2.2. For each item a specific criterion may be formulated for the project case at hand.

The second step is describing the internal project and reaction case information needed for the preselection.

The third step is the actual selection of the reactor type family and perhaps a subselection of a major reactor type class. Here solids processing capacity criterion to select between reactor families is the major element. For cases where large amounts of solids have to be processed a fluid bed (or slurry) should be selected. Some industrial cases are briefly described below. Sections 13.3 and 13.5 and Chapter 14 describe the industrial cases in detail.

10.5.1.2 Industry cases with large-scale solid processing

There are many industry cases where a large solids processing capacity is needed. Examples are biomass conversions to fuels and chemicals, fluid catalytic cracking (FCC), Flexicoker, polyolefins manufacturing, and minerals processing.

The threshold for what can be called large-scale solids processing may be derived from these cases. The smallest solid processing case is the biomass to liquid fuels, BTG–BTL case, with a biofuels production capacity of 30 kt/y [5]. There the heat for the endothermic reaction is provided by hot sand particles. Coke deposited on the sand particles as well as noncondensables are combusted in the second fluid bed combustor. The hot sand is recycled back to the pyrolysis reactor. The pyrolysis reactor has an inverted rotating cone which ensures mixing of biomass with the hot sand and probably ensures some staging of the biomass particles and the produced pyrolysis oil vapor [5].

An alternative set of cases for determining the minimum threshold solids process capacity is the catalyst lifetime in between regenerations. Figure 2.1 of Chapter 2 shows when a fixed bed and when a fluid bed (or slurry) should be chosen.

The classic industrial case is FCC of a refinery stream to produce crude petrol. In this case the overriding cost element was the investment cost of the reactors. Heat transfer to the reactor is very important to maintain the endothermic reaction. Multitubular fixed beds are far more expensive than a fluid bed riser where the heat is supplied by hot particles, obtained in the second fluid bed combustor. Moreover very large production capacity required also meant that a large number of fixed beds would be needed. Rapid coking of the catalyst meant that the fixed bed reactors had to switch from production mode to coke combustion mode, requiring even more fixed bed reactor.

If the catalyst lifetime is in the order of one month then multiple fixed bed reactors may be chosen so that one of the reactors can be in the regeneration mode, while others keep producing.

An interesting case is heavy crude oil upgrading. Section 13.5 shows that three companies selected three different reactor types. Shell chose a three-phase moving bed, trickle-bed reactor, Lummus, in their LC-FINING^{TM} process, chose a three-phase ebullated bed (slurry) reactor and Exxon chose a fluid bed solution. That case shows that reactor selection inside a company also depends on the knowledge and experience of that company. It also shows that knowing all in hindsight still selecting the best reactor is hard.

10.5.1.3 Reactor selection criteria in concept stage

For the concept stage when information on kinetics and phases are generated a more informed selection can be made. In this stage some information can be gathered or is already available. Table 10.2 provides an overview of relevant project information and for which decision this information is usefull. The selection is still preliminary but more informed than the ideation stage.

10.5.1.3.1 Selection criteria

The minimum criteria set for process concept design is the safety, health, environment, economical, technical feasibility, and the sustainable development (including social acceptance) (SHEETS) criteria set [6]. Most of these criteria also hold for reactor concept design, as reactor design often is a major part of a process design. For each criterion we provide some guidelines to decide whether it is relevant for the case at hand and how a subcriterion can be derived to make it applicable.

The **safety criterion** could be the most critical if the chemicals in the reactor can cause a runaway or an explosion or if the feed materials or intermediate components in the reactor are very poisonous. The derived criterion could then be which reactor can be operated outside the runaway area and which reactor would have such a low holdup of poisonous materials at it can be considered safe.

Examples are cooled reactors in which there is such a high cooling capacity that runaway cannot occur, or for poisonous materials a reactor with such a high reaction rate that the poisonous amount is still safe such as a reactor with plug-flow behavior and or also high heat transfer rates so that the reaction rate can be high.

For the **health criterion** the same reasoning as for safety may apply for poisonous materials. In addition a reactor with less connection and requiring less valves would be selected so that diffusive emissions of the poisonous materials are low.

For the **environmental criterion** several subcriteria can be applied. A subcriterion can be the energy quality required for the reactor and or the energy quality provided by the reactor. If the reactor is exothermic then the energy quality provided by the reactor should be maximized for use inside the process or even outside the process. The temperature of the energy provided can then be important. In general a higher temperature is beneficial for usefulness elsewhere.

For the **Economic criterion** the yield of product on feed materials is in most cases the overriding subcriterion. As in most cases feed material constitutes 80% of the total cost, while investment costs are 5–15% of the cost and reactor investment cost are only a fraction of the total investment cost. For cases where the product yield is the overriding cost factor, detailed reactor selection can best be made in the concept stage using all classical reaction engineering methods based on critical performance factors. An initial reactor family selection may be based on Chapter 2.

If the product reacts further to unwanted by-products, then the reactor should be designed such that the by-product formation is minimized. This means a short resident time of the product in the reactor. This can be achieved by having a low conversion, separating the product from the feed components in the separation section after the reactor or inside the reactor, such as in reactive distillation.

For all other reaction paths and kinetics, it is advocated to make reactor models first on residence time distribution effects to find the best RTD. Then mass transfer effects on by-product formation should be modeled to find the best gas–liquid–solids contacting and sizing.

For cases where the investment costs are overriding the feedstock cost the selection will be much harder. The selection method is then probably best carried out in steps.

The first step in the selection is fixed bed, or fluid slurry reactor, depending on the time in between catalyst regenerations, see Figure 2.1.

The second step is selecting the best residence time distribution can be chosen for minimal reactor volume. From that some real reactor types can be chosen.

The third step is then to explore the resulting reactor types on mass transfer and heat transfer and select the reactor with the highest heat transfer coefficient (lowest heat transfer area to install).

The **technical feasibility** criterion is in general not an important item in the concept stage. If the reactor type is not selected from a pool of existing reactors but the reactor type is new and needs considerable research and development, then it is wise to have an existing reactor type as a fallback option next to the new reactor. Later in the feasibility stage the choice for the new reactor is then evaluated in comparison with the existing reactor type. In that stage a choice between reactors with different risks and different development cost and different economic potential is made.

The sustainable development criterion is in general hard to apply for reactor selection in its own. The criterion is to be applied for the whole plant-wide process design. There it is concerned with the choice of renewable feedstock and the choice of product.

In addition to these guidelines, it is advised to involve experienced process development engineers in applying criteria. They may intuitively know when considering the case which criterion is most relevant.

Table 10.2: Internal project information required for reactor family and major class selection.

Information	Accuracy	For decision
Production capacity	Order of magnitude	Batch versus continuous Fluid bed versus fixed bed
Unavoidable phases present		Reactor family choice
Catalyst lifetime before regeneration	1 month or longer	Fluid versus fixed bed
Heat of reaction	Factor 4	Adiabatic reactor or not
Reaction scheme main and by-products	Consecutive or not	Reactor type selection
Reaction rate	Order of magnitude	Reactor type selection

10.5.2 Feasibility stage reactor selection

10.5.2.1 Introduction

This chapter is about selecting between reactor types in the feasibility stage. The purpose of the feasibility stage is to justify the development cost with future benefits of the implemented commercial-scale process, or if the justification cannot be found to stop the innovation project. Reactor type selection can affect both the development cost and the implemented benefit. It is therefore beneficial to have a method that considers reactor types requiring little development cost and reactor types that require a lot of development cost (but with a higher benefit at commercial scale). The purpose of the development effort is to reduce the risks of commercial-scale implementation. So an evaluation method should also include the risks and the effects of the risks to have some semirational approach to justify the development effort.

In Section 10.6.2, we provide a method called value of information (VoI). A modified version of this method facilitates choices between reactor types differing in development effort and facilitates a risk-benefit assessment. In Section 10.3 we explain how to determine a development program for reactor types where a lot of knowledge is available. In Section 10.4 we explain how to make a development program for reactors where little information for commercial-scale design is available.

10.5.2.2 Reactor selection by value of information analysis

The so-called value of information method is used for two purposes. The first purpose is to rationally – based on economic value of the information aimed to be gathered – choose between alternative development programs for a new reactor or process and to justify the selected development program (or come to the conclusion that there is no justification). The second purpose is to economically clarify decisions made with respect to selection between reactors (or processes) with a large difference in estimated investment cost for the commercial scale and a large difference in development cost.

The method has been briefly described by Harmsen [7]. It should be noted that this is a significantly modified and simplified version of the so-called VoI method used within Shell. There it was originally used within "Upstream" to make rational decisions on test drillings based on knowledge before and anticipated knowledge after a potential test drill. Hence the term Value of Information. Harmsen [7] has modified it such that it may be applied for reactor selection as well.

Here the version as modified by Harmsen [7] is explained in more detail.

The reactor selection method consists of three steps: Step 1 is defining the development program for the reactor type for which lot of knowledge for the commercial-scale design is available. The cost of commercial scale as well as the cost of the development program is estimated in this step.

For the presented case this is the fixed bed multitubular reactor. This reactor type is applied for other processes inside the company. Engineering Procurement Construction contractors with experience for these reactors are available to execute the EPC stage.

Step 2 is defining the development program for the reactor type for which little is known about the commercial-scale reactor design so that a number of assumptions for that design have to be made. Those assumptions then need to be tested in the development program. The cost of the commercial-scale process and the cost of the development program are defined for this reactor type.

Step 3 is a systematic comparison for the two reactor types using the modified version VoI method. All economic values are expressed in net present value. By this the development time required is also included in the cost. The modified method is briefly described by Harmsen [7, pp. 65–68]. We here explain the method with the following example case.

10.5.2.3 Example case reactor selection

The example case is about a very large commercial-scale reactor process to be developed. The reaction is a heterogeneously catalyzed exothermal reaction with three phases involved. One of the feeds is a gas. A second feed is in the liquid. The reaction is highly exothermic so that the reactor needs to be cooled. The reactor pressure is 50 bar.

The catalyst lifetime, due to coking, is 1 year so that each year the catalyst needs to be regenerated by tempered oxygenation, which takes 1 week. The irreversible catalyst chemical lifetime is 6 years so that at the regular maintenance stop of every 6 years the catalyst is replaced by fresh catalyst.

The reaction is also so fast that for fixed bed catalyst particles of 1 mm, the Thiele modulus has the value of one. For a slurry reactor the catalyst particles can be reduced to 0.2 mm so that the Thiele modulus becomes 0.2 and no diffusion limitation occurs. So a fixed reactor system is an option, but it will require very high investment cost due to the multiple tube heat exchange reactor. The high Thiele modulus means that the reactor bed volume is also high. The other option is a three-phase slurry reactor system with heat exchanger inside the reactor. Several reactors will be needed given the reactor diameter limitations for pressure vessels. So liquid staging can be obtained by having five reactors in series.

Reaction engineering concept designs have been made for both reactor types for commercial-scale capacity. The fixed bed reactor type required 20 reactors operated in parallel and catalyst regeneration is executed at one reactor at a time, with a 5% lower overall capacity. The slurry reactor required five reactors in series. The catalyst is regenerated continuously in a small regeneration section.

Investment costs have been estimated for both reactor types. The slurry reactor type has 30% lower installed investment cost due to lower heat exchanger area required (higher heat transfer coefficient) and due to much smaller total reactor volume (effectiveness factor = 1).

Table 10.3 contains all elements of the selection method. Each element will be explained here.

Table 10.3: Value of information table [7].

Value of information element	Multitubular fixed bed		Fluid bed	
	Direct implement	Development	Direct implement	Development
1. Design base NPV	100	100	110	110
2. Consequence failure NPV	−125	−125	−120	−120
3. Chance of failure (%)	40	40	80	80
4. Expectation failure NPV	50	50	−96	−96
5. Initial net NPV	50	50	14	14
6. Development NPV	0	−10	0	−30
7. New design base NPV	100	105	120	125

Table 10.3 (continued)

Value of information element	Multitubular fixed bed		Fluid bed	
	Direct implement	**Development**	**Direct implement**	**Development**
8. New consequence failure	−125	−125	−120	−120
9. New chance of failure (%)	40	10	80	20
10. New expectation failure NPV	−50	−12.5	−96	−24
11. Final net NPV	50	82.5	14	71

Example case exothermic heterogeneous catalyst reaction systems. All NPV in M$.

The first term of Table 10.3, NPV of the design base, is determined using investment cost, input cost, and product sales as input. It is the NPV of the commercial-scale reactor, designed with information available to the designer from the concept stage.

The second term, consequences of failure, is the loss of the investment cost and also additional losses such operator salaries, and loss of feed to waste, if the commercial-scale process totally fails after start-up to produce the product at the required quality and quantity. As it is a loss of all investment made, it has a negative sign.

The third term, the chance of failure of the design, is determined. It should be remarked here that this is the chance of failure by directly implementing the design without any development effort. To estimate the chance of failure, an external consultant such as Independent Project Analysis may be involved. An alternative way is to ask experienced process engineers who have witnessed successes and failures to determine the chance of failure. To avoid group bias influences the experts should be asked individually to provide the chance of failure percentage. Given the limited data of their experience available to them this chance of failure figure itself will have an uncertainty range. The range provided by the expert data can be used. The top of the range should be used as the chance figure in the table.

The fourth term is the expectation failure. This is simply by multiplying the second term, consequences of failure with the third term, chances of failure.

The fifth term, the initial NPV, is then obtained by subtracting the expectation failure from the initial NPV.

The sixth term is the development cost NPV. For the first and third column this is zero by definition. For the second column this is the estimated development cost figure, so it has a negative sign and is expressed as NPV. In a section further up in this chapter elements of a development program for reactors are described.

The seventh term, the new design base, is determined by knowing that a development program will be carried out by which for a less conservative design needs to be made, resulting in a higher NPV value. Because of this new design base the net present value can be higher, as indicated in the table.

The eighth term, the consequence of failure, may be the same value of the original value, or for the second column and fourth column slightly different due to a less conservative design.

The ninth term is the new chance of failure after the development has been executed. The figure in the second and fourth columns now will be far lower than the figure in the first or third columns.

The 10th term, the new expectation failure, is the chance of failure (ninth term) multiplied by the eighth term.

The 11th term, the final net NPV, is the value of terms 6, 7, and 10 added up (using the proper signs) for the columns.

That 11th term is then first used to compare columns 1 and 2 and also columns 3 and 4; hence the cases with no development and the cases with development included to see whether a development can be justified. Second, the final results of columns 2 and 4 are compared to the fixed bed and the fluid bed.

For the example case the fixed bed with its development cost included has the highest NPV. So this is the selected option. This option has a higher NPV than the fluid bed with its development cost included. The difference however is small. Other considerations can also be included, such as will in future more processes be constructed and started up? Then the development effort and its cost for these future process designs will not be longer there. Or in other words the development cost may be spread over all foreseen processes to be designed. The fluid bed option is then a better alternative.

10.5.3 Development stage front-end engineering design reactor selection

Reactor selection in the development stage can occur when the experimental part of the development stage is finished, and it appears that the reactor type selected in the feasibility stage and tested in the development stage performs worse than expected.

Then the VoI method should be applied again, now with the information generated by the development. Other experienced engineers can be invited to assist in the risk assessment to avoid tunnel vision. The results of this exercise can be that the lower risk reactor type is chosen for the front-end engineering design (FEED) the major deliverable of the development stage.

10.5.4 Engineering procurement construction (EPC) stage reactor selection

In the engineering procurement construction (EPC) stage reactor selection in general does not take place. However, in procurement of equipment a technology provider may offer a reactor not considered in the feasibility and or development stage, but which looks attractive. Then the VoI method should be applied once again to extract all information about this new reactor and to assess its risks. It is of utmost importance to obtain evidence of the reliability of this new reactor by obtaining information about its commercial-scale performance or pilot-scale performance and in particular in which industry branch this reactor has been applied. Experience in the chemical industry for instance may not be valid for the food industry and vice versa.

10.6 Exercises

10.6.1 Industrial exercise 1: reactor type selection in ideation stage

Given: A start-up company wants to convert waste polymeric PET material into its monomers. Clean the product stream from other components in the waste material and convert the monomers back to PET.

Their chemist has found a solvent to dissolve the polymer. The chemist also found that the depolymerization can be catalyzed by iron oxide magnetic catalyst particle 450 nm. He could withdraw the catalyst particles after the reaction by a magnet.

The depolymerization reaction of polymer to the monomer is

$$\text{Polymer} + n\,\text{water} = n\,\text{bis(2-hydroxyethyl) terephthalate (monomer)}$$

Also, under the reaction conditions a consecutive reaction takes place:

$$\text{water} + \text{bis-terephthalate} = \text{terephthalic acid} + \text{ethyleneglycol}$$

Water is fed as steam vapor to the liquid system at elevated temperatures.

In the ideation stage both the monomer and products of the consecutive reaction are acceptable products. The reason is that after removal of all contaminants PET is made again from the clean monomer as well as the other two products.

In the ideation stage when a few proof-of-principle experiments showed a nearly 100% yield of the three products on the PET feed. You as the newly hired chemical engineer are asked to provide a short list of reactors for the depolymerization and a rationale for the most promising option for further research.

Q1: What is your short list?
Q2: What is your rationale for selecting the most promising option?
Q3: What is your selected option?

10.6.2 Industrial exercise 2: reactor selection concept stage

Given: The chemist and the chemical engineer described in Exercise 1 work together in the concept stage. The chemist has carried out dedicated kinetic experiments in a mechanically stirred reactor operated continuously. It is fed with a liquid, a solution of PET in the solvent. It is also fed with the fine catalyst particles as a slurry. Steam is fed to the reactor. The reactor has a liquid output stream and a vapor output stream. The reactor is furthermore kept isothermal by a heating cooling heat exchanger coil inside the reactor.

The experimental setup was such that the PET depolymerization reaction rate was determined from the produced monomers, the molecular weight distribution of the feed, and the product stream.

The experimental results were analyzed with the following results:

> The depolymerization reaction rate is linearly proportional to the catalyst concentration in the reactor.
>
> The depolymerization reaction rate is proportional to the steam partial pressure in the reactor.
>
> The depolymerization reaction rate depends on the PET polymer units according to a Langmuir–Hinshelwood expression.
>
> The depolymerization reaction rate depends strongly on temperature. Beyond a temperature of 240 °C by-product components are formed. The optimum (safe) temperature is 230 °C. The required liquid residence time in the kinetic lab reactor for catalyst slurry holdup of 10% and a conversion of 95% appears to be 60 min. The depolymerization reaction rate is not affected by a factor 4 higher stirrer speed. It seems from a long duration experiment with catalyst recycle that the catalyst decays in time. It is assumed to be due to contaminants (dyes) present in the waste PET. Additional experiments with the spent catalyst reveal using nitrogen with 1% oxygen for 5 min at 400 °C restore the catalyst activity.

The chemical engineer must make process concept design including a reactor concept design.

Q1: Is liquid-phase residence time distribution a parameter for selecting a reactor?
Q2: Is gas phase residence time distribution a parameter for selecting a reactor?
Q3: Is the solid catalyst phase RTD a parameter for selection a reactor?
Q4: What promising reactor types should be considered?
Q5: What reactor type is ranked top for the concept design?

10.6.3 Industrial exercise 3: reactor family-type selection ideation stage

Given: In a catalyst manufacturing company a new catalyst is made for hydrodesulfurization in refineries. The expected commercial-scale market is 10,000 ton/year. First experiments in a microreactor lab scale are promising. The experimental results on the selectivity and activity of the catalyst are presented to management. The manager asks how should we in the end make this catalyst impregnation? Our existing impregnation facility has a small capacity, so we will have to build a new facility. There are two options: moving belt and a fluid bed. In both cases, the impregnation is by spraying the impregnation solution on top of the moving belt on top of the fluid bed.

Question: Which option is preferred?

10.7 Takeaway learning points

Learning point 1
A qualitative reactor type selection in the ideation stage is provided.

Learning point 2
Reactor type selection in the ideation stage can be enormously beneficial to the project outcome.

Learning point 3
Reactor selection in the concept stage requires reaction schemes.

Learning point 4
Reactor type selection in the feasibility stage includes risks and the cost of de-risking for various reactor types.

References

[1] Van Swaaij WPM, van der Ham AG, Kronberg AE, Evolution patterns and family relations in G–S reactors. Chemical Engineering Journal, 2002, 90(1–2), 25–45.
[2] Krishna R, Sie ST, Strategies for multiphase reactor selection. Chemical Engineering Science, 1994, 49(24, Part A), 4029–4065.
[3] van Helvoort T, van Veen R, Senden M, Gas to Liquids – Historical Development of GTL Technology in Shell. Amsterdam, Shell Global Solutions, 2014.
[4] Wittgenstein L, Philosophical Investigations. Oxford, UK, Basil, 1953.
[5] Harmsen J, Verkerk M, Process intensification – Breakthrough in Design, Industrial Innovation Practices, and Education. Berlin, De Gruyter, 2020.
[6] Harmsen GJ, de Haan AB, Swinkels LJ, Product and Process Design – Driving Innovation. Berlin, De Gruyter, 2018.
[7] Harmsen J, Industrial Process Scale-up: A Practical Innovation Guide from Idea to Commercial Implementation. Amsterdam, Elsevier, 2019.

11 New reaction systems through all innovation stages

The purpose of this chapter is to reveal how a new reactor project can be planned and executed from ideation stage to commercial-scale start-up.

11.1 Introduction

Any reactor with a new feed and/or a new product that has not been implemented at commercial scale for that feed or product before is a new step in a process [1]. This means that the reactor and its surrounding process steps need to go through all innovation stages, if the company wants to make a success out of the new process. To that end, we provide guidelines for reactor design, modeling, and experimental validation for each innovation stage (reactor type selection is treated in Chapter 10).

11.2 Ideation stage (also called discovery stage, or early research stage)

11.2.1 Ideation stage design

In the ideation stage of a new reaction, little design of the commercial-scale reactor will, in general, take place. The focus will be on the proof of principle of the new reaction. However, it is advantageous to also spend some effort on the reactor type selection for the commercial scale, as discussed in Chapter 9, and it is also advantageous to spend some time on reaction engineering design. The design can be a simple block flow diagram showing a reaction function as a block and only showing input and output flows. If several reaction steps are involved, then several function blocks can be drawn. Even the complete process can be designed using the functions necessary to get from the process input to the process output. Function combinations such as reactive distillation or reactive extraction can then also be drawn. The resulting design may result in a temperature window and, perhaps, also in the concentration window of the reaction system.

The advantage of having some information about the envisaged commercial scale is that the proof of principle experiments can be better defined, and the reporting of the results will be more suited for passing the stage gate to the concept design.

Examples of these advantages are shown in the Shell shale retorting and combustion reactor system, Section 13.8, and in the Eastman reactive distillation cases of Section 13.7.

https://doi.org/10.1515/9783110713770-011

11.2.2 Ideation stage modeling

In ideation stage, very simple modeling of the reaction engineering design sketch can be beneficial. For instance, simple input and output streams can be calculated assuming stoichiometric reactions with complete conversions.

Also, a simple heat balance can be determined assuming no heat exchange, hence, just determine the reaction heat and the adiabatic temperature change due to the reaction. From that calculation, it may be concluded that the reactor can be designed adiabatically or will need heat exchange.

11.2.3 Ideation stage proof of principle experiments

Guidelines for proof of principle experiments for a new reactor type are hard to provide. It requires the ingenuity of the researcher to design and make a simple laboratory-scale test unit that will convince him and his supervisor that the idea may work. It will also depend on what the technical service department can construct.

11.3 Concept stage (also called research stage)

11.3.1 Concept design

11.3.1.1 Optimization target for concept design

The reactor concept designer should focus on by-product (waste) reduction, as in the majority of design cases, feed cost is by far the larger part of all costs. In chemical processes, feed cost is 70–80% of all cost. Investment cost, energy, and labor are the other cost items. In reaction engineering terms, this means that the highest selectivity should be obtained by reactor design. In the sections below, this design target is worked out into guidelines.

11.3.1.2 Reactor concept design in relation to process concept design

In the concept stage, the process concept is defined. This means, amongst others, that the reactor type is to be selected. This reactor type selection is treated in Chapter 10 in detail, for each stage including the concept stage.

The reactor conditions and the reactor size have also to be determined in this stage. This determination, however, should be done in combination with the unit operations surrounding the reactor section. We follow the sequence of process concept design advocated by Douglas in his book concept design of chemical processes [2]. The sequence is to start at the input and outputs of the process, so start at the process conditions and then work from outside to the inside. However, when the sequence has ended at the reactor and its conditions, we advocate reexamining the process concept design and the reactor concept design from the reactor outwards to the process design boundaries.

Let us then consider the most important elements for reactor design in relation with its surrounding unit operations, using the Douglas sequence.

11.3.1.3 Feed composition design

We start with the question, should the feed composition to the reactor be the same as initially decided in the ideation stage, or should it change? In the ideation stage, a solvent may have been chosen by the chemist in the ideation stage to obtain the reacting component in the liquid phase and not as a solid. The process concept designer, however, may review this initial choice. He will check whether he can select one of the process streams of the process concept to fulfil this solvent function. This means that he has already made a rudimentary process concept design, so he knows what streams are available to him as options to use them as solvent for the feed component. Here, you also see that making a process concept is not a linear set of steps from beginning to end, but that it comprises one or more iterative steps.

The feed may be diluted stream, meaning that the reactant for the reaction is diluted with an inert. An example is oxygen diluted with nitrogen because air is taken as the feed stream for the oxidation reaction. The process concept designer may calculate the additional cost of using pure oxygen and the reduction cost by this higher oxygen concentration. If the selectivity in the reactor is increased by using pure oxygen, then, in most cases, the feed cost reduction of the other feed component by this higher selectivity will compensate for the additional cost of pure oxygen.

11.3.1.4 Single-pass reactor conversion design

If the reaction kinetics show that the desired product reacts in a consecutive reaction to an undesired by-product, then it is, in most cases, better to have a low single pass conversion of the feed and separate the unconverted feed component from the product in a separator after the reactor and recycle the unconverted feed back to the reactor inlet.

In some cases, a process intensified option can be designed, where the reaction and the separation are combined in a single piece of equipment such as in reactive distillation or, more general, in multifunctional reactors. This subject is worked out in more detail in the section on process intensification.

11.3.1.5 Residence time distribution design

When a consecutive reaction of the product to an undesired component plays a role, then the residence time distribution (RTD) of the reactor should be close to plug flow, because in that case, the product component concentration in the reactor is always the lowest.

11.3.1.6 Product removal from reaction phase

For undesired consecutive reactions of the product component, rapid removal from the reaction zone is an alternative option to be considered. If the reaction phase is a liquid or a liquid with a heterogeneous catalyst, then removal of the product to the gas phase is an interesting option to consider. Options available are reactive distillation and a horizontal bubble column. In the latter, the gaseous product can directly go the gas phase and then quickly leave the reactor.

If the reactor is a gas–solid system, then a horizontal fluid bed can be an attractive option. The product leaves the reactor quickly at the top of the reactor. This option was applied in the Shell shale process design, as described in Chapter 13.

11.3.1.7 Reactor conditions, pressure, and temperature selection for highest selectivity

The reaction kinetics has to be examined first, in order to find the optimum pressure and temperature for highest selectivity. If the desired main reaction is favored over by-product formation by a high pressure (high concentration of gaseous reactant) then choose that even if the investment costs of the reactor increase enormously. In the economic evaluation of the process concept with the reactor, the high pressure–high yield option can be compared with the low pressure–low yield option. The total cost of the high pressure–high yield option will convince the designer and management that it is, indeed, the right choice.

The reactor temperature should also be chosen such that the maximum yield is obtained. Reactor modeling and simulation will show the optimum temperature for highest yields of product on feedstock.

11.3.1.8 Operation mode batch versus continuous

The choice of operation mode batch versus continuous should not be taken for the reactor in isolation, but should be taken for the whole of the new process. In a batch reactor, often, several functions such as heat up, reaction, and separation (crystallization) can be fulfilled. For process capacities >5,000 ton/year, Douglas recommends the continuous operation [2].

For small capacities when only gases and liquids are involved, however, often, process-intensified continuous operation is also attractive. The next section treats this subject.

11.3.1.9 Process intensification as a concept design method

Process intensification (PI) is an innovative approach to process concept design, which is, in some respect, also useful for reactor concept design. The definition of PI reads:

Process Intensification is a set of radically innovative process design principles, which can bring significant benefits in terms of efficiency, cost, product quality, safety, and health over conventional process designs, based on established unit operations [3].

Process intensification has developed into a large academic field [4] with a comprehensive textbook on how to research these radically innovative process principles and what has already been achieved [5]. How to educate PI is also reviewed and described [6]. Also, practical industrial process design and innovation methods are available [3, 7]. Specific PI designs have been implemented at commercial scale. Those with a reaction involved are amongst others: reactive distillation, reverse flow combustion, and fluid bed rotating cone reactor. Descriptions of these and others are provided by Stankiewicz [5] and Harmsen [3].

The field is too large to be discussed here, extensively. However, the main connections between PI and multiphase reactors can be shown using the functional design step (FDS) process design method [3, 7]. The steps involved are:

Step 1: Define functions to transform process input streams to output streams.
Step 2: Integrate functions.
Step 3: Select dynamic building blocks.
Step 4: Select thermodynamic building blocks.
Step 5: Select equipment.

Now. the steps are briefly described in view of their relevance to multiphase reactor design with a focus on byproduct minimization.

Ad Step 1: Focus on defining process functions such that by-product formation is avoided, and only essential functions are taken into consideration to transform the input stream to the output stream.

Reaction systems with no by-product formation, first of all, call for finding and developing highly selective catalyzed reactions. Second, is to remove the product from the reaction phase rapidly, such as in reactive distillation. Third, keep the residence time short. Fourth, have a reactor with excellent heat transfer capability, such as in a high heat exchange reactor, so that the optimum temperature can be obtained everywhere in the reactor.

Ad Step 2: Integrate functions, such as reaction and separation, as much as is possible into a process unit. Examples are the Eastman methyl acetate process in which seven reaction and separation functions are integrated into a single column [8] and the BTG-BTL biomass pyrolysis process, in which in a rotating cone fluid bed reactor, heat transfer by hot sand, mixing, and reaction takes place in single vessel [3].

Ad Step 3: Select dynamic operations such as a reverse flow packed bed reactor, in which the flow direction changes every few minutes by switching the feed from one side to the other side. The center part of the reactor remains at a hot temperature, while the input and output streams have a low temperature.

Ad Step 4: Consider alternative energy supply to the reactor such as microwaves for ultra-rapid and uniform heating.

Ad Step 5: Select equipment with high heat and or mass transfer performance such as a rotating packed bed reactor for high gas-liquid mass transfer rates, or as micro-channel reactor for high heat transfer rates.

11.3.2 Concept modeling

In the concept stage, reactor modeling starts with simple flow sheet modeling in which the selected reactor type may be modeled as a plug-flow or backmixed model. Heat control modeling may be chosen as an adiabatic or an isothermal reactor. Optimum reactor conditions such as temperature and pressure for minimum by-product formation can be determined once the reaction kinetics has been determined.

When the reaction conditions window for pressure and temperature have been established from the kinetics obtained and a reactor type has been selected, a mini-plant process can be designed as a downscaled version of the commercial-scale plant. Reactor modeling can now be more elaborate and closer to the real reactor, including now mass transfer between the phases and also heat transfer and temperature profiles in the reactor. Experimental mini-plant results can then be used to validate the model.

If the reactor type selected is a fluid bed or slurry reactor, then modeling will include 3-D CFD modeling for mass transfer and RTD behavior of the commercial scale. Specific internals for the commercial-scale reactor may be designed and optimized, using CFD modeling.

11.3.3 Experimental validation

The reactor concept design and its model should be validated experimentally. In most cases, this can be done by a mini-plant or pilot-plant process, including the real reactor type chosen. If the reactor is a fixed bed, then a small diameter reactor of a few cm can often be chosen. If the commercial-scale reactor type is an adiabatic reactor, special precautions have to be taken to insulate the small-scale mini-plant reactor. If it is a multitubular reactor, then a single tube reactor can be chosen, if possible, with the same tube diameter and length as anticipated for the commercial scale.

If the commercial-scale reactor is a fluid bed or slurry-type reactor, then a small-scale reactor is harder to design. Also, not all features of the commercial-scale design can be maintained in the small-scale design. The large-scale circulations of the commercial-scale fluid bed or the slurry bubble columns will not occur in the downscaled version.

Experimental validation of the CFD model by a mock-up cold model will, in general, be carried out in the development stage. Designing the mock-up model with its costing will, in general, be executed in the feasibility stage. However, having the cold mock-up model prior to the hot pilot plant is advantageous. The commercial-scale plant and the pilot plant can then be designed with more confidence, and development time is shorter. More information is provided in Section 11.4.

11.4 Feasibility stage design (also called first part of development stage)

11.4.1 Introduction

The purpose of the feasibility stage is to determine the feasibility of the new commercial-scale design and whether its benefits are sufficient to warrant the execution of the development stage to reduce the risks of commercial-scale failure. To that end, the feasibility stage treats two main subjects:
- a commercial-scale design and its economics
- a development plan content for risk reduction and its cost

In the feasibility stage, a commercial-scale reactor (and process) design is needed for three reasons. The reasons are obtaining a commercial-scale design that
- is safe, healthy, environment-friendly, economic, technically feasible, and sustainable;
- facilitates a downscaled pilot plant version with a test program; and
- evaluates the need for a development program.

If no feasibility stage with a commercial-scale reactor design is included in the development plan, often a small-scale pilot plant is designed, constructed, and a test program is carried out. After that development, a commercial-scale design is made based on the pilot plant. The designer of the commercial-scale reactor (and process) often then struggles. The pilot plant equipment has no commercial-scale equivalent, or the scale-up knowledge is absent for the pilot plant reactor. Hence, reversing this sequence makes sense.

11.4.2 Reactor development plan overview

The reactor development plan is to be derived from the critical performance factors of the commercial-scale reactor. This means that, first of all, the commercial-scale reactor has to be selected and designed. The reactor selection is described in Chapter 10 feasibility stage 10.5.2. The reactor designed in concept stage can be taken as the starting point for the commercial-scale reactor design, in this feasibility stage. The difference is that, now, an experienced process engineer is designing the commercial-scale reactor (and process).

The development plan can be made using the following elements; first, the critical performance factors are analyzed for the case at hand. Then, a scale-up method is selected. Furthermore, reactor modeling for design, a final validation of the design, and its modeling are in the plan.

11.4.3 Critical performance factors for commercial-scale reactors

We identify four key critical performance factors:
- Feed distribution
- RTD and mixing
- Mass transfer
- Heat transfer

All these factors are critical to the commercial-scale performance and are dependent on the configuration, dimensions, and scale of the reactor. Hence, these factors need special attention in the reactor design, scale-up, and scale-down. Each of these factors is discussed in the following sections.

11.4.3.1 Feed distribution

Feed distribution is the phenomenon that the feed entering the process equipment will be needed to be distributed over its cross-sectional area to obtain a good reactor performance, in terms of conversion and selectivity. This critical performance phenomenon is hardly treated in reaction textbooks. We deal with it, here, extensively, because it is often neglected during the design phase and then creates enormous problems in the new reactor at commercial scale.

Feed distribution at small scale: this distribution is often easily obtained by single point feed in equipment with a small diameter. Often, the phenomenon at small scale is not even considered to be important. Upon scale-up, the phenomenon is nearly always very important for the process equipment performance. If no information has been gathered in the concept stage on the effect of feed distribution parameters on the performance, then only the brute force scale-up method (see this section further ahead) is available for reliable scale-up. The feed distributor to be used for the commercial scale should be tested in a cold-flow test facility, or information from other industrial-scale applications should be obtained about its feed distribution quality.

11.4.3.2 Feed distribution fixed bed reactors

The feed distribution over the top of the fixed bed should be even. In our industrial experience, we have seen several commercial reactors where the initial distribution of the feed was very poor. In one case, the liquid feed entered the reactor via a central inlet of 0.5 m diameter, while the reactor diameter was 5 m, so the entrance velocity was 100 times higher than the averaged superficial velocity. This acted as a jet whirling the top layer of catalyst. Catalyst particles broke when hitting each other and the wall.

A special feed inlet device (e.g., a "Schoepentoeter") should be installed to distribute the feed evenly on the bed. In addition, inert large particles (ceramic balls) are often placed on top of the catalyst bed to increase even feed distribution.

In a trickle-bed case, the initial liquid distribution had 300 feed points per square meter. This means that the distribution points are 6 cm away from each other. This means that the bed has a very uneven gas and liquid flow. Gravity is the main force acting vertically on the liquid, so radial liquid distribution is very small. For exothermic reactions, this also can mean locally different temperatures, which then propagate through the whole bed. In the particular case, product selectivity was not reached, and the technology provider had no clue why this happened. He proposed a new catalyst bed to solve the problem. The plant operator did not like this proposal and asked for help to analyze the problem. We went to the trickle-bed specialist; he immediately concluded that the liquid distributor was not correct. With a new distributor; the selectivity problem was solved.

Technology providers may claim to have propriety inlet designs. The client; however; should ask for large-scale experimental proof of the performance. An example of published inlet device for single phase distribution is the Shell "Schoepentoeter." It is used for gas and for liquid distribution. An entry into the device information is given in [9].

11.4.3.3 Feed distribution fluid bed and slurry reactors

For fluid bed and slurry reactors, the feed distribution should be properly designed, not only for a good distribution but also for mechanical integrity. Commercial-scale fluid bed and bubble columns have very large vertical internal circulations of several meters per second. These cause enormous forces on the distributor. This happens in particular when the distributor is a set of tubes inside the vessel. Special precaution should therefore be taken in distributor design.

11.4.3.4 Closing remark on feed distribution

Although feed distribution is related to both macro mixing and micromixing, it should not be confused with these latter two scale-dependent phenomena. We will describe these in the next two sections.

11.4.3.5 Residence time distribution and mixing

Reactor selection in view of RTD and mixing is discussed in Chapter 10. Design choices in relation to RTD and mixing are discussed in Chapter 4. One general remark is made. If some of the reactions involved are fast and take place in the gas or liquid phase, then mixing intensity can be very important and Chapter 4 should be consulted.

11.4.3.6 Mass transfer

If the reactor involved is multiphase, then mass transfer is an important subject. Chapter 10 deals with the subject in relation to choosing the best reactor type. Chapters 5 and 6 can then be used to design the reactor for optimum mass transfer.

11.4.3.7 Heat management and transfer

Reactor type selection in relation to heat management is discussed in Chapter 10. Reactor design in relation to heat management and transfer is discussed in Chapters 5 and 6.

11.4.3.8 Critical performance factors for easy to scale up reactor types

In this feasibility stage, an estimate has to be made about the content and cost of the development program. This section is about reactors (such as fixed bed reactors) where most information for making a commercial-scale design is known. Given the information generated in the concept stage on the reactions and the physical data and the derived knowledge on what is the best reaction engineering concept in terms of RTD, mass transfer, heat transfer, heat control, catalyst deactivation, and reactivation, a commercial-scale design can then be easily made without the need for big assumptions.

11.4.3.9 Critical performance factor for hard to scale up reactor types

To determine the development program for reactor types, such as three-phase slurry reactors, the critical performance factors may be useful. Table 11.1 lists them for a three-phase slurry reactor in which the scale-up aspects are highlighted.

Table 11.1: Critical performance factors and their scale-up aspects for three-phase slurry reactors.

Performance factor	Gas phase	Liquid phase	Solid phase
RTD	Wide, uncertain	Wide, staging feasible	Wide, staging feasible
Mixing	Uncertain	Low, uncertain	Segregated flow
Mass transfer	Limited, uncertain	High, Sh = 2 certainty	High for fine solids
Heat transfer	External loop option	Medium, minimum	Medium
Shear rate distribution	Wide, uncertain	Wide	
Impulse transfer	Wide high, uncertain,	High on internals	Low

The development program for the three-phase slurry reactor will then include:
- Redesign of the commercial-scale reactor, to minimize the uncertainty.
- Modeling of all critical performance factors and simulating the performance.
- Simulations including optimization and finding design parameter value combinations, such that the performance uncertainty is reduced for input uncertainties.
- Model validations by a hot pilot plant.
- Model validations for RTD and mass transfer in cold-flow experimental setup.

11.4.4 Reactor scale-up methods and applications

In the feasibility stage, a scale-up method has to be selected. The following scale-up methods are available unit operations, hence also for reactors. These are, brute force, model based, empirical, hybrid empirical-model, dimensionless numbers, and no-scale-up [10]. Some guidelines are presented here to choose the best scale-up method for the case at hand.

11.4.4.1 Brute force scale-up method

In this method, all critical performance factor values are kept the same in the pilot plant and the commercial-scale plant. This means that the commercial-scale plant dimensions and conditions have been decided, so that these dimensions and conditions, and feed quality, can be kept, and are kept the same in the pilot plant, except for the production capacity. The method is called "scale-out," and "numbering up" if the commercial-scale reactor is a multiple of reactors in parallel, equal to the single reactor of the development stage.

This method is, for instance, applied for fixed bed reactors and reactive distillation reactors [10]. For those applications, the following parameters are kept the same for the pilot plant and the commercial-scale design: feed qualities, reactor height, superficial velocities, catalyst, catalyst particles shape and size, pressure, and temperature.

The method is applicable for all cases where the hydrodynamics are governed by a small-scale structure and not by the outside reactor dimensions, that is, packed bed reactors, structured packing reactors, and multitubular reactors. If the commercial-scale reactor is a multitubular reactor, then the pilot plant may contain three tubes to test the even feed distribution and the even catalyst loading procedure over more than one tube.

The advantage of the method is that it is very reliable. The only uncertainty is the initial feed distribution of the commercial scale.

The disadvantages of the method are: a) that the cost of the pilot plant is higher than for other scale-up methods; and b) that it is essential that the commercial scale be designed with confidence. If that is not the case, then the pilot plant test results will show the design failure, and so the high cost of the pilot plant is lost.

11.4.4.2 Dimensionless number scale-up method

The dimensionless numbers scale-up method has two variants. The first variant is for reactors where RTD, mass transfer, and heat transfer performance are predicted by dimensionless numbers Péclet, Sherwood, and Nusselt as correlations with Reynolds and relevant other dimensionless numbers, Schmidt for mass transfer, and Prandtl for heat transfer. Correlations are found in Chapters 4, 6, and 7.

The scale-up method then consists of testing the reactor performance case at hand, based on those correlations and validating the correlations for their quantified predicted effects, and whether fluid velocities effects on the performance are correctly predicted.

This variant scale-up method is applicable for all reactors where dimensionless number correlations are available, so this applies to many tubular reactors, packed bed reactors, and specific structured packing reactors.

The second variant is called dimensional analysis and scale-up and is described below.

11.4.4.3 Dimensional analysis and scale-up method

We briefly discuss here, a generic method called Dimensional analysis and scale-up. The first step is listing all dimensions and all variables of the phenomena involved. Then following the method, all dimensionless numbers for the case are generated. This first step is called Buckingham pi (π) theorem. The second step, the scale-up step, is keeping each dimensionless number of the pilot plant unit operation and the commercial-scale unit operation at the same value. The method and its applications for fluid process unit operations are described extensively by Zlokarnik [11].

Many researchers have investigated this scale-up method for gas–solid fluidized bed. Anderson generated dimensionless numbers for the fluid mechanical description already in 1967 [12], and subsequently, many academic researchers have tried to validate his dimensionless numbers of variants of his numbers, experimentally or by modeling. All proved that the method is not reliable. A recent paper by Rudisuli shows this in some detail [13].

Van Deemter, in his review article [14] on basics of process modeling, questions the generic method on two aspects. His first question is how far one can go in reducing the number of independent variables (and thereby increasing the number of dimensionless groups), and his second question is, whether there are any rules which dictate what variables should be included in the dimensional analysis.

This second variant of dimension analysis and scale-up is not applied in any process industry, as far as we know. This is probably due to the fact that the other scale-up methods described here have been proved to work and also can be easily communicated to management, while this dimensionless analysis and scale-up is hard to communicate. That it has been proven to be unreliable for the fluid bed case does not help. Our conclusion is that this second variant is unreliable.

11.4.4.4 Model-based scale-up method

Computational fluid dynamics (CFD) models can now be used to predict local flow patterns, mixing rates, and reactions in complex geometries. The models can also predict turbulent dynamic behavior. For single-phase reactors with fast reactions and complex configurations, such as in furnaces, these models are very useful to optimize the commercial-scale configuration. Also, for gas–liquid reactors such as horizontal bubble columns with baffles, CFD models are very useful to optimize gas distributor and baffle geometries. Hence, this is model-based scale-up.

The CFD model needs, of course, to be validated for the most critical elements. The geometry effect predictions on flow pattern and mixing can be validated in large-scale cold mock-up experimental setups. It may also be validated in an existing reactor using tracer experiments, such as in Section 13.5.

The advantage of model-based scale-up is that complex geometry optimization can be quickly carried out, and a far better optimum design can be found than with a pure experimental approach. The disadvantage is that the model may not be valid for the problem at hand, but this is unknown and may not appear in the experimental validation.

Chapter 13 shows several applications of CFD modeling for commercial-scale reactor optimization.

11.4.4.5 Empirical scale-up method

The empirical scale-up method is a step-by-step approach. Each next step is at a larger production capacity. The effects of parameters such as stirrer speeds, temperature, and recipe components are studied at the first step and then used in the next scale step to get the desired product quality again. Often, 4 or 5 capacity steps are used to get to the final commercial-scale production.

The method is mostly applied in pharmaceuticals and fine chemicals, with a mechanical stirred tank reactor mostly operated batchwise. The five reactor scales are often already available, so that only the experiments at each scale have to be executed. In this way, often, the scale-up is carried out within a year. Hence, it is also a rapid scale-up method.

The advantages of the method are that, first of all, it is fast if the reactors are already installed. For companies in pharmaceuticals, this is nearly always the case. Second, it does not need a large research program to study all phenomena, such as kinetics, emulsion formation, mass transfer, and heat transfer.

The disadvantage is that the existing reactors are used, and the best reactor type for a case is not chosen. Because of the heat exchange limitations of the mechanical stirred reactor, often fed-batch operation is applied with a slow addition of a limiting reactant. The whole reaction time then takes far longer than in a dedicated high heat exchange reactor. This long reaction time often causes significant by-product formation by consecutive reactions of the product.

11.4.4.6 Hybrid empirical-model-based scale-up method

Hybrid empirical-model-based scale-up is a powerful method for cases where the empirical scale-up model is costly and takes a long time. By making a model that captures some critical aspects of the reactor performance and also predicts, to some extent, the effect of scale-up, the scale-up cost and time can be reduced. As it is hard to write down a completely generic method, a specific example is worked out.

An example is fermentation for the production of pharmaceutical intermediates, which are called Active Product Ingredients in that industry. The reactor type is mechanically stirred reactor operated in fed-batch mode.

The empirical scale-up method is carrying out the production in a 1 L fermenter called scale 1. This comprises optimizing the parameters temperature, stirrer speed, pH, and batch time, and also determination of the effects of these parameters on the production rate and product quality. Then the production is carried out in the next scale 50 L fermenter called scale 2. Again, the conditions are optimized to obtain the desired product quality and to optimize the production rate by using the trends for parameter effects of the 1 L scale. This method is then applied at 3 m^3 scale called scale 3 and, finally, at a 200 m^3 scale, the commercial production scale called scale 4.

The hybrid model consists of determining some simple lumped kinetics and some model for the oxygen mass transfer by CFD modeling. The observed experimental trends are used to validate the model for some predictions. In addition, some measurements of the oxygen concentration in the liquid are used to fit the CFD model to the experiments for scale 1. Then, in step 2, the model is used to predict the performance of scale 3 and the required parameter values. A limited number of experiments are carried out to validate the CFD model and also to adjust parameter values to optimize the product quality and production rate. Then, in step 3, the model is used to predict the commercial-scale performance and optimized parameter values. The commercial-scale production is started, and some parameters will need some adjustments. The method reduces the scale-up time and effort and also a better combination of parameter values may be obtained for product quality and production rate.

11.4.4.7 No-scale-up method

The no-scale-up method has two variants. The first variant is that, in the concept stage, a reactor is selected and then used for a few hours to produce samples for testing and later, the same reactor is used for commercial-scale production, but now run continuously. The reactor itself can be a process-intensified reactor, and hence has high heat and mass transfer capacities and near plug-flow behavior.

We have heard in corridors of congresses that this method is used for pharmaceuticals development and commercial-scale production for continuously operated tube reactors, also called flow chemistry. Corning also advocates this method on their website [15].

The second variant is that directly after lab-scale experiments, the commercial-scale reactor is designed and constructed. In the design, all risks are minimized by selecting unit operations and equipment, which the designer believes will work certainly and by oversizing to account for lack of precise information of kinetics, mass transfer, and heat transfer.

Independent Project Analysis (IPA) has analyzed over 10,000 implemented process projects and has come to the conclusion that this scale-up method can be successful only if the project is a single reactor with no process recycle stream involved. In all other cases, the method failed [16, 17]. The statistical definition of failure by IPA is that over 30% additional cost was added to the investment during start-up and over 38% more time was needed than scheduled for [18]. In many cases, no significant amount of product was produced at all [16].

11.4.4.8 Avoiding scale-up uncertainties by design

In some cases, the critical scale-up phenomena, RTD, mass transfer, and heat transfer can be avoided by changing the reactor design. The effect of RTD is, for instance, avoided if the reactor is designed such that it has a uniform concentration of reactants, for instance, by a low single pass conversion (and consecutive product reactions do not take place). Mass transfer limitations may be avoided by designing for surplus of mass transfer capacity. Heat transfer limitations in the reactor may be avoided by low single pass conversion allowing for adiabatic reactor design and having an external heat exchanger in the external recycle stream.

The development program can then be much shorter. An example case of such a design and scale-up approach is found in Chapter 14.

11.4.4.9 Modeling plan

Guidelines for a modeling plan are that:

First, the purpose of the modeling should be defined. Options for a modeling purpose can be: Optimization of the reactor system, be it the reactor internal shape, the number of reactors in series, optimizing the reactor conditions such as pressure, temperature, residence time, residence time distribution, or other aspects.

Second, the modeling tool should be selected.

Third, a modeler, or a team of modelers, should be selected.

11.4.4.10 Reactor test facilities: various scales

New reactors or new reaction systems need to be tested on their performance. Often, the test results are also used to validate reaction engineering models. To limit the cost of operation and feedstocks, the smallest scale is preferred. In most cases, the new reaction systems, and/or new reactors are part of a new process. This means that the reactor is often tested in an integrated small-scale plant. There are many names and many scales used in the industry. To clarify the nomenclature, Table 11.2 is made using Vogel from BASF [19], de Haan, DSM [20], and Illg, IMM [21], so from three different sources. Therefore, the table is not internally consistent. The values listed can easily vary one order of magnitude. Each test facility is briefly described.

Table 11.2: Names and purposes of test plants.

Names	Volume 10^{-3} m^3	Reactor type	Purpose
Flow chemistry microstructured reactor	0.01	Microchannel,	Catalyst screening, proof of principle, no-scale-up commercial production
Laboratory tests	0.1	Stirred	Proof of principle
Micro-plant, bench scale	0.1	Fixed bed	Proof of concept
Mini-plant	1	Fixed bed	Design and model validation
Pilot plant	10	Fixed, fluid bed	Design and model validation
Demo plant, test plant	10,000	Fixed, fluid bed	Risk reduction commercial scale
Cold-flow, Mock-up	Large	Complex new	Validate CFD model or design

11.4.4.11 Flow chemistry microstructured reactor

Flow chemistry is a name used in fine chemicals and pharmaceutical industry for reaction and separation experiments at laboratory scale in special microstructured equipment operated continuously. The equipment is designed to ensure plug-flow behavior plus very high mass and heat transfer. In this way, rapid and deep conversion can be reached and by-product formation is reduced. There is a whole catalogue of these microstructured reactors, mixers, heat exchangers, and separators. The reader is referred to work by the group of Hessel [21, 22] for an overview and detailed description of many microsystems. The equipment is very suitable for chemical route selection and rapid catalyst screening.

For two-phase flow, gas–liquid and liquid–liquid, the microchannel Taylor flow reactor is very interesting. It is a co-current flow of blobs of gas and a continuous phase liquid through a narrow channel. The blobs have a high mass transfer with the liquid phase due to the thin film between the blob and the channel wall. The gas (or second liquid) circulates inside the blob, also causing higher mass transfer. A good entrance to the information is provided by Etminan [23].

Schrimph describes all relevant correlations on RTD, mass transfer, and heat transfer for reaction engineering concept design. He also describes classes of potential industrial applications [24]. Technology provider LPR Global also provides a list of potential industrial applications [25].

The Taylor flow reactor is particularly useful for lab-scale kinetic studies of two-phase and three-phase reactions. Paunovic, for instance, studied kinetics of the reaction to hydrogen and oxygen with a wall-coated catalyst [26]. This Taylor flow reactor is also useful for determining kinetics of photocatalyzed reaction systems, as light can easily by supplied via the transparent wall, as Su showed this in her study on gas–liquid photocatalyzed reactions [27].

11.4.4.12 Laboratory test setup

There are many types of laboratory reactors and methodologies for determining reaction rates and kinetics. These have pros and cons. This field is full of pitfalls and, in our experience, the effort to conduct an accurate kinetic study is nearly always underestimated. This topic is too broad for an in-depth discussion here. A valuable source of information is provided by the Eurokin organization [28].

Most of the literature focuses on one- or two-phase systems only. For three-phase heterogeneously catalyzed reaction systems with particle sizes of a few millimeters or smaller, the laboratory tests can be carried out in a small fixed bed, provided they are done with great care. Special design rules to avoid wall flow and wide RTDs have to be taken [29]. Another type of reactor that has been successfully developed over the last two decades is the so-called Robinson-Mahoney reactor, see e.g. Lauwaert et al. [30].

11.4.4.13 Micro-plant, bench scale

A micro-plant is the smallest version of the most important elements of the process. Its purpose is to quickly establish which process elements may work [19].

11.4.4.14 Mini-plant

A mini-plant is in fact a small-scale integrated pilot plant. For gas–liquid processes with fixed bed reactors, this is the preferred scale for the integrated pilot plant.

11.4.4.15 Pilot plant

A pilot plant is an integrated downscaled version of the commercial-scale design. Its purpose is to validate the process design and to validate the flow sheet model of the commercial scale process.

11.4.4.16 Demo plant

A demo plant is, by the Technology Readiness Level definition, the first commercial-scale production plant of the innovation project [10].

In various companies, however, the term is loosely used. In some cases, a pilot plant is called demonstration plant. Also, it is mentioned that no demonstration plan will be built, but the commercial-scale plant will be built directly.

11.4.5 Cold flow test rigs

Cold flow test rigs are large configurations through which water and nitrogen or air is flowing, and for three-phase systems, also solids. The configuration is the same as the commercial-scale reactor. Often, the geometric size is close to the commercial-scale reactor size. Often, the walls are transparent, so that flows can be visually observed.

The purpose of the cold flow test rig is to find optimum configurations for the reactor and its internals to validate a CFD model for gas, liquid, and solids flows.

Cold flow test rigs are chosen for complex reactor geometries to study flow patterns and for mass transfer rates. Often, the geometric size is close to the commercial-scale size, as the feeds in most cases are water and air, or nitrogen; so the feed costs are low. Often, the test rigs are made of transparent material, so that flow patterns can be seen, filmed, and photographed using color traces. RTD and mixing behavior can also be studied using tracer injections and measuring the response. These test rigs are also very useful in determining the best configurations and design details of complex reactor shapes and internals.

Due to complex geometries, the investment costs are considerable. The investment cost estimate should therefore be part of the feasibility stage. The detection technique should also be selected, as these can also be costly. Detection techniques for flow patterns can be simple color trace injection and a fast camera. However, the detection is then only for a pattern close to the wall and not a 3-D detection.

Cold flow test rigs may be not a proper name for all test rigs with these purposes. Prins, for instance, had a 2-D fluid bed model of transparent material operating at 500 °C [31]. Some call these test rigs mock-up models.

11.4.5.1 Flow pattern detection methods

Nowadays, there are many multiphase flow pattern detection methods available. It is beyond the scope of this book to describe all these methods. Here are a few methods with references.

For gas–liquid flow patterns, capacity, resistive, pressure fluctuations and electromagnetic methods are described by Keska [32]. Classic methods using dye injection and filming the flow behavior also still provides a lot of insight. This holds, of course, only for transparent mock-up rigs.

For gas–solid fluid beds, direct visualization, tomography, optical probes, capacitance probes, and pressure fluctuation measurements are available. A review is presented by van Ommen [33].

For three-phase flow systems, Shakya describes X-ray tomographic methods [34].

Section 13.3 shows detection methods applied in a commercial-scale bubble column reactor.

11.5 Development stage

11.5.1 Introduction

In the development stage, all major risk items (identified in the feasibility stage) are addressed. Most of them are reduced by experimental setups such as a pilot plant and or a cold flow test rig, as described in Section 11.3. The test results are used to validate the design, to validate models, and to validate scale-up theory. The results of all those investigations are put into a front end engineering design (FEED).

11.5.2 Pilot plant and test program execution

In the feasibility stage, a preliminary pilot plant and costing has been made. In the development stage, the pilot plant is to be designed in detail, constructed, and commissioned, followed by a test program to validate the commercial-scale design, and to validate models.

11.5.3 Front-end engineering design

Front-end engineering design (FEED) is the most important deliverable of the Development stage [17]. This deliverable has two purposes. First, it provides all information on which the decision can be taken to pass the development stage-gate into the engineering, procurement, and construction (EPC) stage. Second, it is the document used by the engineers in the EPC stage as input for all their detailed design.

The FEED contains, according to the authors experience and supported by Bakker [18] and Vogel [19], the following documents:

- Development results and evaluation
- Basic design engineering package
- Economic potential
- Safety
- Risk register updated
- Planning

The development evaluation document contains all risk items of the feasibility stage and all results of the development stage addressing these risk items. As multiphase reactors have, in general, gone through an extensive development addressing all main identified risks, it means that this development evaluation will be elaborate. It will contain reactor performance results in terms of yield, selectivity, capacity, and product quality. It will also contain catalyst fouling, deactivation, and reactivation. It will contain heat transfer performance as also long-term behavior of construction materials, such as corrosion, erosion, and fouling.

The basic design engineering package will contain a lot of detail on the reactor shape, sizing, and also details on reactor internals shape and sizing. It will also contain mass and energy balances, and Pressure and Temperatures, and profiles. It will also contain a flow sheet computer model containing for each stream the chemical composition. It may also contain other computer models for dynamic behavior and runaway behavior.

The economic potential determination needs various inputs. The major items are product market prices and feed market prices. For these, the marketing and sales department will be involved to forecast the economic potential. Often scenarios are used to determine the risks involved and how these risks can be reduced, for instance, by contract models with suppliers and clients.

A 10% accurate investment cost estimate will be made for the whole plant. For multiphase reactors such an accurate cost estimate is not easy to make, as a large part of the cost will be in the engineering of the reactor with its internals. Often, the reactor investment cost estimate is made using information from an engineering construction firm that specializes in detailed reactor design and construction.

Safety in reactor design and operation are always very important aspects. For multiphase reactors, the devil is in the detail, hence special safety studies; a detailed HAZOP and a Dow Explosion index analysis will be executed and reported. Also, experimental runaway studies will be reported.

A health statement is also needed, relating to components used in the process and components made in the reactor section, which can affect the health of the process operators.

Environmental aspects of the new process are also addressed in the FEED. For multiphase reactors, several items need to be reckoned:

- By-product destinations
- Diffusive emissions
- Global warming gas emissions due to utilities used

For by-products formed in the reactor, destinations have to be reported. In our experience, we have come across a case, where the process engineer stated that the by-product destination will be solved at a later stage. In the end, at the process start-up, the destination still had not been found and the plant had to stop production until a very expensive waste treatment solution was constructed.

Diffusive emissions by valves and connections: these should be minimized by reducing the number of process equipment with connections and by reducing external control valves. Instead, flow control by variable speed pumps should be asked for.

Global warming gas emissions caused by the use of utilities based on fossil fuels should be avoided. Utility use can be minimized by heat integration in which the reaction heat is utilized.

All risk types such as economic, technical, and operational, are all reported in a risk register.

Planning contains a project implementation plan, an execution schedule for the EPC and start-up stages, and an activity planning: who will do what in the next stages.

11.5.3.1 Critical aspects for commercial-scale design to be addressed in EPC stage

The authors of this book have often experienced mechanical failures of reactor internals such as gas distributors in fluid bed reactors, poor distribution of gas and liquid in reactors, catalyst breakage, consequences of that breakage for fluid flows through the reactor, and corrosion in certain unit operations such as heat exchangers.

It is beyond the scope of this book to expand on this subject, in detail. Here is just a short list of critical aspects to deal with, in the detailed engineering stage of the EPC stage:

- Feed distribution of gas and liquid in reactor
- Momentum transfer and pressure drop forces on internals and catalyst
- Local shear rates on catalyst particles causing breakage and attrition
- Local corrosion
- Local erosion due to high fluid (including slurry) velocity

The authors have seen many industrial start-up cases where the feed distribution designed by the EPC contractor was not good enough. Often, available design items such as a gas or liquid distributor constructed directly below the inlet pipe are not installed.

Experiences and observed problems in the pilot plant should be communicated to the EPC contractor.

11.6 Engineering, procurement, and construction (EPC) stage (also called execution stage)

11.6.1 Contractor choice and co-operation

The first important decision in the EPC stage is choosing the main EPC contractor for the whole process, and also subcontractors for important complex process equipment such as the multiphase reactor. The selection criterion for the main contractor is experience in the particular industry branch and even better experience in the specific process engineering design, procurement, and construction. The sub-contractor should have experience in the multiphase reactor, experience in the industry branch, and as a wish, experience in the particular application. The consequences of not selecting an EPC contractor experienced in the industry branch are described in detail for actual cases by Harmsen [10].

The second important decision is to involve the reactor development engineers of the manufacturer in the detailed engineering design. Harmsen describes methods to enhance this interaction between the development engineer and the EPC. He also describes that, for a complex multiphase reactor design of BTG-BTL, one of the BTG-BTL development engineers was temporarily moved to the EPC contractor organization. This constituted one of the reasons for the success of that process [3].

11.6.2 Reactor procurement and construction

The multiphase reactor should be procured from a construction firm specialized in that reactor type. The finalized construction should be checked in detail on whether it is in agreement with the detailed design.

11.6.3 Commissioning

Commissioning is the physical final check on the constructed installed process. Nitrogen and water flows are mostly used for checking and cleaning all pipes and most vessels. For the multiphase reactor often containing a catalyst, special commissioning tests are designed and executed. For multitubular reactors, special catalyst loading technique and procedure will have to be in place. Also, even packing density and even gas flows over each tube will need to be checked.

Also, all controls will be checked for a multiphase reactor with heat exchange; all flow, temperature; and pressure controls will be checked on the reactor side as well as on the coolant (or heating) side.

Coordination between the EPC contractor personnel and the manufacturer operators will be essential for a successful commissioning.

11.7 Start-up and normal operation (also called demonstration stage)

Guidelines and methods for start-up and normal operation for a new process also holds for a conventional process with a new reactor, as the new reactor makes it a new process. Not applying these guidelines and methods leads to start-up disasters. The methods that were successful and the disaster cases, including those with multiphase reactors,

are described by Harmen [10]. The critical success factors for start-up preparation are listed below.
- Potential problem analysis
- Precaution measures
- Complete start-up team and organization
- Experienced start-up leader
- Start-up support team
- Trained operators
- Start-up plan
- Documentation

11.8 Exercises

11.8.1 Industrial exercise: glucose to ethylene glycol

11.8.1.1 Context

A bulk chemicals company presently producing ethylene glycol from ethylene, which, in turn, is produced via steam cracking from a crude oil fraction, wants to develop a renewable feedstock based process. The first commercial-scale process should have a capacity of 100,000 ton/year.

The company sets up a project team to develop a process based on this chemical route. It has found literature that ethylene glycol can be produced from a biomass source glucose. A promising route appears to be a one-reactor option presented by Murillo [35]. The paper is open access type, so it can be freely downloaded and contains kinetic information and experimental results.

The reactions involved are:

1. Retro-aldol condensation of glucose to erythrose and glycolaldehyde	$C_6H_{12}O_6 = C_4H_8O_4 + C_2H_4O_2$
2. Glucose hydrogenation to hexitols	$C_6H_{12}O_6 + H_2 = C_6H_{14}O_6$
3. Glucose side reactions	$C_6H_{12}O_6 = xByP$
4. Erythrose conversion to glycolaldehyde	$C_4H_8O_4 = 2\ C_2H_4O_2$
5. Glycolaldehyde hydrogenation to ethylene glycol	$C_2H_4O_2 + H_2 = C_2H_6O_2$
6. Glycolaldehyde side reactions	$C_2H_4O_2 = yByP$
7. Methane production	$C_2H_6O_2 + 3\ H_2 = 2\ CH_4 + 2\ H_2O$

11.8.1.2 Experimental setup

The reactions have all been carried out in the presence of a porous catalyst. It is likely that the hydrogenation reactions are catalyzed and that the others are thermal reactions. However, specific information is not provided. The experiments have been carried out in a mechanically stirred reactor with hydrogen bubbling through the liquid. The reactor is operated in batch and semi-batch mode. In the latter mode, glucose is continuously fed to the reactor.

The chemical reaction engineer in the team has to answer following questions for each innovation stage.

Exercise: ideation stage

Q1: What is the maximum achievable yield of ethylene glycol on glucose using this chemical route?

Exercise: concept stage

Q2: What parameters will be investigated in the experimental program?
Q3: What lab-scale reactor type is to be used for determining the kinetics?
Q4: Which phases are involved in the reactor system?
Q5: Which critical performance factors play a role for this reaction system?
Q6: What reactor types should be considered?
Q7: What RTDs would you like to have for the gas phase, liquid phase, and solids phase?
Q8: ould mass transfer limitations of gas–liquid affect the selectivity?
Q9: Can the reactor be operated adiabatically?
Q10: Which reactor type would you select?

Exercise: feasibility stage

Q11: What unit operations are surrounding the reactor system for the commercial-scale process?
Q12: Would you integrate the reactor with a downstream separator and recycle unconverted feed components?
Q13: Which components would you recycle?
Q14: What are risks of the reactor type chosen by you?
Q15: What is the capacity and major dimensions of the pilot plant reactor?
Q16: What are the main elements of the pilot plant reactor development program?

Exercise: development stage

Q17: If the development reveals that the catalyst decays in less than 10 days, would you change the reactor type selection?

Exercise: EPC stage

Q18: What criteria would you apply for selecting the EPC contractor?

11.9 Takeaway learning points

Learning point 1

In ideation stage: Maximum yield of product on feedstock can be derived from stoichiometric reaction equations.

Learning point 2
In concept stage: Kinetics and physical properties have to be determined to be able to select and design a reactor concept.

Learning point 3
In feasibility stage: Designing a commercial-scale process helps integrate the reactor and optimize it.

Learning point 4
In development stage: Catalyst decay rates may change the reactor type selected in previous stages.

Learning point 5
In the EPC stage: An important criterion for EPC contractor selection is experience in industry branch where the reactor is to be applied.

References

[1] Merrow EW, Estimating startup times for solids-processing plants. Chemical Engineering, 1988, 24, 89–92.

[2] Douglas JM, Conceptual Design of Chemical Processes. New York, McGraw-Hill, 1988.

[3] Harmsen J, Verkerk M, Process Intensification: Breakthrough in Design, Industrial Innovation Practices, and Education. Berlin, De Gruyter, 2020.

[4] Keil FJ, Process intensification. Reviews in Chemical Engineering, 2018 Feb 23, 34(2), 135–200.

[5] Stankiewicz A, Van Gerven T, Stefanidis G, The Fundamentals of Process Intensification. Hoboken, NY, Wiley, 2019.

[6] Rivas DF, et al., Process intensification education contributes to sustainable development goals. Part 1 and Part 2. Education for Chemical Engineers, 2020 Jul, 32, 1–4, 15–24.

[7] Harmsen J, Verkerk M, New A, Approach to industrial innovation. Chemical Engineering Progress, 2021, 117(3), 50–53.

[8] Siirola JJ, An industrial perspective on process synthesis. AIChE Symposium, 1995, Series 91, 304, 222–233.

[9] Bravo JL, Distillation technology: What's next? Chemical Engineering Progress, 2019 Aug 1, 115(8), 56–60.

[10] Harmsen J, Industrial Process Scale-up: A Practical Innovation Guide from Idea to Commercial Implementation, 2nd ed. Amsterdam, Elsevier, 2019.

[11] Zlokarnik M, Dimensional analysis, scale-up. In: Flickinger MC, Drew SW, eds, Encyclopedia of Bioprocess Technology: Fermentation, Biocatalysis and Bioseparation. Hoboken, NY, John Wiley & Sons, 2002.

[12] Anderson TB, Jackson R, A fluid mechanical description of fluidized beds. Industrial and Engineering Chemistry Fundamentals, 1967, 6, 527–539.

[13] Rüdisüli M, Schildhauer TJ, Biollaz SM, van Ommen JR, Scale-up of bubbling fluidized bed reactors – A review. Powder Technology, 2012, 217, 21–38.

[14] van Deemter JJ, Reactor modelling-effect of scale. Chemical Engineering, London, United Kingdom, 1983, 390.

[15] Corning, Advanced Flow reactors, accessed 26 Sept. 2022,https://www.corning.com/worldwide/en/innovation/corning-emerging-innovations/advanced-flow-reactors/Corning-Advanced-Flow-reactors-take-continuous-flow-process-production-from-lab-to-full-scale-manufacturing.html
[16] Merrow EW, Commercialising new technologies in the chemical process industries. Independent Project Analysis Report 1991. Referred to in Harmsen, GJ, Kritische succesfactoren bij het ontwerpen en opstarten van chemische fabrieken. NPT, June 1996.
[17] Merrow EW, Industrial Megaprojects, Concepts, Strategies, and Practices for Success. New York, NY, John Wiley and Sons, 2011.
[18] Bakker HL, Kleijn JP, eds. Management of Engineering Projects: People are Key. Nijkerk, Netherlands, NAP-The Process Industry Competence Network, 2014.
[19] Vogel GH, Process Development: From the Initial Idea to the Chemical Production Plant. Weinheim, Germany, Wiley VCH, 2005.
[20] De Haan AB, Process Technology: An Introduction. Berlin, de Gruyter, 2015.
[21] Illg T, Löb P, Hessel V, Flow chemistry using milli-and microstructured reactors – From conventional to novel process windows. Bioorganic & Medicinal Chemistry, 2010 Jun 1, 18(11), 3707–3719.
[22] Hessel V, Renken A, Schouten JC, Yoshida JI, eds. Micro Process Engineering, 3 Volume Set: A Comprehensive Handbook. Hoboken, NY, John Wiley & Sons, 2009.
[23] Etminan A, Muzychka YS, Pope K, A review on the hydrodynamics of Taylor flow in microchannels: Experimental and computational studies. Processes, 2021, 9(5), 870.
[24] Schrimpf M, Esteban J, Warmeling H, Färber T, Behr A, Vorholt AJ, Taylor-Couette reactor: Principles, design, and applications. AIChE Journal, 2021 May, 67(5), e17228.
[25] Global LPR, Taylor Flow Chemical Reactor Brochure, Accessed September 6, 2021, at https://www.uskoreahotlink.com/wp-content/uploads/Taylor-Flow-Chemical-Reactor-Brochure-.pdf
[26] Paunovic V, Schouten JC, Nijhuis TA, Direct synthesis of hydrogen peroxide using concentrated H2 and O2 mixtures in a wall-coated microchannel–kinetic study. Applied Catalysts A: General, 2015 Sep 25, 505, 249–259.
[27] Su Y, Hessel V, Noël T, A compact photomicroreactor design for kinetic studies of gas-liquid photocatalytic transformations. AIChE Journal, 2015 Jul, 61(7), 2215–2227.
[28] Eurokin, accessed August 11, 2022, at www.eurokin.org
[29] Gierman H, Design of laboratory hydrotreating reactors: Scaling down of trickle-flow reactors. Applied Catalysis, 1988 Jan 1, 43(2), 277–286.
[30] Lauwaert J, Raghuveer CS, Thybaut JW, A three-phase Robinson-Mahoney reactor as a tool for intrinsic kinetic measurements: Determination of gas-liquid hold up and volumetric mass transfer coefficient. Chemical Engineering Science, 2017, 170, 694–704.
[31] Prins W, Siemons R, Van Swaaij WP, Radovanovic M, Devolatilization and ignition of coal particles in a two-dimensional fluidized bed. Combustion & Flame, 1989 Jan 1, 75(1), 57–79.
[32] Keska JK, Williams BE, Experimental comparison of flow pattern detection techniques for air–water mixture flow. Experimental Thermal and Fluid Science, 1999, 19(1), 1–12.
[33] van Ommen JR, Mudde RF, Measuring the gas-solids distribution in fluidized beds – A review. International Journal of Chemical Reactor Engineering, 2008, 6, 1.
[34] Shakya S, Munshi P, Behling M, Luke A, Mewes D, Analysis of dynamic bias error in x-ray tomographic reconstructions of a three-phase flow system. International Journal of Multiphase Flow, 2014, 58, 57–71.
[35] Murillo C, et al., Modeling of ethylene glycol production from glucose in a semi-continuous reactor. Chemical Engineering and Technology, 2020, 43(3), 950–963.

Part D: **Education**

12 Education guidelines

The purpose of this Part D on education is to provide guidelines, hints, and industrial cases for teaching chemical reaction engineering (CRE).

12.1 Introduction

Both authors have had no formal training in education, so we do not claim to provide solid education methods. Our knowledge on educating CRE is obtained from our practice of teaching industrial practitioners and students. It is also derived from observing fellow students struggling to get to terms with CRE. We have searched literature on teaching CRE but did not find the reason why some students find the subject hard while others find it easy. In the next section we present a hypothesis for this difference between students and some education proposals to test this hypothesis.

We intend to provide content and cases for teaching students at BSc, MSc, PDEng, PhD, Post-doc level, and in industry.

12.2 Challenges in chemical reaction engineering education

Teaching CRE to students as a first course is very difficult. A literature search on teaching agreed on this difficulty. The difficulty was linked to the problems of solving differential equations and to having to master a large and diverse number of subjects.

However, we could not find information about some of our own observations when entering our first courses on CRE, respectively, 38 and 50 years ago.

12.2.1 From Jan's recollection

I still remember my first encounter with the course CRE in my third year of the bachelor program in 1972. I was warned by older students that this was a very difficult course. Only 30% passed the exam the first time. In the first lesson our teacher, the late Professor Ton Beenackers, explained that the CRE discipline was a very young discipline. It started in the 1950s. Because of this not all subjects were well-established but the subjects treated in this course were the well-established basics. He then explained one item by making a mass balance over a reactor. I found it all easy to understand. After this first lecture, I discussed the content with fellow students. To my astonishment the majority found the subject very hard to understand. I asked them: "What is so hard?" They said: "how do you make such a model of a reactor?" I said:

https://doi.org/10.1515/9783110713770-012

"You just draw a box and put an arrow as inlet stream and an arrow as the outlet stream and then the equations simply follow." That answer did not land at all with them and I was unable to find out their real problem.

The students that struggled with reaction engineering also struggled with the course physical transport phenomena. For the same reason, they could not start to draw a picture about a problem to solve. Not even a picture of a bath with a tap for the water input and a sink outlet for the output. It seemed to me that the students had a strange mental block to draw any simple picture of the real thing. After years and years of trying to remove this blockage, I found that it remains hard to remove but practicing and practicing making drawings help some. A step-by-step interactive lecture using the Levenspiel wastewater problem (see Chapter 1) and asking the course participants to draw a box and then draw bubbles and then draw a liquid level inside the box help. Seeing simplified pictures of real reactors as shown in this book may also help.

I also remembered that I was very poor in organic chemistry, in particular, very poor in experimental organic chemistry synthesis practice. But I also observed that students very good in laboratory organic chemistry often found reaction engineering and physical transport phenomena very difficult. This made me think about this difference. I understood finally a key difference in learning approach, when reading the book of Petroski [1] on the importance of failure in design. Petroski [1] explains that designers define a hypothesis, which they call a design. This design is experimentally validated (years) later by transforming the hypothesis in a real process and testing it. The designer has no problem of this very delayed testing of his hypothesis. He even has no problem in that it is never tested. Organic chemistry-oriented students however may have a problem with designing a theoretical artifact, for which they do not know themselves directly whether it will work. This may even frighten them. This fear may block their thinking.

There are few more observations about people having mental blocks to make an abstract picture of a real problem. One was in the course of thermodynamics about a fridge in a kitchen. The question to answer was: What happens to the temperature in the kitchen when you leave the fridge door open. I had learned to draw a picture of the system and systematically put in all inputs and outputs. So I made a picture of the kitchen and a picture of the fridge. It took me some time to find the energy source for the fridge. It appeared to be the electricity cable to the fridge. Then giving the answer was easy. Only a few students had drawn a picture. Most students however gave an answer without drawing system boundaries. They said that the kitchen would become cooler, which is obviously the wrong answer.

The most striking example however of not making simple balances was inside Shell. In a one-day course for advising engineers on client focus, the following problem was given. An employee of Shell in the Netherlands buys a house for 200,000 guilders. After 5 years he moves to the UK and sells his house for 300,000 guilders. After again 5 years he moves back to the Netherlands and buys the same house he has had before for 400,000 guilders. After again 5 years he sells the house for 500,000 guilders. The question to answer is: How much money will he have earned from his house buying and selling. The remarkable thing is that a large number of Shell engineers got the wrong answer. Even more remarkable, I was the only one who made a balance of all incoming and all outgoing money. The others tried to solve the problem without first making balances and mixing things on earning and losing money by buying and selling. So making balances is an acquired habit and not a natural habit.

My conclusion is that making an abstract picture of a problem and also connecting balances to the picture is the real stumbling block for students to acquire knowledge about CRE. By practicing and practicing making simple pictures of the problem and making balances the mental hurdles to do so may be overcome.

My personal hypothesis is that I found all courses involving balances (mechanical forces on objects, thermodynamics, physical transport phenomena, and CRE) easy because when I was a kid, I always played with water flows. When it rained, I would go outside and make small rivers and lakes on a slope to see how levels in lakes dropped and raised and how water flowed in the bent rivers.

I also was a part-time student assistant for the group of van Swaaij and had to prepare experimental setups to determine mass transfer. Two setups I still remember. One setup as a small vessel with a simple blade stirrer. The vessel was partly filled with an aqueous sulfite Na_2SO_3 solution to which cobalt salt was added. The space above the liquid was filled with oxygen and a gas burette with a piston was connected to the vessel top. When the stirrer was started at a very gently speed (a few rotations per minute) the piston would slowly drop as the oxygen reacted with the sulfite in the water. The mass transfer rate could be seen. Increasing the rotational speed increased the lowering speed of the piston. So the stirrer speed affected the oxidation rate. I found it fascinating that a physical action could influence the chemical reaction rate.

A different setup involved a mechanically stirred reactor filled with water through which air could be bubbled or nitrogen. A probe measured the oxygen content in the water. By feeding nitrogen the oxygen concentration would drop with an exponential decay. By feeding air the concentration would increase. Then the mass transfer coefficient $k_L a$ could be determined from the curves. For different rotation speeds $k_L a$ could be related to the stirred speed.

I believe that doing these experiments I learned that mass transfer was a real phenomenon that could be observed. Perhaps CRE should not only be taught as a theoretical discipline but also be practiced in the laboratory. So that people who want to see that the theory is real can see this reality.

12.2.2 From René's recollection

Only building on Jan's last remark: some 15 years after Jan I had my first encounters with CRE, at the very same University and still Prof. Ton Beenackers, next to Prof. Wim van Swaaij and Prof. Geert Versteeg. By then, there was an experimental part included in the course. Visually I could see that a truly perfect CISTR does not exist in real life, no matter how hard you tried.

But also in my time as CRE student, the course was a clear "divider". For me it was like "coming home". Everything coming together. But many others hated it and had to do the exams two, three, or four times before passing. One aspect that played a role in my experience was that when one just reads the theory and reads the math this is far from sufficient to also put it to practice, even when one memorizes it all. One has to struggle with the problems by oneself. Almost reinventing it. For me, the math was actually very helpful. After reading the concept basics and all the math involved, I tried to do it all by myself, without "peeking". Then inspecting the starting points and sometimes surprisingly simple end-equations, made me think about the fundamental concepts again. My hypothesis at the time was: many fellow students did not pass the exam the first time – or with a mediocre grade – because the exams required *both* a relatively high degree of math skills and a proper understanding of the concept fundamentals. And indeed, making proper drawings and balances as Jan explained in the previous section.

12.2.3 CRE as a language game linked to teaching

The problem of teaching CRE may also be understood by seeing CRE as a language game as defined by Wittgenstein. Wittgenstein considers language not to be a one-on-one representation of reality but as meaningful reference act in a social cultural context. The meaning of a word is the use of the word (meaning is use). According to Wittgenstein language is instrumental; it creates a reality. People who speak a different language also experience reality differently. Language is part of a form of life (community) in which worldview, culture, and language are interwoven. The specific community is formed in history by and inside its language practice.

Wittgenstein calls a specific language a language game governed by rules that have to be understood and applied. The rules create the game, legitimate acts, and determine what is meaningful [2]. This all means that a certain language has only meaning by a specific use inside a specific context. The specific language of a group has to do with its perspective on a reality.

CRE can be seen as such a language game. Its words such as "reaction rate" and plug-flow reactor have only their specific meaning inside the CRE community. The words are used for reactor design acts. The design received by another engineer is understood and transferred to detailed process engineering design from which equipment can be purchased and from which the process can be assembled. The operator of the process has then to be trained with the CRE language to facilitate the plant operation.

Once CRE is understood as a language game which has its use inside a CRE community it is also understandable that learning this language is not easy, as learning a language is always hard for grownups. Learning this language and specifically its perspective on reality may be speeded up by the experimental practice approach where important phenomena can be experienced such as in the mass transfer setups described.

12.3 Guidelines to use this book in academic education

This section was written when the book was nearly finished. When we started, we were not sure whether or not the book could be used as a textbook on its own for teaching purposes. We were afraid that we would treat complex theories for multiphase reactors, so only folks that already have a good knowledge and understanding of CRE might benefit from our industrial experiences. On the other hand, we never intended to write another CRE textbook, as there are already many very good and excellent ones. Now that the book is nearly complete, we realize that we spend a large effort on explaining basics of CRE in the first six chapters including many exercises. So we think that the book can indeed be used in teaching.

This book is different from other textbooks in two ways:

- It is focused on understanding virtually without any math.
- It contains many real-life examples from the authors' own experience in industry and teaching to students and practitioners alike.

The book can therefore be used in introduction courses for students with no prior knowledge of CRE using the first six chapters. We also think that the book will be very useful for universities of applied sciences because of the many industrial cases in Chapters 13 and 14 where theoretical items are applied.

For students who are sufficiently proficient in math, we see huge potential of using our book as companion to the classic books by Levenspiel [3], Westerterp et al. [4], Fogler [6], and Froment et al. [7]. We could envision the teaching, per topic, to start with our book where we tried hard to introduce clearly and concisely the various topics without any math and from an industrial experience perspective. Then the teacher may revert to one of the classic textbooks or own course note, where the same topic is introduced and explained, but virtually always with a lot of math involved. Others may prefer it the other way around and start with Levenspiel and use our book for those struggling with the math. In other words, use our book to take away the math barrier towards understanding. The added value lies in applying the two different approaches.

We could also see value in (only) adopting a number of our real-life exercises and industrial cases, of Chapters 13 and 14, by giving them of students and have them work on it in groups of three to five students for 30–60 min per exercise.

The book can also be used for advanced courses using Chapters 6–8.

It could well be that a combination of practical experience with phenomena such as residence, time distribution, and mass transfer with theoretical classroom teaching would be beneficial to students.

A simple transparent vessel with a water inlet and an outlet in which a dye pulse can be injected in the inlet and then observing the coloring inside the vessel could help to understand residence time distribution and mixing. Mass transfer measurements in a mechanically stirred vessel, partially filled with water, and an oxygen probe, with air and nitrogen gas feed options could also help.

12.4 Guidelines to use this book in industry

The book is also intended for use by practitioners in industry, be it for education/relearning or application in their own work. This is not only because of the many real-life examples. In our experience, for many Chemical Engineers who have been taught CRE, the concepts, theory, and application thereof have become dormant or rusty. They are not the reactor engineer, but in their projects or work within the plant he or she recognizes CRE aspects playing a role. We aimed for a book that provides a low barrier reference book for those practitioners. It covers many multiphase reactor topics and covers both the fundamentals (without the math barrier) as well as applications.

We believe Part C to be very relevant for an industry practitioner and hence also for the students who may later become one.

12.5 Education options for industry practitioners

The authors of this book have set up several reaction engineering courses inside the company we worked for. Here is a short description of the courses with topics treated.

12.5.1 Learning course: industrial chemical reaction engineering and process concept design for nonchemical engineers

This course treated CRE and process concept design for nonchemical engineers. Most participants were chemists with several years of industrial practice in chemistry and all had witnessed laboratory-scale reactors.

It was set up as a learning house with interactive lecture sessions of 1.5 h, with homework in between. The lectures were given every 14 days. Only one topic was treated per session.

The books used were Levenspiel [3], Westerterp [4], and Douglas [5]. Each participant had these books.

The following 13 subjects were covered:

- Heterogeneous reactions: Levenspiel chapters 17 and 18
- Packed bed catalytic reactor: Levenspiel chapter 19
- Fluidized bed reactors: Levenspiel chapter 20
- Deactivating catalysts: Levenspiel chapter 21
- Reactions on solid catalysts: Levenspiel chapter 22
- CISTR with heat exchange: Westerterp, chapter VI.4
- Autothermal reactor operation: Westerterp chapter VI.5
- Maximum temperature and dynamics: Westerterp chapters VI.6 and VI.7
- Flow sheet input and output structure: Douglas chapter 5
- Flow sheet recycle structure: Douglas chapter 6
- Reactor selection: Douglas 6.6
- Reactive distillation: handout by Harmsen
- Safe design methods: in-house handouts

This course can now also be given using this book in combination with the book by Douglas [5].

12.5.2 Hands-on course: industrial reaction engineering and conceptual process design

This course was for graduated chemical engineers who just entered our company to enhance their knowledge on industrially relevant subjects not covered formal academic courses on CRE and conceptual process design. This was a 2-day course with highly interactive lectures – almost without PowerPoint – and group exercises.

12.5.3 Course program

Day 1: Industrial reaction engineering
Introduction to course and its purpose
Introductions and overview course steps and purpose

Block 1: Reaction rates
In groups of 3: discuss solution of the "Levenspiel homework"
Presentation of group results and feedback

This is a highly interactive session in which the course conductors asked questions about the problem they had been asked to solve upfront. It is the only "academic" case of the whole course. Chapter 1 provides all details of how this case study is performed and the learning points.

Block 2: Residence time distribution, mixing, and reaction
RTD theory
Case study 2: Shell's "O-reactor" scale-up to commercial scale
Group Exercise case 2
Presentation of group results and feedback Shell's solution in practice

This is essentially exercise 1 of section 4.7.4: RTD of one of Shell's reactors.

After a 30–60 min plenary introduction/recap of RTD theory, the scale-up problem handed out to the students. They were divided into small groups of three to four students and asked to discuss the problem, summarize, and capture their findings/answers. Timing for group work 45–60 min. Then reconvene in a plenary session where all groups presented their work to the other groups and the course leader(s), typically averaging 5 min per group. Finally the latter presented the company's solution and the reasoning behind it (20–30 min, depending on questions from the groups).

Block 3: Mass transfer and chemical reaction
Theory mass transfer with reaction for heterogeneous catalysts
Case study 3: Catalyst size and shape selection
Group Exercise case 3
Presentation of group results and feedback Shell's solution in practice

This is essentially the exercise of section 5.8.2: catalyst particle size and shape for the dehydration of MPC. Same way of interactive, group wise, teaching as for the above case 2. Typically, 45–60 min plenary theory, 60–75 min working in groups, and 20–30 min "company solution and reasoning."

Block 4: Reactor system design for catalyst decay
Types catalyst decay and design remedies
Case study 4: Oxidation reactor system for catalyst decay
Group Exercise case 4 (presentations next morning)

This case involves conceptual design for the epoxidation reaction section of the Shell's SM/PO process where, as in many cases in industry, catalyst deactivation, and the factors governing catalyst effective lifetime dominate reactor selection, reactor line-up, and operational strategy decisions.

Day 2: Conceptual process design

Block 1: Conceptual process design methodology
Theory: Smith Onion and Douglas Chapter 6:
Description of the overall case; the commercial process in context of Shell
Group exercise 1: Reaction analysis to identify key design issues
Report results and reaction analysis + feedback

The course conductors provide first an overview of two process concept design methods: the Smith "onion" inside-to-outside sequence method and the Douglas outside-to-inside method. Then the course conductors guide the students during the whole day through our mixed method: an iteration between outside–in and inside–out sequence. So, first output and input streams, then a reaction section with input and output and then required other unit operations to connect all input streams to all output streams. Often then more input and output streams are needed to complete the design.

Then details on a complex industrial process are given; some five pages of info on the reaction scheme, physical properties, and safety aspects. The group task is to analyze the design sequence with the onion method. Each group then presents their results plenary.

Block 2: Block flow diagram input, output recycle
Theory: Douglas Chapters 5–7: (Tables 5.1.1–5.2.2)
Description of Exercise 2
Group exercise 2: Input, output, recycle of the Shell process
Reporting results and feedback

Same setup: first, the course conductors provide some theory. In this section, it involves the Douglas outside–in method. In the group sessions, they will make a block flow diagram of the input, stream output streams, and recycle streams and main process blocks.

Block 3: Reactor and separation system selection
Theory: Douglas Chapter 6: Table 6.6.1 and Chapter 7, Tables 7.3.3–7.3.6
Description exercise 3
Group exercise 3:
Reporting and feedback actual Shell process
Feedback attendants on course

In this last and longest block we focus on how to select the sequence of reactors and separators as block-flow diagram of the given case. In groups they further build on what they had done in the previous two blocks and finally come up with a complete conceptual design. After the group presentations, the real process design is discussed.

12.6 Position of reaction engineering in chemical engineering curriculum

The authors of this book have noticed in their long industrial career that lot of elements of reaction engineering, such as defining first the required output in amount and purity and then the input in combination with a first reactor concept, and also learning the concept of single pass conversion, external recycle of a separator, are all very useful to learn process concept design, or in fact to learn to design, as a different thinking and acting mode from learning theory, or doing experiments. Also the role of mass balances in design is learned. One could therefore consider placing basic reaction engineering in the first year of chemical engineering programs. Chapters 2 and 3 of this book could be used for this purpose.

12.7 Takeaway learning points

Learning point 1
Drawing pictures of the reaction engineering object is an essential part of solving a problem.

Learning point 2
Teaching to draw pictures needs interaction with students to overcome their blockage to draw an abstract picture of the real object.

Learning point 3
Drawing pictures with inputs and outputs and then making mass balances are excellent ways to understand and communicate the essence of the problem to be solved.

Learning point 4
Practical simple experiments showing mass transfer and mixing can help some students to acquire the awareness that these concepts are real.

Learning point 5
Doing exercises are essential to acquire CRE knowledge.

References

[1] Petroski H, To Engineer is Human. NY, Random house, 1992.
[2] Wittgenstein L, Philosophical Investigations. Oxford UK, Basil, 1953.
[3] Levenspiel O, Chemical Reaction Engineering, 3rd ed. New York, John Wiley & Sons, 1999.
[4] Westerterp KR, van Swaaij WPM, Beenackers AACM, Chemical Reactor Design and Operation. John Wiley & Sons, Chicester, UK, 1984.
[5] Douglas JM, Conceptual Design of Chemical Processes. McGraw-Hill, NY, 1988.
[6] Fogler SH, Elements of Chemical Reaction Engineering, 6th ed. Pearson, 2020.
[7] Froment GF, Bischoff KB, De Wilde J, Chemical Reactor Analysis and Design, 3rd ed. UK, Wiley, 2010.

13 Industrial cases

The purpose of this chapter is to describe industrial cases for educational purposes. To that end, cases containing all major theoretical elements of this book have been described.

13.1 Introduction

Table 13.1 gives an overview of which theoretical subjects are treated in which cases.

A lot of attention in these cases is spent on reactor selection. The cases show that a selected reactor type for a certain project can change during the project stages of ideation, concept, and feasibility. Cases in Section 13.4 also show that reactor selection between companies for the same application function can differ widely.

Table 13.1: Overview of subjects treated in industrial cases.

Theory	Sections
Residence time distribution	13.3, Chapter 14
Micromixing	13.3
Mass transfer of gas–liquid	13.3
Mass transfer of heterogeneous catalysts	13.2, 13.4
Heat transfer and heat control	13.2, 13.6, Chapter 14
Reactor type selection: ideation stage	13.6, 13.7
Reactor type selection: concept stage	13.2, 13.5
Reactor type selection: feasibility stage (scale-up risks included)	13.2
Reactor type selection: development stage	13.2
Reactor type selection: fixed bed versus fluid bed or slurry reactor	13.2, 13.7
Modeling	13.2, 13.3, 13.3

13.2 Gas-to-liquid (GTL) Shell case

13.2.1 Introduction to GTL case

The purpose of this first industrial case is to show how the theory of our book has been applied for all stages – from idea to commercial scale. The entire 50-year history of this case is described by van Helvoort in enormous detail in his book of 333 pages, *Gas to Liquids – Historical Development of GTL Technology in Shell* [1]. So, obtaining a lot of information is easy. We have very concisely summarized that highly recommended book in Section 10.4.

https://doi.org/10.1515/9783110713770-013

Here, we not only summarize some aspects of that book but also add some of our own hands-on experiences.

The heart of the GTL process is the very well-known Fischer–Tropsch reaction as depicted in Figure 13.1:

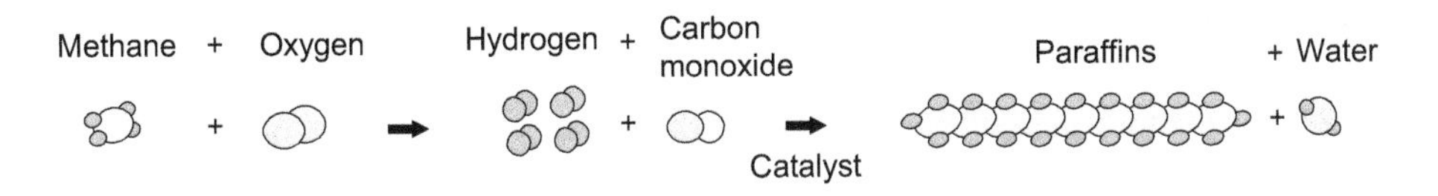

Figure 13.1: The Fischer–Tropsch reaction.

The reaction is quite exothermic, with a reaction enthalpy of −165 kJ/mol, corresponding to an adiabatic temperature rise of several hundreds of degrees, depending on the exact feed composition and reaction conditions. The reaction itself can be regarded as an oligomerization on the catalytic surface. A very simplistic picture of the chain reactions is given in Figure 13.2.

13.2.2 A consecutive or a parallel reaction?

The very first thing to notice is that from a catalytic point of view this chain growth is clearly a set of consecutive reactions. However, from a reaction engineering point of view, one can argue that it is a set of parallel reactions!

In a classic reaction engineering model of such a heterogeneously catalyzed gas phase reaction, rates are described from the point of view of the gas phase molecules. In the simplified picture, gas phase reactants CO and H2 react on the surface to form gas phase molecules C2, C3, C4, C5, and so on, via surface species intermediates, in parallel. There would only be consecutive reactions if, for example, a C2 molecule in the gas phase, previously formed from CO and H2 via surface reaction and desorption,

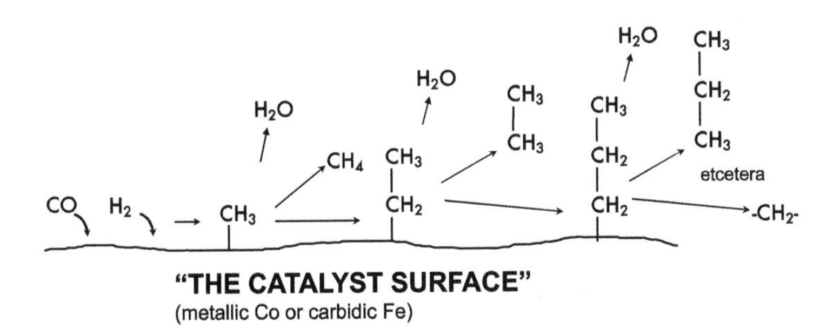

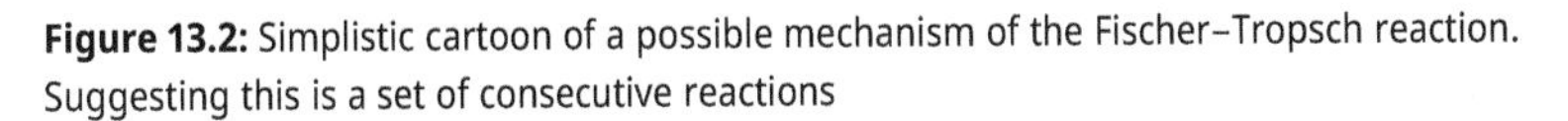

Figure 13.2: Simplistic cartoon of a possible mechanism of the Fischer–Tropsch reaction. Suggesting this is a set of consecutive reactions

would re-adsorb on the surface and react further to C3, C4, C5, and so on, via surface reactions and the corresponding intermediates.

Figure 13.3 depicts the classic view on this chain growth mechanism and how using a chain growth probability parameter α can describe some, if not all, of the basic features of the product distribution. This is referred as the Flory–Schulz FT kinetics or also Anderson-Schulz-Flory (ASF) kinetics. So, where Figure 13.2 suggests a set of consecutive reactions, Figure 13.3 makes it clear that, in fact, we should treat it as a set of parallel reactions (unless all surface reactions are going to be modelled in detail!).

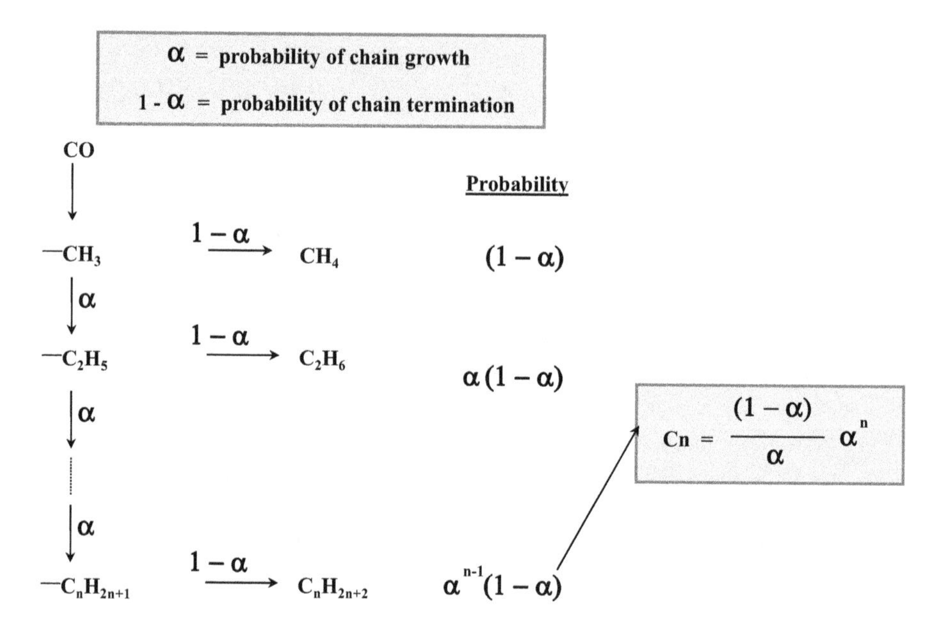

Figure 13.3: The Flory–Schulz kinetic model. (suggesting this is a set of parallel reactions)

13.2.3 Flory–Schulz distributions

Figure 13.4 shows how the product distribution depends on this simple chain growth probability parameter α. In the Shell GTL process, catalysts have been developed with a high α of 0.9 or higher. In order to make a commercially more attractive final product slate, the Fischer–Tropsch reactor is followed by, among other steps, a catalytic hydro-cracking reactor section.

Figure 13.5 shows such a Flory-Schulz distribution as a function of the chain-growth probability in a different way. It also indicates how this would also drive reactor selection, or at least one aspect of that the selection. For a relatively low α, say 0.7, the Fischer–Tropsch reaction produces mostly hydrocarbons in the diesel and gasoline

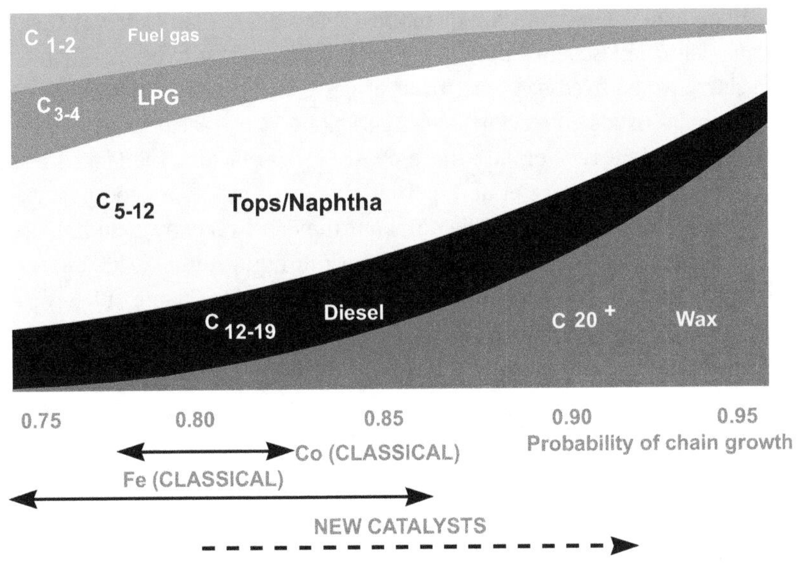

Figure 13.4: Flory – Schulz product distributions (ASF).

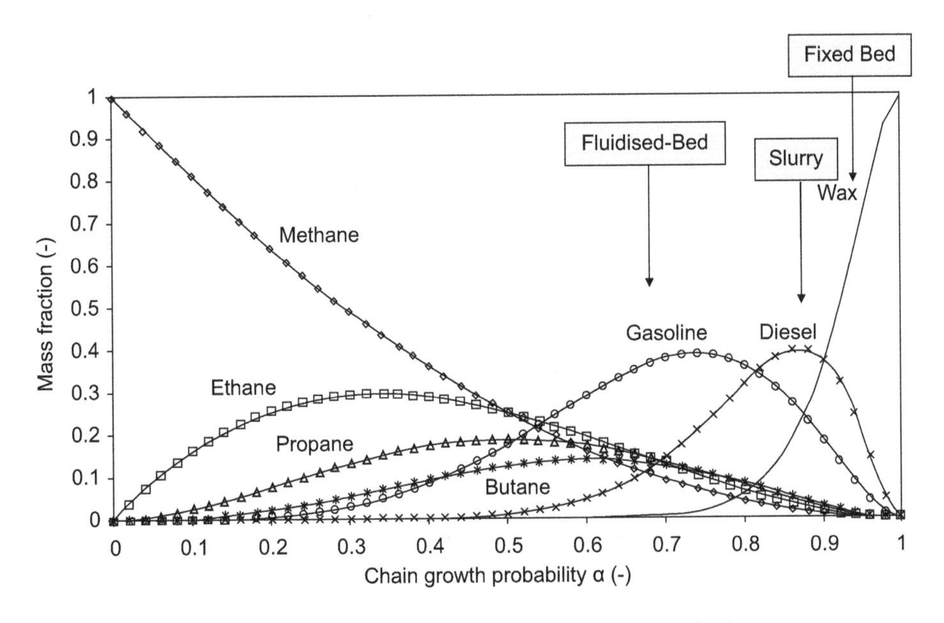

Figure 13.5: Flory – Schulz product distributions (ASF) and tentative reactor selection.

range, next to lighter ones. Very few hydrocarbons are made that, under reaction conditions, are in the liquid phase. Hence, a fluidized bed becomes a real option. Of course, the disadvantage of a low α is that 10% or more methane is made. And methane was, of course, the original feedstock from which the syngas was made; so that is one product you do not want to make in significant quantities. This is one explanation why the major large commercial-scale Fischer–Tropsch reactors that are currently in operation are either of the slurry or the fixed bed type. These are essentially Sasol's slurry reactors and Shell's multi-tubular fixed bed reactors, see also Section 10.4 on reactor selection.

The Sasol GTL plant reactors are three-phase slurry reactors, see Figure 13.6 and the book chapter by Steynberg et al. [2]. That chapter provides a good overview of slurry reactors and provides lots of detail about the Sasol reactor, including cooling and catalyst separation inside the reactor from the liquid outlet. This shows that knowledge and selection also depend on the company, see also Chapter 10 .

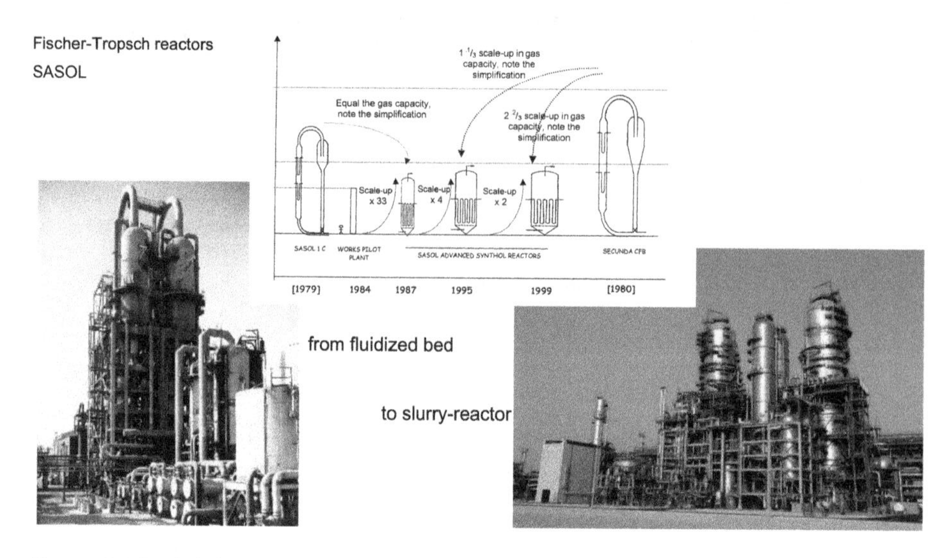

Figure 13.6: Sasol Fischer–Tropsch reactor technologies; see also [2].

13.2.4 Why Shell experts "like" fixed bed reactors for GTL?

In Chapter 10 we already discussed in quite some detail the history of the reactor selection and development within Sasol's main competitor, Shell. Here, rather than making an extensive list of pros and cons of each of the two main reactor types, we summarize the list – what Bos and Hoek [3] called in 2009: "Fixed bed GTL reactors, . . . and why we like them", based on years of experience in Bintulu (the Pearl plant in Qatar was still under construction).

First some basics: each of the 12 reactors per GTL "train" has tens of thousands of tubes, see Figure 13.7 for a snapshot of a part of the tube-sheet and 2 pictures, giving an impression of the size of these reactors. The mechanical design has been optimized and the reactor operation is totally stable. Also, the catalyst has quite a good stability,

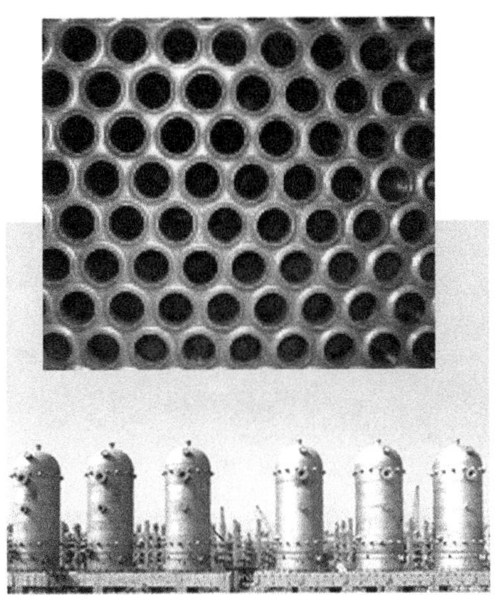

Figure 13.7: Shell gas-to-liquids: Fischer–Tropsch reactors.
Left: on transport on the Danube; top right: part of the tube-sheet;
bottom right: reactors during the construction phase of the Pearl plant.

allowing for effective lifetimes of years, also thanks to the repeated in situ regeneration procedures.

Obviously, with a fixed bed, there are no catalyst/wax separation issues. Shell could build on decades of experience with the design and operation of wall-cooled multi-tubular reactors, not only for the production of ethylene oxide, but also for a number of other processes, including three-phase systems. So, the scale-up is well understood, not only with respect to the process side -scaling up by numbering up, but also with respect to the coolant side. Moreover, the years of operational experience had shown that the design and operation of the Bintulu reactors was still quite far away from the limits of heat removal. A combination of this experience, clever engineering solutions, and improved catalysts has led to quite some improvements over time, supported by a variety of multiscale models. Figure 13.8 illustrates this improvement for the scale of a single reactor tube – relating to the reactor design for stability topics treated in Chapter 7. As this specific example shows, the conditions at the top can be quite different from the conditions at the bottom. This has also induced some patent applications wherein different catalysts are loaded at the top and the bottom of the reactor tube, see [4].

Figure 13.9, also taken from Bos and Hoek [3], shows these improvements from the start-up of the Bintulu plant and just before start-up of the Pearl plant, that is, excluding all improvements after start-up.

One classic advantage of fixed bed reactors is, using our preferred phrasing, "you always know where your catalyst is".

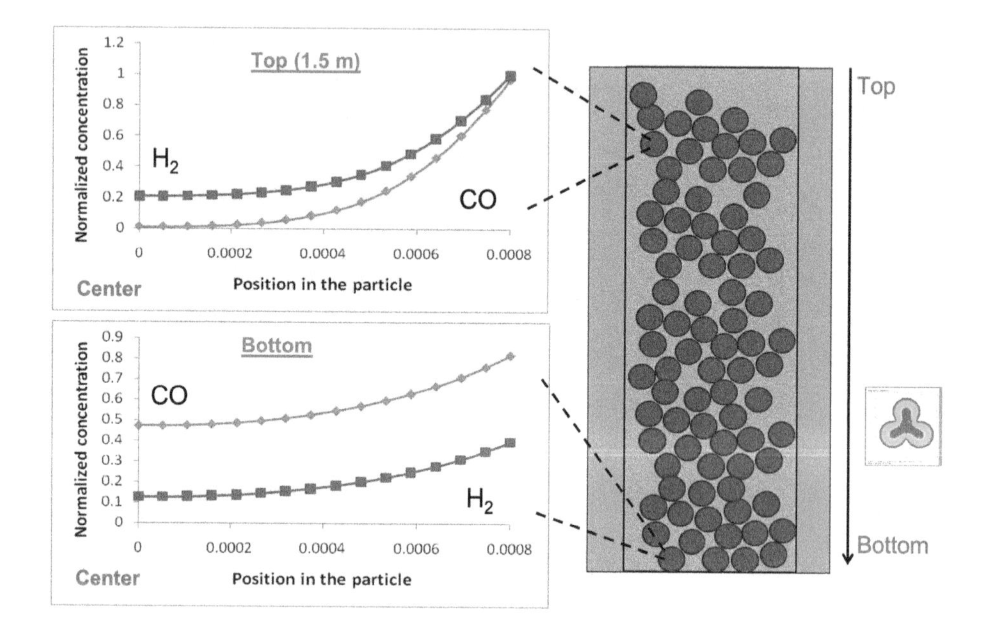

Figure 13.8: GTL tubular fixed bed reactor and catalyst-scale modeling.

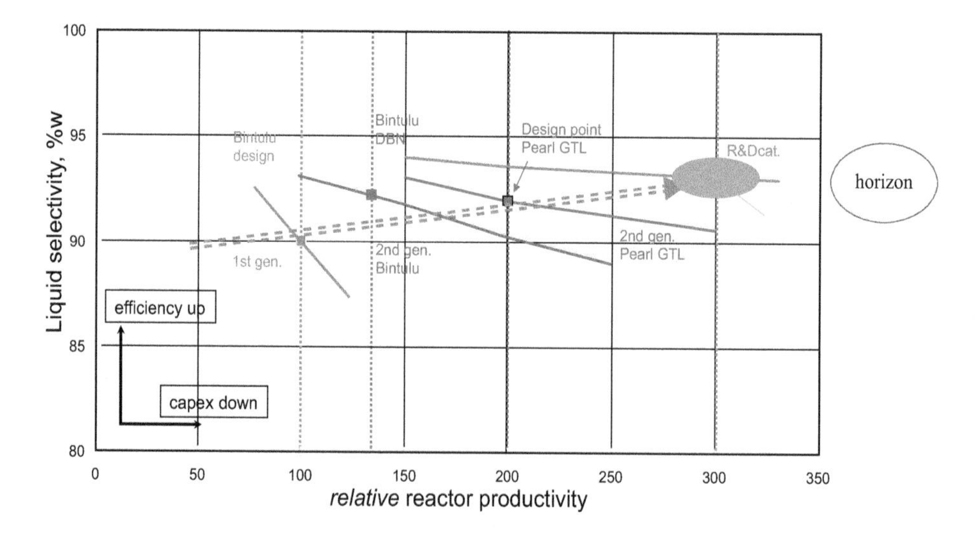

Figure 13.9: Progress in Fischer–Tropsch catalyst and reactor productivity.

Finally, an additional advantage – stemming from the disadvantage of having to have as many as 12 reactors per train – is that the "$N+1$" design/operation strategy becomes relatively affordable. Here, N is the number of reactors needed to reach the total production capacity and the "+1" reflects the fact that, regularly, one of the reactors needs to be taken off-line in order to replace the deactivated catalyst. With, for example $N=2$, one loses 1/3 of the capacity during the exchange – or one would have to let the remaining reactor work harder, typically at the cost of selectivity loss. With $N=11$, the impact if far less. Moreover, it allows for a lot of scope for optimization with respect to operational strategies.

In our book, we of course focus on the reactor, but we would like to stress the importance of true collaboration between experts from a huge number of disciplines in order to make a complex project like a GTL plant a success; see Figure 13.10.

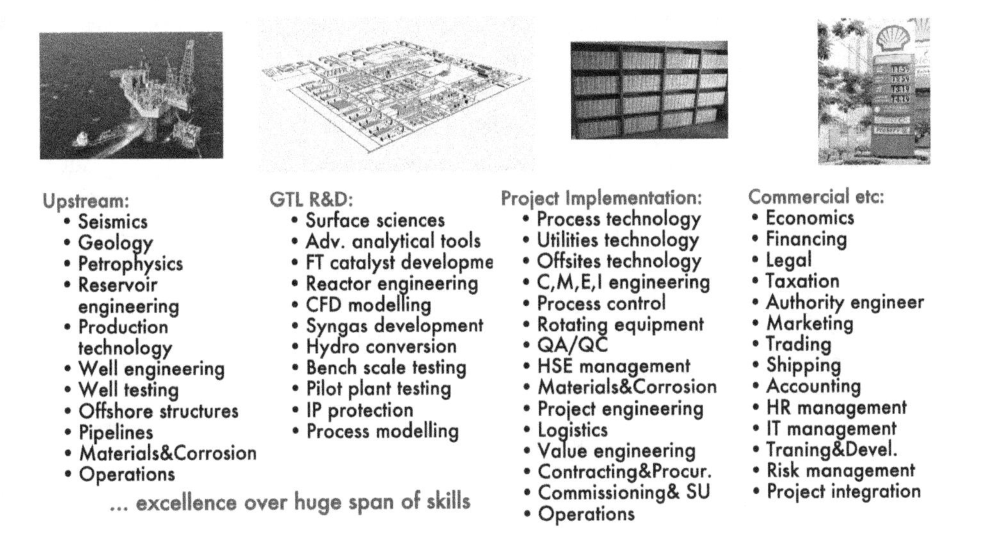

Figure 13.10: Overview of the disciplines involved in the GTL plant design.

References: gas-to-liquid (GTL)

[1] van Helvoort T, van Veen R, Senden M, Gas to Liquids – Historical Development of GTL Technology in Shell. Amsterdam, Shell Global Solutions, 2014.

[2] Steynberg AP, Dry ME, Davis BH, Breman BB, Fischer-Tropsch reactors, in studies. In: Steynberg A, Dry M, eds, Surface Science and Catalysis 152. Elsevier B.V, 2004.

[3] Bos ANR, Hoek A, Continuous improvement of shell's GTL technology. Invited lecture WCCE-8, Book of abstracts, pp. 512 f, Montreal, 2009.

[4] Bos ANR, Van den Brink PJ, Remans TJ, Stobbe ER, Unruh DJM, Wisman RV, Stacked catalyst bed for Fischer-Tropsch. US 8980194 B2, 2015.

13.3 Ethyl benzene peroxidation reactor (EBHP)

13.3.1 Introduction to the case

The purpose of this case is to show a commercial-scale gas–liquid bubble reactor type with special features:
- Cross-flow gas–liquid for low gas feed pressure drop; so, low compressor costs
- Liquid horizontal flow baffled to limit RTD, to reduce consecutive reaction by-product formation
- Cooling by heat exchange bundle in the reactor

We will show computational fluid dynamics (CFD) modeling of liquid micro mixing to optimize reactor performance, commercial plant scale residence time distribution experiments (RTD) both in the liquid phase and gas phase, and finally, a concrete reactor improvement implemented in the real plant. This relatively comprehensive example thus illustrates the application of the theory and other aspects of quite a number of different sections of this book, that is Chapters 2, 4, 5, 8, and 11.

Propylene oxide is a versatile chemical intermediate used in a wide range of industrial and commercial products. Current world production is well over 10 million metric tons a year. While other processes exist, Shell Chemicals' companies have derived a strong competitive advantage by using and continually developing their proprietary SM/PO technology, see Buijink et al. [1]. In this process, propene and ethylbenzene (EB) are converted into propylene oxide (PO) and styrene monomer (SM), respectively. Worldwide, there are now six world-scale SM/PO plants based on Shell Technology; the most recent one started up in China in 2021.

The first step in this multi-step process is the air-oxidation of ethylbenzene to ethylbenzene-hydroperoxide (EBHP). This is performed by Shell in a cross-flow operation in a series of large horizontal bubble column reactors with a very low aspect ratio: height 4–6 m, length 15–25 m, see Figures 13.11 and 13.12 taken from Klusener et al. [2].

Typically, a reactor train consists of 4–5 of such horizontal columns, which are equipped with baffles and heating/cooling coils. Air is introduced via separate middle and side sparger systems. The gas outlet stream contains a very significant amount of EB from evaporation/stripping, besides unconverted oxygen, and this EB is recovered in a condensing column and recycled to the reactor train.

The literature on horizontal cross-flow bubble columns is rather limited; see [3–5]. Pohorecki et al. [6] did mention tracer experiments in similar commercial reactors but did not disclose any details. Within Shell, we have developed computational fluid dynamics (CFD) models as well as integrated reactor models, with complex reaction kinetics, for these rather unconventional bubble columns. To validate some of the model predictions, it was decided to perform radioactive tracer experiments in one of the commercial plants during normal operation. Two completely different types of tracer

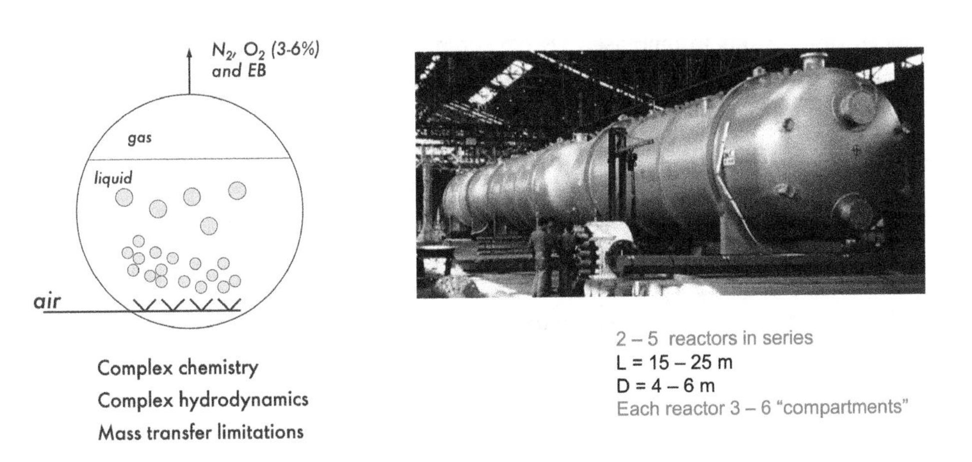

Figure 13.11: The horizontal cross-flow reactor for peroxidation of ethylbenzene.
Left: Simplified picture of gas flow in cross-flow bubble column
Right: Photo of commercial reactors.

experiments with completely different purposes have been performed: (1) liquid phase and (2) gas phase tracer experiments.

13.3.2 Reaction description

The peroxidation of EB to EBHP is autocatalytic, implying that a minimum concentration of the EBHP product is necessary to initiate the reaction, and that the reaction goes faster on increasing the EBHP concentration. On the other hand, a higher EHBP

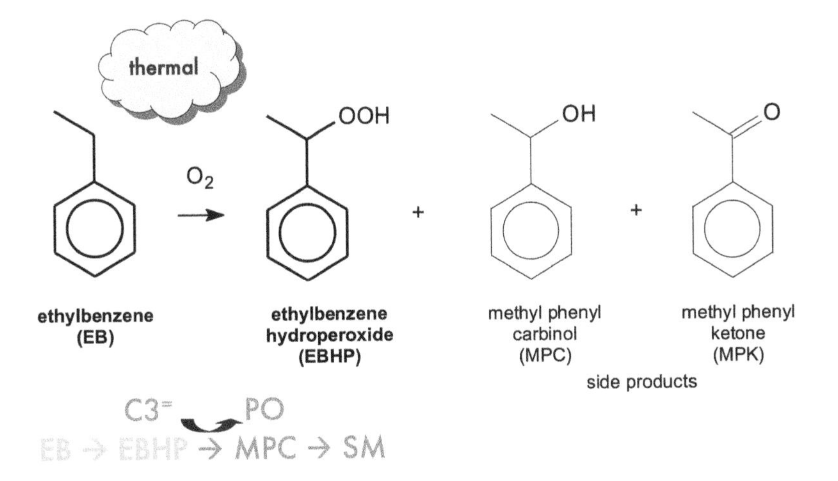

Figure 13.12: Simplified chemistry of the peroxidation of ethylbenzene.

concentration leads to increased EBHP decomposition, increased by-product formation, and lower selectivity. For this reason, using the theory presented in Chapter 4, backmixing is desirable in the front-end of the first reactor, whereas plug flow behavior is desirable in the remaining reactor train. To facilitate this behavior, the reactors are segmented by a few transversal baffles. However, in the first reactor of the train, a lower number of such baffles are installed.

13.3.3 The liquid-phase RTD experiments

The two main incentives for investigating the actual liquid phase mixing in the reactor train are to determine:

- The effectiveness of the transversal baffles in facilitating plug flow ("staging" by forming compartments in series) and, subsequently, the scope for improved selectivity by further reducing the deviations from the plug flow
- The effectiveness of installing a lower number of such baffles in promoting backmixing in the first reactor and hence the scope for improvement by (further) enhancing backmixing

Figure 13.13 illustrates where the tracer injections and detections in the actual plant were positioned.

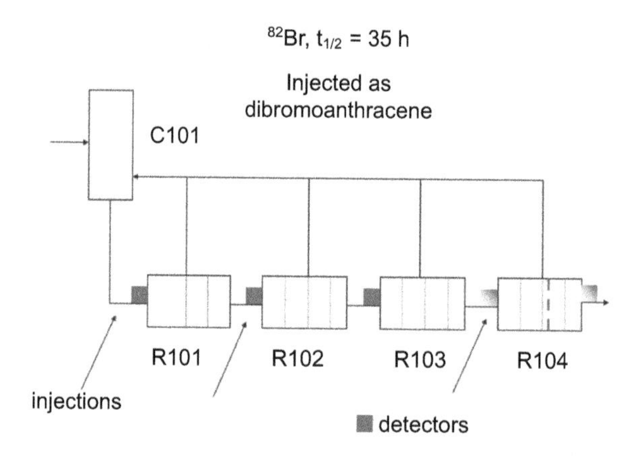

Figure 13.13: Tracer injection points and detection points in the reactor train.

The experimentally obtained RTD curves (count rate vs. time) were fitted with both the tanks-in-series model and the axial dispersion model, see Chapter 4. Both models have basically two parameters: the mean residence time and the parameter describing the spread in residence time, that is, the number of tanks N and the Péclet number, respectively. In principle, the residence time could be calculated for (accurately) known flow rates, volumes, levels, and gas holdups. However, here the mean residence time is considered to be an unknown model parameter and is one of the two fit parameters. This is not only because of the presence of significant EB evaporation but also because the plant measured or calculated values in practice are often not very accurate. The time axis was set to zero at the moment the injected bolus of activity passed the detectors positioned at the reactor inlet.

13.3.4 Results of the liquid-phase RTD experiments

Figure 13.14a shows the results of fitting the RTD data of R102 with the tanks-in-series model for $N = 3$, 4, 5, and 6. For each N, the average residence time was optimized. It can be clearly seen that this model cannot quite fit the data: either the left-hand or the right-hand side of the measured RTD curve can be modelled well, but not the full curve with one set of parameters.

Figure 13.14b shows the result of fitting the data of R102 using the axial dispersion model, using open–open boundary conditions to allow easy fitting in the time domain. A remarkably good fit was obtained and the resulting Péclet number was 6.6. In Figure 13.14c, the tracer had been injected into the inlet of R101 with measurement at the outlets R101 and R102, respectively. Unfortunately, no detector was present at the outlet of R104 during these experiments, and for R104, we only have the signal recorded by the detector placed at the circumference of R104, that is, mounted on the reactor vessel itself rather than on the outlet piping; so, this is not really the RTD over the whole reactor train. Consequently, this R104 signal can only be used qualitatively for interpretational purposes.

For the full train of four reactors, only an estimate of the RTD can be made because no measurement results are available for injection into R101 and measurement at the outlet of R104. However, using other data available, it was estimated to be approximately equivalent to 13 tanks-in-series.

R101 contains fewer baffles and hence fewer compartments; therefore, it might be expected (and actually was hoped for) to have more backmixing in R101.

The Péclet number of R101 is slightly higher – that is, slightly less backmixing – than the number found for R102: Pe = 6.7 versus Pe = 5.7. This would suggest that the transversal baffles do not significantly influence the backmixing, that is, no significant staging is achieved with these baffles.

However, to assess the effect of the number of baffles, one should also look at the dispersion coefficient, D_{ax}, rather than only at the Péclet numbers, because the residence times of the reactors increase significantly from R101 to R104 due to EB evaporation.

The variation of the Pe and D_{ax} values is of the same order as our estimated experimental accuracy. However, if we do assume the differences to be statistically significant, two lines of reasoning may be applied, leading to different conclusions. On the one hand, D_{ax} for R101 is slightly higher than for R104 and R101 + 102, and is also

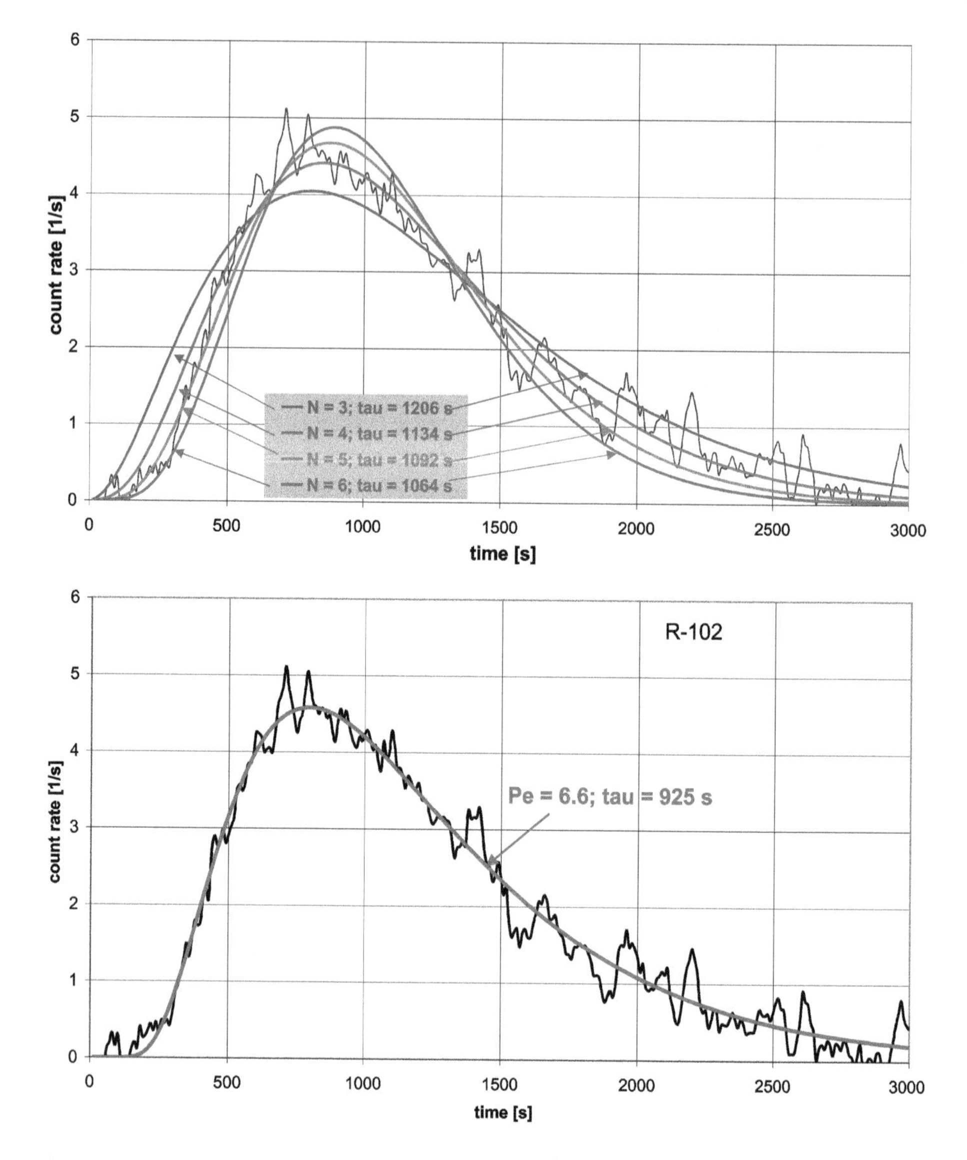

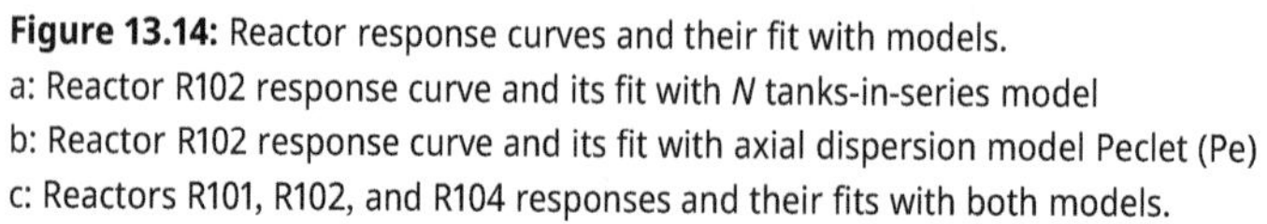

Figure 13.14: Reactor response curves and their fit with models.
a: Reactor R102 response curve and its fit with *N* tanks-in-series model
b: Reactor R102 response curve and its fit with axial dispersion model Peclet (Pe)
c: Reactors R101, R102, and R104 responses and their fits with both models.

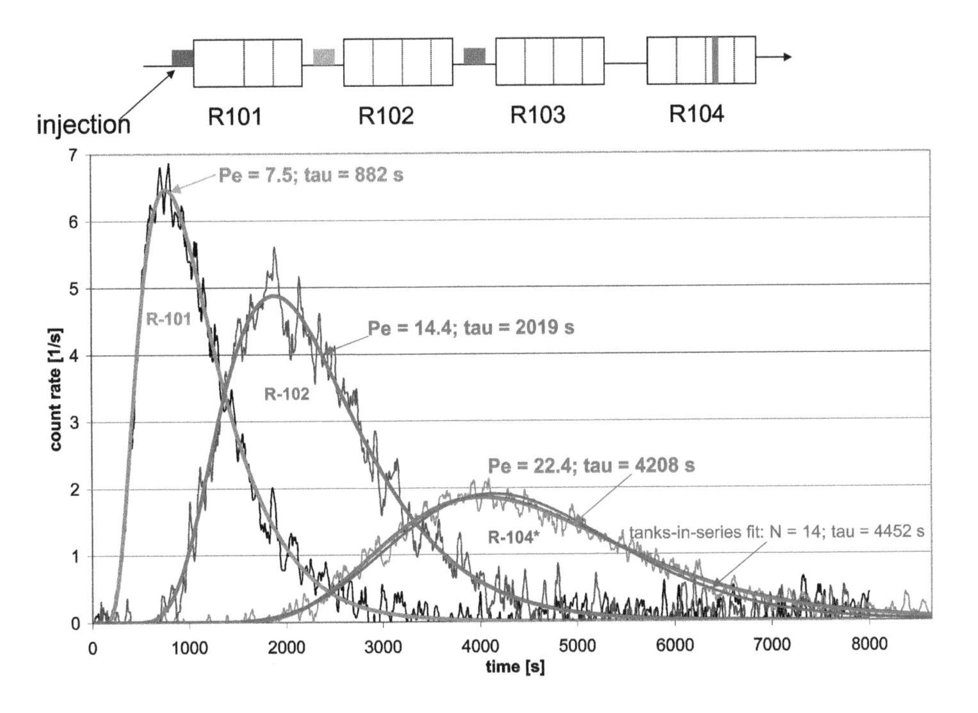

Figure 13.14 (continued)

slightly higher than for the complete reactor train, indicating a slightly higher dispersion for R101. So this would suggest that higher R101 will have slightly higher backmixing, but this leaves the higher D_{ax} for R102 unexplained. On the other hand, one could assume that due to the decreasing liquid velocity, D_{ax} would have a decreasing trend from R101 to R104, if these vessels were the same. In that case, a higher value for R101, than for R101, could be expected if the lower number of baffles did not have effect. The fact that the D_{ax} for R101 is found to be lower, rather than higher, suggests that in R101, the lower number of baffles has slightly reduced the backmixing.

This second line of reasoning is supported by the fact that if one assumes D_{ax} linearly increases with the liquid velocity, the results of R102 and R104 are fully consistent (Equal Pe, despite differences in residence time). In that case, one would conclude from the higher Péclet number for R101 that reducing the number of baffles in R101 has, if anything, decreased the backmixing rather than increased it.

13.3.5 Results of the gas phase RTD experiments

It is important that oxygen is well-distributed, as a main side reaction of EB oxidation is a decomposition of the product EBHP, which is enhanced under the so-called "oxygen starved" conditions. Therefore, air is introduced along the bottom of the reactors via the middle and side spargers. The airflow through each of these can be regulated separately.

Because the selectivity observed in the commercial plants falls up to a few percent short of what is achieved in ideal laboratory conditions, it was hypothesized that this may be due to "starvation" in poorly aerated regions of the large reactors. Starvation is defined as the presence of reaction zones in which the dissolved oxygen concentration – and thus the EB oxidation rate – is virtually zero.

For a better understanding and quantification of this starvation phenomenon, we modelled the complex hydrodynamics including mass transfer and chemical reaction with CFD. For this exercise, we used the discrete particle model of the commercial CFD modelling package FLUENT along with the so-called k-epsilon turbulence model. To reduce the computational effort, a 2D geometry model was applied. A uniform bubble size was used because preliminary simulations with different bubble size distributions had showed no significant effects on the hydrodynamics. In-house derived mass transfer correlations were used to describe the gas-to-liquid oxygen transfer. Simplified reaction kinetics for O2 consumption was adapted. An instantaneous EB flash was taken into account.

The key CFD predictions are illustrated in Figure 13.15, showing the presence of very strong liquid circulation patterns in the transverse direction and a fast upward flow of gas bubbles along quite narrow paths, with very high gas hold up.

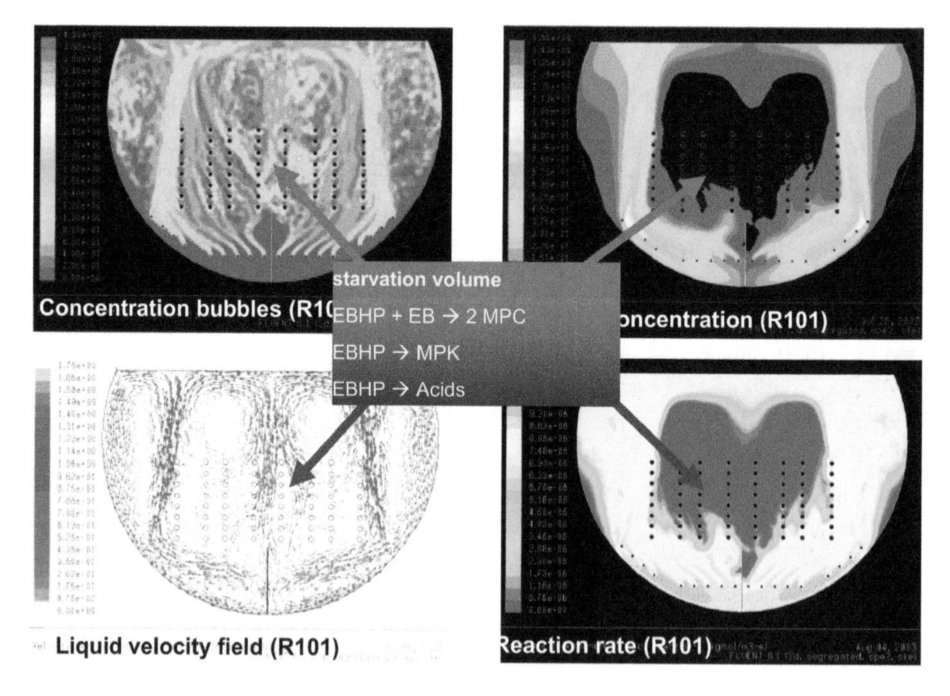

Figure 13.15: Reactive CFD predictions.
Gas bubbles concentration (local gas holdup), liquid phase oxygen concentration, liquid velocity field, and reaction rates.

For this specific case, the gas is predicted to mainly rise through two relatively narrow paths near the walls, and gas injected in the middle spargers is predicted to predominantly rise sideways, that is, the liquid in the center of the reactor is not effectively aerated and is predicted to suffer from starvation.

To validate the model predictions and to assess the scope for improvement, we performed gas phase radioactive tracer experiments in one of our commercial plants during normal production. Tracer injections (41Ar) were done separately in either the air feed at the side or the middle spargers.

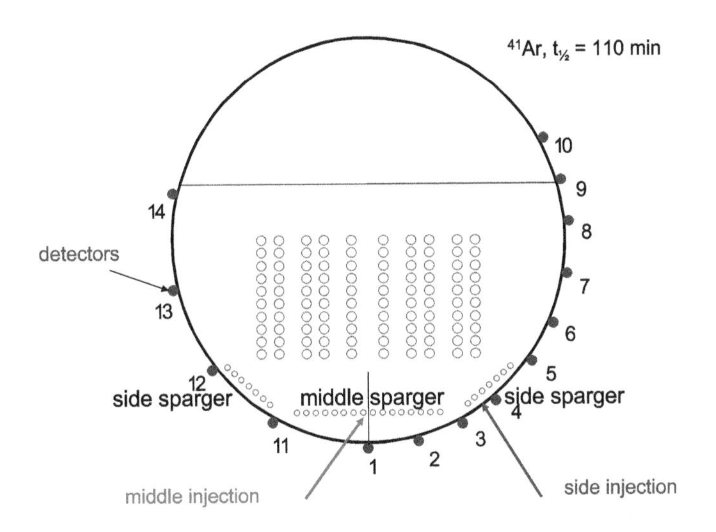

Figure 13.16: Positioning of the detectors and the 3 inlet spargers.

Using 14 detectors mounted on the reactor wall, see Figure 13.16, we could obtain a tracking of the gas tracer (from which, among other things, the gas rise velocities and the circulation patterns were determined).

First, and maybe foremost, these results confirm the key prediction of the CFD model and the assumed cause of the "starvation": gas injected in the middle sparger is detected flowing upward near the side of the vessel.

From the data, average gas velocities of 2.5 and 0.8 m/s were calculated for side and middle injection, respectively. For reference, the natural single bubble rise velocity is only of the order of ~ 0.1 m/s. The significantly different velocities found for the side and middle injected gas tracer pulses indicate separate, and largely segregated, paths for these gas flows. Remarkably, without realizing it, this was also predicted by our CFD model before we had done the experiments, see Figure 13.17. This shows the results of particle tracking of the gas injected from the side and middle spargers indicated in red and blue, respectively. Quantitatively, the experimentally determined breakthrough times at the gas/liquid interface, 2 and 4 s for the side and middle injection, respectively, matched very well those modelled with the CFD model.

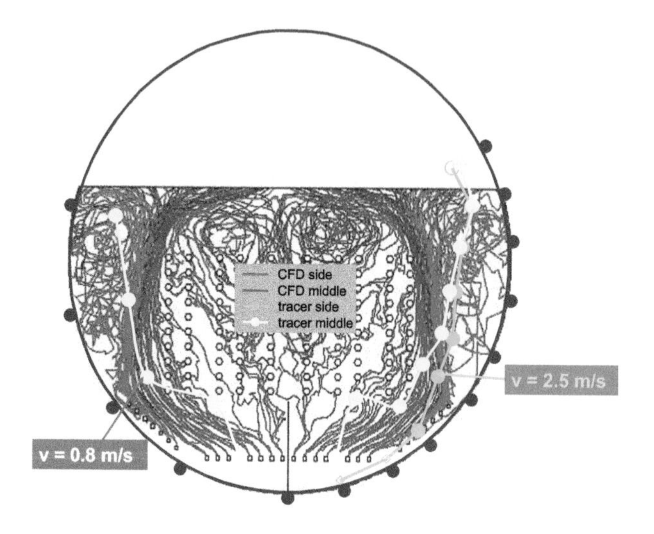

Figure 13.17: CFD particle tracking predictions and experimental results.
red and blue lines: CFD predictions
yellow and green: tracer experiments
CFD predictions generated before the plant tests!

The detectors at the side of the vessel – opposite to where injection took place – show much higher "breakthrough times" – 13–15 s. The responses of detectors 9 and 14, located at opposite positions at the gas/liquid interface, confirm the prediction that the gas from the side predominantly flows towards the interface along the side of the reactor and not via the middle. In the latter, detectors 9 and 14 would have observed the tracer more or less at the same time. This is furthermore confirmed by the fact that the tracer released from the (right-hand) side sparger is observed by the detectors 4–10 in increasing order of time.

The response of detector 12, located opposite to detector 4, shows a curve with a maximum that is similar to the second maximum of the response curve of detector 4, suggesting a recycling period of ~ 18 s, see Figure 13.18.

In order to better comprehend the recorded signals, a simple data-acquisition model accounting for solid angle of acceptance and attenuation (homogeneous fluid density) effects for a point-source (representative for all injected activity being present in one 'air bubble') in the reactor has been set up. Relative signal responses have been computed, starting from detector 4 for a vertical and an along-wall source trajectory.

The computed experimental flow paths are depicted in Figure 13.17 along with the CFD prediction of the bubble tracks. The flow path for side sparger injection indicates that after initially flowing somewhat towards the reactor wall, the air flows more or less vertically towards the level interface. The computed velocities are high near detector 5 (~ 5 m/s), slowing down to ~ 2 m/s in between detectors 5 and 8, while

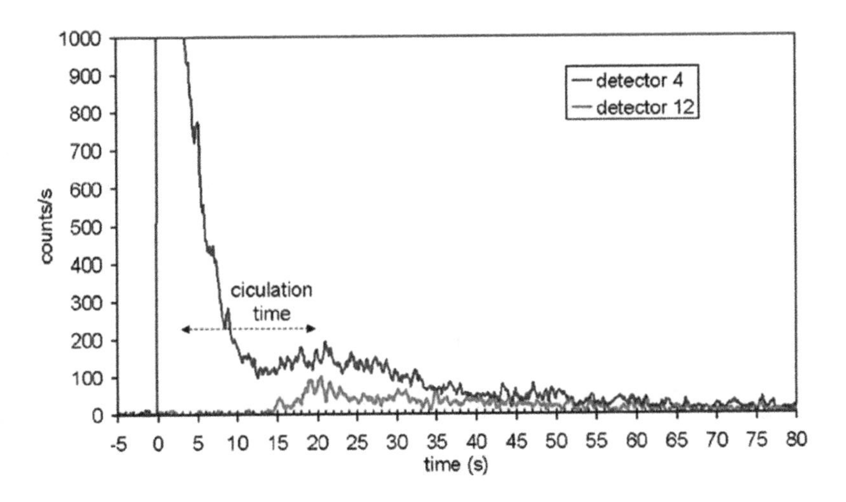

Figure 13.18: Tracer response of detectors 4 and 12 (locations provided in Figure 13.16).

increasing to 3–6 m/s in the gas cap (detector 9 to 10). The small fraction, flowing from the injection point into the bottom section of the reactor, moves very close along the reactor wall. The computed velocity in traversing from detector 4 to 2 is ~ 0.5 m/s. The shape of the recorded signal for detector 2 supports the flow, closely along the wall.

The flow path for middle sparger injection initially has a substantial horizontal component; above the side sparger, it starts to flow more vertically. The velocities between detectors 2 and 3 are high (~ 6 m/s), slowing down in the middle part to ~ 1 m/s, and increasing to ~ 6 m/s in the gas cap. The average gas velocity for side sparger injection is significantly higher than for middle sparger injection under normal operating conditions, approximately 2 and 1 m/s, respectively. This again strongly suggests that air bubbles from the side sparger and the middle sparger follow distinct paths through the reactor: air bubbles emerging from the side sparger predominantly do not mix with air bubbles from the middle sparger; each largely following their own, distinct path towards the liquid/gas interface. In comparison with the "jet" stream predicted by CFD, the positions of the experimental flow path points are somewhat skewed toward the reactor wall.

13.3.6 Commercial plant improvements

As with most tangible results of this work, we developed a number of potential measures to reduce the "starvation" predicted to occur for certain configurations and severe operating conditions. These options included repositioning of the gas inlets and installing additional baffles at specific locations, see also the patent application by Hollander et al. [7].

After implementation of the optimum configuration in a number of commercial reactors, the reaction temperature could be lowered significantly (when compared with equal oxygen consumption rates), see Figure 13.19. This translates to improved yields and/or increased production rates.

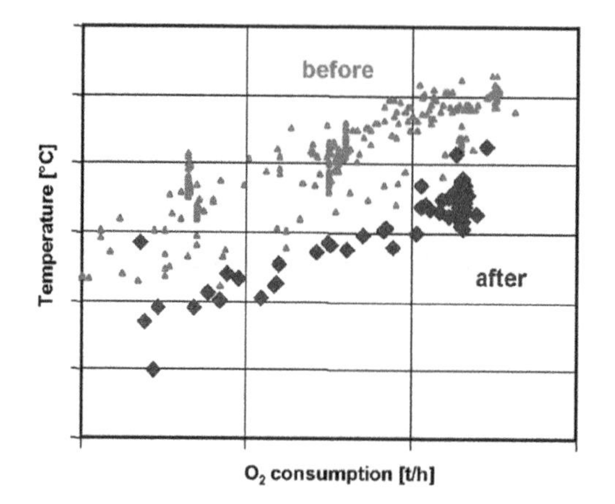

Figure 13.19: Reactor temperature vs oxygen consumption. Comparison before and after reactor modifications.

13.3.7 Takeaway learning points

Learning point 1

A well-designed set of RTD tracer experiments in a real commercial-scale reactor can be a very powerful diagnostic tool and a justification for reactor design modifications.

Learning point 2

RTD tracer experiments at commercial scale, with detectors before and after the reactor(s), can be expanded with detectors on the reactors themselves to gain even more insights with respect to multiphase flow patterns inside the reactors.

Learning point 3

In case the RTD of a reactor can be well described with an axial dispersion model, it can also be well described by a tanks-in-series model (provided the best fit is 3 or more tanks-in-series). The opposite is also true: if a tanks-in-series model does not fit the data well, an axial dispersion model will also not do well.

References: ethyl benzene peroxidation reactor

[1] Buijink JKF, Lange JP, Bos ANR, Horton AD, Niele FGM, Propylene epoxidation via shell's SMPO process: 30 years of research and operation. In: Oyama ST, ed. Mechanisms in Homogeneous and Heterogeneous Epoxidation Catalysis. Elsevier Science, The Netherlands, 355–371, 2008.

[2] Klusener PAA, Jonkers G, During F, Hollander ED, Schellekens CJ, Ploemen IHJ, Othman A, Bos ANR, Horizontal crossflow bubble column reactors: CFD and validation by plant scale tracer experiments. Chemical Engineering Science, 2007, 62(18–20), 5495–5502.

[3] Tilton JNRTWF, Designing gas-sparged vessels for mass transfer. Chemical Engineering, 1982, 89 (24), 61–68.

[4] Zuiderweg FJ, Bruinzeel C, Liquid mixing and oscillations in sparged horizontal cylindrical vessels. Proceedings of the European Symposium of Chemical Reaction Engineering, Fourth Meeting, 1968, pp. 183–189.

[5] Krzysztoforski A, Wojzit Z, Pohorecki R, Baldyga J, Industrial contribution to the reaction engineering of cyclohexane oxidation. Industrial Engineering Chemistry Process Design and Development, 1986, 25(4), 894–898.

[6] Pohorecki R, Baldyga J, Moniuk W, Podgorska W, Zdrojkowski A, Wierzchowski PT, Kinetic model of cyclohexane oxidation. Chemical Engineering Science, 2001, 56, 1285–1291.

[7] Hollander ED, Klusener PAA, Ploemen IHJ, Schellekens CJ, Horizontal reactor vessel. Patent WO2006024655, 2006.

13.4 A new catalyst shape: pressure drop and packing density

13.4.1 Introduction

This case is about a new catalyst charge of a fixed bed reactor. The catalyst in this multi-tubular reactor is expected to reach the end of its economic lifetime 10 months after loading. The traditional catalyst supplier has offered a larger particle with a smaller bore size that enables us to put more mass of catalyst into the reactor. The previous charge consisted of cylindrical rings measuring 8.5 mm height, 8.5 mm external diameter, 3 mm bore diameter, and with a packing density of 970 kg catalyst/m^3. We will assume that the bed porosity ε_b is 0.55 m^3/m^3. The offered larger particles have the dimensions of 9.5, 9.5, and 2.5 mm, respectively. A batch of the new catalyst has been tested and exhibited a packing density of 1057 kg catalyst/m^3. All other catalyst properties have remained the same. So, in effect, around 9% more mass of catalyst can be put into the reactor. The recycle compressor is operated near the declining end of its curve, that is, a pressure drop increase of more than 5% would be unacceptable.

We will now evaluate whether the proposal of the catalyst supplier is a good one and check whether the existing compressor can handle the expected pressure drop over the tubular reactor. Further data are that under normal operation, the Reynolds numbers varied between Re = 600 and Re = 1600. Under typical operating conditions, diffusion limitations can be neglected. The Thiele modulus of Chapter 5 is around 0.3. This example is based on information provided in a series of patent applications [1–3].

13.4.2 Initial evaluation

Without dedicated experimental data, we used the classic Ergun equation for the evaluation of the pressure drop ΔP, which can be rewritten as

$$\left(\frac{\Delta P}{\rho_f u_s^2}\right)\left(\frac{d_p}{L}\right)\left(\frac{\varepsilon_b^3}{1-\varepsilon_b}\right) = 150\frac{1-\varepsilon_b}{\mathrm{Re}} + 1.75 \tag{13.1}$$

where ρ_f is the gas density, u_s is the superficial velocity, and L is the bed length. For the Reynolds number, different equivalent diameters should be used, depending on the particle shape (see Table 13.2).

Table 13.2: Equivalent diameters for particles of different shapes.

Shape	Equivalent diameter (d_p)
Sphere, diameter d_s	d_s
Cylinder with length H equal to diameter $d_c = 2R_u$	d_c
Long extrudates with radius R_u	$3R_u$
Ring with length H, outside radius R_u, and inside diameter R_i	$\frac{3(R_u + R_i)H}{R_u - R_i + H}$

For the given range of Reynolds numbers, the value of the right-hand side of eq. (13.1) ranges from 1.91 to 1.81: the first part of the right-hand side, reflecting the laminar contribution, is small compared to the second 1.75 part, which is the turbulent flow contribution. Because the catalyst particle density has not changed, the bed porosity can be calculated from the ratio of the packing densities. This yields a bed porosity of 0.51 for the new catalyst. To calculate the equivalent particle diameters, we use the expression given in Table 13.2 for a ring and find that the equivalent particle diameter for the old catalyst is 6.2 mm and for the new catalyst, it is 7.7 mm. For the same production level and gas recycle flow rates – the same value of ρu_s^2 – we then have the ratio of the pressure drops:

$$\Delta \mathrm{P}_{\mathrm{new}}/\Delta \mathrm{P}_{\mathrm{old}} \approx \mathrm{d}_{\mathrm{p\,old}}/\mathrm{d}_{\mathrm{p\,new}}\left(\varepsilon_{\mathrm{b\,old}}/\varepsilon_{\mathrm{b\,new}}\right)^3\left(1-\varepsilon_{\mathrm{b\,new}}\right)/\left(1-\varepsilon_{\mathrm{b\,old}}\right)$$

Introducing the numbers, we find that the pressure drop increases with a factor of 1.11. So based on the Ergun correlation, the 9% extra mass of catalyst is expected to yield a 11% increase in pressure drop. As this is more than twice the maximum of 5% increase that was stated to be acceptable, without further data, we would advise against the proposal.

13.4.3 Experimental results

However, since the Ergun equation is a semi-empirical correlation, with empirical constants and shape-dependent “correction factors”, it was decided to experimentally determine the pressure drop of the proposed new catalyst particle size.

This showed that in fact the observed pressure drop was 8% lower than the old catalyst (instead of 11% higher than the old catalyst as predicted by the Ergun correlation, without experiments).

This example illustrates on the one hand how the Ergun equation can be used to evaluate catalyst design proposals, but on the other hand that it remains highly recommended to carry out representative measurements. This is particularly true for changes in the geometry of the catalyst, for which the Ergun equation is not very suited to make a priori predictions.

The evaluation of the proposal made by the catalyst supplier can now be positive: the 9% increase in loaded catalyst will not cause pressure drop problems. The extra mass of catalyst inside the tube can be exploited by lowering the temperature, which – in general– will yield both a better selectivity and a longer lifetime, which justifies the costs of the extra amount of catalyst. Therefore, his proposal may in fact backfire the catalyst supplier: he had proposed it hoping to sell 9% more catalyst per batch. But, in practice, the lifetime of the catalyst may increase by more than 9%, resulting in a reduction in catalyst sales. Of course, ultimately, the proposal represents a value increase and the added value could be shared between the supplier and the customer via an increase in catalyst sales price.

13.4.4 Takeaway learning points

Learning point 1
The Ergun expression can be used for engineering estimates (±30%) for pressure drop in fixed beds of (catalyst) particles with regular shapes like spheres and cylinders, provided an appropriate hydraulic or equivalent diameter and shape factor is applied.

Learning point 2
For relatively subtle changes in the shape of a particle, such as changing the bore size in a ring-shaped particle or the length over diameter ratio, the Ergun expression can fail to predict, even directionally, whether the pressure drop will increase or decrease.

Learning point 3
If the estimated value should be accurate within 10%, the pressure drop should be checked experimentally using the actual particles. For low ratios of tube diameter and particle size, the same tube diameter as in the commercial-scale reactor must be used, and the loading (speed) should not be too different from the commercial-scale loading.

Learning point 4
Relatively subtle changes in catalyst particle size and shape can bring a lot of value in case a balance needs to be found between catalyst holdup, pressure drop, heat transfer, mechanical strength, and diffusion limitations.

References: a new catalyst shape

[1] Richard MA, McAllister PM, Coleman AT, Syrier JLM, Bos ANR, Process for selecting shaped particles, a process for installing a system, a Process for reacting a gaseous Feedstock in such a system, a computer Program, a computer program product, and a computer system. WO2006036677 A3, 2006

[2] McAllister PM, Bos ANR, Richard MA, Rekers DM, Reactor system and process for the manufacture of ethylene oxide. US 2004/0225138, 2004.

[3] McAllister PM, Bos ANR, Richard MA, Rekers DM, Reactor system and process for the manufacture of ethylene oxide. US 2005/0019235, 2005.

13.5 Heavy residue oil upgrading: reactor type selections and development

13.5.1 Heavy residue upgrading introduction

The purpose of this case study is:
- To show the different reactor type selections that different companies made
- To compare their commercial-scale successes
- To draw lessons *for reactor type selection methods, in general*

First, the reaction systems for heavy residue upgrading – conversion to lighter and cleaner hydrocarbons – will be described, followed by three reactor types, with their development history to commercial-scale implementation.

The three three-phase reactor types are:
- Trickle-flow moving-bed bunker flow reactor of Shell
- Ebullated bed slurry reactor of LC-Fining
- Fluid bed reactor of Exxon (Flexicoker)

These three-phase reactor types have been developed and implemented in the same period: 1960–1980. So the three companies had the same general information available to them. However, the companies made very different decisions in the selection of the reactor type.

13.5.2 Heavy residue upgrading reaction chemistry

Physically, heavy residues at room temperature look like asphalt on the road. They are solid black matter. At temperatures above 350 °C, they become an easily flowing liquid. Chemically, they have the following characteristics. Their molar hydrogen/carbon ratio is around 1. Their chemical composition is made up of organic sulfur components, organic heavy metal (mainly vanadium and nickel) components, poly-aromatic components, and heavy aliphatic components.

There are two chemical routes to convert this feedstock to hydrocarbon streams that can be treated in conventional refinery processes to market products. One route is to add hydrogen to get a higher hydrogen/carbon ratio, and convert the organic sulfur and organic metal components to H_2S and metal sulfides. The other route is to thermally crack the feed into coke and to lighter oils. Metal components stay in the coke. The lighter products are treated in the refinery units to transport fuels.

The reactions involved are many, but they can be lumped into four main reactions:

(1) hydrogen (g) + organometallic components (l) = metal sulfides (s) and hydrocarbons (l)
(2) hydrogen (g) + organic Sulphur components (l) = H_2S (g) + hydrocarbons (l)
(3) hydrogen (g) + hydrocarbon large molecules (l) = hydrocarbon small molecules (l)
(4) hydrocarbons (l) = C (s) + hydrocarbons (g)

Reactions 1–3 are heterogeneously catalyzed. Reaction 4 is a thermal reaction (no catalyst). This reaction takes place mainly in the FLEXCOKING™ process of Exxon, but also takes place in the local absence of hydrogen in the other hydrogenation reactors in local areas depleted of hydrogen by mass transfer limitations. If this happens at or in the porous catalyst, then this is called "catalyst black death".

The reaction enthalpy of the hydrogenation reactions for alkane hydrogenations is about – 40 kJ/mol H_2 [1] and for aromatic hydrogenations, it is −70 kJ/mol H_2 [2]. So these reactions are highly exothermic. The reaction enthalpy of the coking reaction (4) is around +80 kJ/(mol bonds broken); so this reaction is endothermic. Its free enthalpy $\Delta G = -60$ kJ/(mol bonds broken) [3]. So the reaction needs energy from outside but the reaction itself runs due to the increase in free entropy caused by the many smaller molecules formed. A thermodynamic analysis of catalytic cracking reactions is provided by Nazarova [3].

The temperature and pressure for the hydrogenation reaction to run are high. For the hydrogenation reactions, the temperature range is 350–450 °C and the reaction pressure is between 100–200 bar. All three hydrogenation reactions are catalyzed by a porous heterogeneous catalyst. The metal sulfides that are formed deposit inside the pores. This reduces the number of active catalytic sites and reduces the diffusion rates inside the pores. The catalyst life is therefore limited to 10–30 days. For the coking reaction, the temperature is even higher, 450–500 °C [4], but the pressure is 1 bar.

13.5.2.1 Reactor type options

The known reactor type options for these reaction systems in the 1960–1980 period were for the hydrogenation reactions:
- Fixed bed trickle-flow swing operation type
- Three-phase slurry bubble column reactor type

And for the coking reaction, a fluid bed reactor system similar to fluid catalytic cracking (FCC) was known.

The fixed bed trickle-flow reactor is operated adiabatically. To avoid reaction run-away behavior, the adiabatic temperature rise is limited. This is achieved by operating with a large surplus of hydrogen gas, which after cooling and H_2S separation is recycled to the reactor inlet.

The RTD of the gas and liquid flow is nearly plug flow. So the adiabatic temperature rise can be set for a plug flow reactor (see Chapter 7). The adiabatic temperature rise is limited.

The fixed bed trickle-flow reactor with swing operation means that the system has at least two reactors in parallel. One reactor is in normal operation and the other reactor is in catalyst refreshment mode. The latter means that at ambient temperature and pressure, spent catalyst is removed from that reactor, for instance by a liquid slurry operation or by a sucking hose, after which, fresh catalyst is loaded to the reactor and the reactor is brought to reaction conditions. The catalyst lifetime is about 1 month, as indicated by van Swaaij for this reactor type in Chapter 2. In that time, a reactor can be un-loaded and re-loaded. The reactor is operated at a high pressure; so the reactor vessels are very capital intensive. Needing two reactors of which one is not producing, is therefore not attractive.

The slurry-ebullated-bed reactor allows for withdrawal of the spent catalyst and feeding fresh catalyst under process conditions, and so looks attractive. It, however, has a backmixed liquid RTD and also the catalyst phase has a backmixed RTD.

The fluid bed coking reactor has a reasonably narrow RTD for the oil fraction in the riser fluid bed reactor part; so reasonably deep conversions can be reached.

In the next sections, the history of the three reactor type selections and their developments in 3 different companies will be described.

13.5.3 Shell bunker flow selection and the development to commercial scale

13.5.3.1 Shell hydrogenation trickle-bed moving-bed reactor selection

Shell had developed three-phase trickle-flow hydrogenation reactors in the fifties of last century for upgrading oil fractions. Their knowledge on scale-up of this reactor type for these applications steadily grew, mainly in the Shell Research Laboratory Amsterdam, and was transferred to the design office in the Hague. There was also a steady trend of treating more ‘heavy’ oil fractions containing larger molecules and also containing more organic sulfur and organic metal components; called: 'whitening of the barrel'. The latter trend caused shorter catalyst lifetimes. Special rapid catalyst loading and unloading procedures with reactor adaptations were developed to cope with these shorter catalyst lifetimes. However, when very heavy residual oil fractions were considered, such as Venezuelan crude oil with high metal content, fixed bed reactors with swing operation were no longer economically feasible.

Then, a conscious decision was made by Shell engineers in selecting one between the fluid bed coking option and the moving-bed trickle-bed hydrogenation reactor type. The Shell engineers preferred the hydrogenation option because they loved the trickle-bed hydrogenation technology and its superior product composition and also because they considered the coking option as wasteful and dirty [5]. The process was called Hydro-DeMetallization (HDM) and the reactor was called a bunker reactor. Later, the whole process, with the reactor, was called HYCON (HYdroCONversion) [5].

13.5.3.2 Shell hydrogenation trickle-bed moving-bed (bunker) reactor development

Figure 13.20 show the moving-bed reactor for hydrogenation of heavy residues, obtained from [6]. The reactor is also called bunker reactor. The bottom part has a very special design. The top conical section ensures that the catalyst moves as in nearly plug flow, called mass flow in the solids flow engineering field. In the next cylindrical section, a screen is placed through which oil and gas flow radially outwards, while the spent

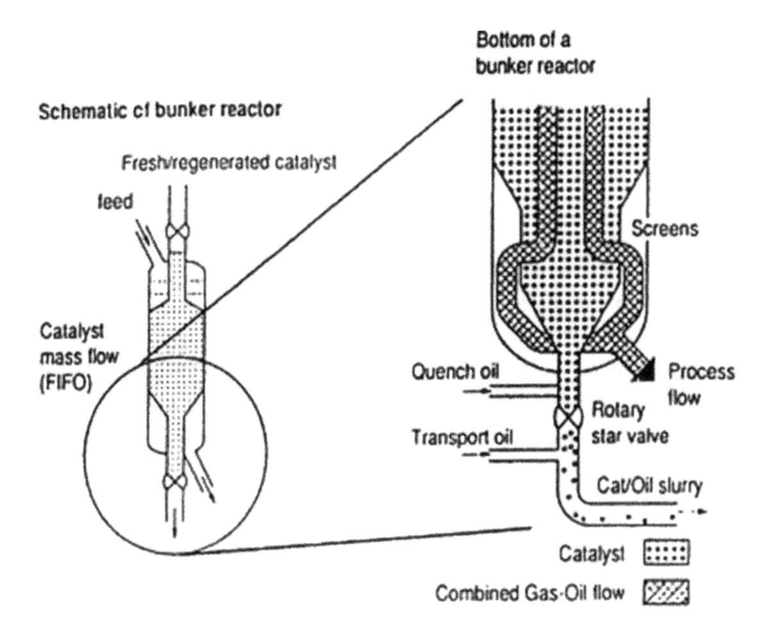

Figure 13.20: Bunker reactor for heavy residue upgrading of Shell [6].

catalyst stays inside. In the central bottom section, the spent catalyst leaves the reactor via a rotary star valve.

13.5.3.3 Bunker reactor development

Table 13.3 provides an overview of the of the HYCON bunker reactor development. We will briefly describe each step.

Table 13.3: Innovation steps of bunker flow reactor for the hydro-conversion of heavy oil residue.

Period	Innovation step	Scale (t/day)
1960–1977	First demonstration of HDM bunker reactor Gothenburg refinery [7]	400
1979–1985	Second demo-reactor Cardon refinery Venezuela [7,8]	400
1977–1989	Development, design, and start-up commercial-scale reactor [9,10]	4,000
1995	Steady-state operation bunker reactor Pernis refinery [8]	4,000

13.5.3.4 First demonstration reactor, Gothenburg

Early in the development, Shell central office decided to design and construct a HDM bunker flow reactor at the Gothenburg Refinery for treating heavy residue oil by hydrogenation. The bottom part of the reactor, obtained from Furimsky, is shown in Figure 13.20 [6]. It contains a cone section, a vertical screen to allow gas and liquid to radially flow out of the reactor, while keeping the catalyst in the reactor. The very bottom section allows the spent catalyst to leave the reactor at process conditions. Fresh catalyst can be fed at process conditions at the top of the reactor. The reactor operates in trickle-flow mode, meaning that gas and liquid co-currently flow downwards over the catalyst particles [6].

13.5.3.5 Development

In 1978, a Venezuelan company showed interest for the residue upgrading capability of Shell and a full research plan was defined for developing a process for Venezuelan heavy short-residue, containing up to 500 ppm metal (as organic components). The research involved a hot pilot plant to study the catalyst lifetime and validate kinetics for the hydrogenation reactions and for the catalyst decay rates. It also involved a large-scale cold-flow transparent model (the size of the Gothenburg reactor) to study the gas, liquid, and solids flow, and the RTD by modeling the gas and liquid flow, and also of the solids flow under gas and liquid flow. Research showed that the radial gas and liquid flow near the screen could hamper the catalyst flow. This initiated a special bottom section design for the commercial scale, which was patented [11, 12].

13.5.3.6 Second demonstration process

A demonstration process was designed, constructed, and started up in Venezuela at PDVSA's Cardon refinery, consisting of a series of four bunker-flow HDM reactors integrated with a Demetallizing Catalyst Recovery (DCR) unit. The capacity was 400 t/day. The Demonstration resulted in the conclusion in 1985 that a commercial-scale design was feasible [8].

13.5.3.7 Commercial-scale design and operation

In 1989, "A HYCON unit at the Shell refinery in Pernis (Rotterdam, the Netherlands), with a throughput of 4,000 ton/day of vacuum residue per day, included five reactors, of which three are the "bunker" demetallization reactors and the remaining two are fixed bed hydrodesulfurization/hydroconversion reactors." [9, 10].

"The process performance (in terms of sulfur/ metals removal and conversion into distillates) is in line with predictions, and the catalyst bunkering system works well. During the start-up phase of the Pernis unit, equipment and materials problems were encountered. These problems were solved by repairs and replacement of the affected equipment. At present, the throughput has been increased to well above the design value and a high conversion of vacuum residue into distillates has been consistently achieved" [7]. In 1985, 6 years after the start-up, an extended stable run could be achieved [8].

This start-up time of 6 years is longer than the industry average for a commercial-scale process, as provided by the Merrow correlation [13]. Even if we assume that the crude feedstock in the Pernis is new (different from the Venezuelan feedstock); the two process units (HDM and HDS) are also new; and we take the feedstock as raw solids, the Merrow correlation provides an averaged start-up time of 18 months, while the actual start-up was 72 months; a factor of 4 longer.

13.5.3.8 Conclusions – Shell bunker reactor selection and commercialization

Having the first demonstration plant at Gothenburg refinery without having any knowledge of the gas, liquid, and catalyst flow behavior, and with no knowledge of design and construction details that are essential for the solids flow and the gas and liquid flow is little value. Over ten years, it never worked properly, without providing insight into what caused the problems. It proved however, that designing, constructing, and operating a bunker reactor for hydro residue upgrading is not easy.

If you enter a new knowledge field, such as a moving bed, in combination with 2-D gas-liquid flow, a cold flow (mock-up) test unit can be very useful in quickly identifying flow problems and for optimizing the reactor geometry.

The start-up time of the commercial-scale bunker reactor of 6 years; a factor 4 longer than the industry correlation estimate of Merrow [13], indicates that the technology was not easy to scale-up. Since 1995, no other bunker reactor process was reported to be implemented at commercial scale; so it cannot be considered to be a mature technology.

13.5.4 LC-FINING™ residue hydrocracking in three-phase slurry-ebullated-bed reactor

The reactor is a three-phase slurry-ebullated-bed reactor. Gas and liquid flow up through the reactor. Liquid is recycled inside the reactor by a pump and recycle tube. The catalyst bed is in the liquid-expanded (semifluidized) condition. The catalyst size is around 1 mm [14]. The reactor is cooled inside by a tube bundle and is nearly isothermal. The spent catalyst is withdrawn from the reactor and fresh catalyst is added online [15] at a frequency of once a day [14].

The reactor conditions are [15]:

Temperature:	410–440 °C
Pressure:	110–180 bar
Conversion heavies:	55–85%
Desulfurization:	60–85%
Demetallization:	65–85%

13.5.4.1 Reactor type selection

No explicit reactor type selection considerations for the LC-FINING™ process could be found. Gillis [15] mentions the following features, which may have been taken as selection criteria for this reactor type by Lummus for their LC-FINING™ process:

- Low pressure drop
- Isothermal
- Online catalyst feed and withdrawal
- Flexible to feedstocks

13.5.4.2 Reactor development and commercial-scale applications

Gillis reports the existence of a pilot plant, but a description of the development could not be found. The process has been continuously improved, from one commercial-scale application to the next.

To quote Gillis: "Key to the application of RHC is the assurance that this process is reliable, which has been confirmed through continual feedback from the commercial units, and advances in equipment and designs." RHC stands for Residue Hydro-Cracking. So far, 8 commercial-scale units have been built, with a total capacity of 400 kbd [16]

The LC-FINING™ (Lummus Cities reFining) process was developed by Lummus Crest and Cities Service Research and Development Company in the 1960s, initially for upgrading bitumen from tar sands [17].

13.5.4.3 Conclusion on three-phase ebullated-bed reactor

The three-phase slurry-ebullated-bed reactor has proved to be commercially very successful with its eight commercial-scale implementations.

13.5.5 Heavy oil upgrading by coking with Exxon Flexicoker fluid bed

13.5.5.1 Reactor system

The flexicoker fluid bed reactor of Exxon upgrades heavy oil fractions to lighter oil fractions, which are also lower in metal and sulfur content, but part of the feed ends up as coke, which is burnt to provide energy. The reactor system works as follows. Vacuum residue feed enters the scrubber for "direct-contact heat exchange" with the reactor overhead product vapors. The higher boiling point hydrocarbons (525 °C) present in the reactor product vapors condense in the scrubber and return to the reactor, in a mixture with the fresh feed. Lighter overhead product vapors in the scrubber go to conventional fractionation and light-ends recovery. The feed is thermally cracked in the fluidized-bed reactor to a full range of gas and liquid products, and coke [18].

13.5.5.2 Reactor selection

No publication could be found on the reactor type selection. It is likely that the obvious choice for the Exxon engineers was a fluid bed coking reactor because of the Fluid Catalytic Cracking (FCC) success in Exxon.

13.5.5.3 Reactor development to commercialization

The first FLEXICOKING™ unit began operation in 1976, and since then five others have come online; the most recent one in 2012. Since 2010, Exxon has licensed three new units with a combined capacity of 114 kilo barrels per day (kbd). These are currently in different stages of construction. ExxonMobil has licensed the fluidized bed coking technology since the 1950s, including the largest fluid bed coker, which is located in Syncrude, Fort McMurray, Canada. Their experience ranges from 5 to 105 kbd unit capacities [19].

It is now a commercially proven technology for over 38 years in ExxonMobil. It has also been licensed to third-parties. All FLEXICOKING™ units are operating today at capacities greater than the original design, for a combined operating capacity of 227 kbd [18]. The Flexicoker fluid bed reactor has been implemented at least 8 times [18].

13.5.6 Reactor type comparison – heavy petroleum upgrade

Hydro-upgrading of heavy petroleum with the Shell moving-bed bunker reactor in trickle flow operation has been implemented once at commercial scale. After a start-up period of 6 years, it reached steady-state production capacity of 4 kbd in 1995.

Hydro-upgrading of heavy petroleum with the LC-Fining ebullated slurry reactor with internal cooling has been implemented 8 times, with a total installed capacity of 4000 kbd; so this technology has reached maturity.

This means that for hydro-upgrading of heavy petroleum, the ebullated slurry reactor is, for the time being, superior to the trickle-bed moving-bed reactor.

A comparison between a coking versus hydro-up grading cannot be made here as it depends on the feedstock and the desired products. It requires a dedicated evaluation for each given case.

It is clear from the number of commercial-scale implemented cases that the slurry reactor and the fluid bed reactor type have reached technical maturity, while the moving-bed bunker reactor has only been applied once, and so has not reached maturity.

The reason for not having more commercial-scale implementations for the bunker reactor technology, while competing technologies have many implementations, is not disclosed by Shell or others.

13.5.7 Exercises

13.5.7.1 Exercises on reaction systems

Task 1: Draw a functional mass flow block-flow diagram for the hydro-upgrading of heavy oil, with heavy oil and hydrogen as input streams, product stream as light oil, and hydrogen di sulfide and hydrogen, free of H_2S, as recycle stream.

Task 2: Add functional blocks for heat generation and transfer.

Task 3: Draw functional block flow diagrams for all flexicoker reactors and their connections.

13.5.7.2 Exercises on thermodynamic calculations

Task 4: Determine the superficial hydrogen mass flow (kg/m^{2}s) to limit the adiabatic temperature rise to 10 °C, neglecting the liquid flow heat balance for the hydrogenation of heavy oil, with a reaction enthalpy of 50 kJ/mol H_2.

Task 5: Calculate the hydrogen conversion for this 10 °C temperature rise.

13.5.8 Takeaway learning points

Learning point 1

Reactor selection for a novel three-phase system, with rapid catalyst decay, strongly depends on the existing knowledge and experience inside the companies.

Learning point 2

Fluid bed and slurry reactors for cases with rapid catalyst decay have been implemented many times.

Learning point 3

For times in between two catalyst regenerations of less than one month, fluid bed or slurry reactors should also be considered.

References: heavy residue oil upgrading

[1] Jaffe SB, Kinetics of heat release in petroleum hydrogenation. Industrial & Engineering Chemistry Process Design and Development, 1974, 13(1), 34–39.
[2] Moore WJ, Physical Chemistry, 4th ed. London, Longmans, 1962.
[3] Nazarova G, Ivanchina E, Ivashkina E, Kiseleva S, Stebeneva V, Thermodynamic analysis of catalytic cracking reactions as the first stage in the development of mathematical description. Procedia Chemistry, 2015, 15, 342–349.
[4] Jahnig CE, Campbell DL, Martin HZ, History of fluidized solids development at EXXON. In: Grace JR, Matsen JM, eds. Fluidization. Boston, MA, Springer, 1980, 3–24.
[5] Howard S, Jonker J, Geschiedenis van de Koninklijke Shell, deel 2. Amsterdam, Boon, Amsterdam, 2007.
[6] Furimsky E, Catalysts for Upgrading Heavy Petroleum Feeds. Amsterdam, Elsevier, 2007.
[7] Scheffer B, et.al, The shell residue hydroconversion process: Development and achievements. Catalysis Today, 1998, 43(3–4), 217–224.
[8] Ramirez-Corredores MM, The Science and Technology of Unconventional Oils: Finding Refining Opportunities. Bottom of the Barrel Upgrading Technologies. Cambridge, MA, USA, Academic press, 2017, Chapter 5.
[9] Magomedova RN, Popovaa AZ, Maryutinaa TA, Kadievc KM, Khadzhi SN, Current status and prospects of demetallization of heavy petroleum feedstock, review. Catalysis Today, 1998, 43, 217–224.
[10] Oelderik JM, Sie ST, Bode D, Progress in the catalysis of the upgrading of petroleum residue: A review of 25 years of R&D on shell's residue hydroconversion technology. Applied Catalysis, 1989, 47 (1), 1–24.
[11] Wijffels JB, Pronk GJ, Den Hartog AP, Moving catalyst bed reactor. United States patent US 4,568,523. 1986 Feb 4.
[12] Wijffels JB, Pegels AA, Wezenberg A, inventors; Shell Oil Co, assignee. Apparatus for the catalytic treatment of hydrocarbons. United States patent US 4,229,418. 1980 Oct 21.
[13] Merrow EW, Estimating startup times for solids-processing plants. Chemical Engineering, 1988, 24, 89–92.
[14] Kressmann S, Boyer C, Colyar JJ, Schweitzer JM, Viguie JC, Improvements of ebullated-bed technology for upgrading heavy oils. Oil & Gas Science and Technology, 2000, 55(4), 397–406.
[15] Gillis D, Residue Hydrocracking Solutions for Refinery Sustainability. Chevron Lummus Global 2019. Accessed December 20, 2021, athttps://refiningcommunity.com/residue-hydrocracking-solutions-for-refinery-sustainability/
[16] Arora A, Mukherjee U, Refinery configurations for maximising middle distillates. PTQ, 2011, 2–11, 16(3), 9.
[17] Steegstra J, High Conversion of vacuum residue. Chevron Lummus Global, 2019. Accessed December 20, 2021, at http://www.aiche.nl/images/presentations/2019-4-11-ldm.pdf.
[18] Exxon, Resid heavy oil conversion. 2021, Accessed May 27, 2021, at https://www.exxonmobilchemical.com/en/products/technology-licensing-and-services/resid-heavy-oil-conversion.
[19] Exxon, Flexicoking. 2021. Accessed May 27, 2021, at http://file:///C:/Users/Gebruiker/Downloads/flexicoking_webpagepdf.pdf

13.6 Reactor stability in an adiabatic trickle-bed reactor

Scientists at the Institute of Catalysis in Novosibirsk in Russia were the first to give full details of the properties of an unstable adiabatic bed reactor, where they analyzed the temperature variations in the outlet plane of the catalyst bed, see Klenov et al. [1]. As the most probable reason for the existence of hot spots in this outlet plane, Matros [2] mentioned a possible fluctuation of the porosity in the local regions of the bed. The real cause for the instabilities was found by Benneker et al. [3], which will now be discussed.

To analyze an adiabatic bed reactor, we always assume plug flow in our calculations and check our assumption by calculating the Re number, based on the particle diameter to check whether we operate in the turbulent regime, that is, Re > 10–100. This means that our assumption was justified and that we work well in the turbulent flow region, so that plug flow may be expected, or not.

In reactors with high heat effects, and especially with high pressure reactors, we must also check whether the temperature gradient over the bed induces free convection, and how important is its contribution to the flow. Benneker et al. carefully checked the influence of the various length variables for free convection in a packed bed and found that the following modified Grashof number describes the phenomenon:

$$\mathrm{Gr} = (\Delta\rho/\rho)\, g\, \left(d_p d_{\mathrm{bed}}\right)^2/(H_{\mathrm{bed}}\nu^2)$$

Here, $\Delta\rho$ is the driving force for the free convection, and has to be taken as the density difference over the entire bed. Neglecting deviations of the ideal gas law, $\Delta\rho/\rho$ is equal to $\Delta T/T_0$. The diameter of the reactor is 1.2 m. After the liquid flow stops, the catalyst particles will start to runaway, and during the process of evaporating the methanol remaining in the catalyst pores, the temperature of the particles will gradually increase to the value of $T_0 + \Delta T_{\mathrm{ad}}$. At a certain moment during the start of the evaporation, we assume that the temperature difference with the surrounding gas has reached a value of 10 K. Under that condition, the Grashof number is already equal to 140,000; so the important ratio of Gr/2.5Re^2 is already equal to 6.2. This means that the free convection is already 6.2 times stronger than the forced flow of the gas. Moreover, the free convection is directed upward whereas the forced flow is directed downward. It is evident that as soon as the particle runaway starts the flow through the reactor, it becomes really chaotic and the reactor is highly unstable!

We note that in case free convection plays a role next to forced convection, one has to look at the Reynolds number $\mathrm{Re}_{\mathrm{res}}$ that includes the free convection contribution as well the forced convection part $\mathrm{Re}_{\mathrm{forced}}$ [4]. In that case, we would have had to calculate the square root of $\mathrm{Re}^2_{\mathrm{forced}}$ + Gr/2.5. Doing that, we would find the value to be in the range of 9025–56,000. This would lead to an imaginary value of $\mathrm{Re}_{\mathrm{res}}$, which of course is absurd. The sign of the Gr number is negative because the forced flow stream and the free convection flow stream are in opposite directions. Nevertheless, the absurd outcome is a serious warning that the flow conditions in a reactor are unstable.

However, the question remains: Is the reactor unstable under normal trickle-flow operation? In that case, we may estimate that the maximum temperature difference between the particle and the liquid flowing over it is 12 °C. In reality, because the catalyst efficiency at the top of the reactor is 0.33, this difference is already reduced to

4 °C. But, now the driving force for free convection is the temperature difference at the interface of the liquid and the gas phase, which will practically be zero. So, the trickle-flow reactor will be stable as long as we can be sure that the particles are fully wetted.

References: reactor stability in an adiabatic trickle-bed reactor

[1] Klenov OP, Matros YS, Lugovskoy VI, Lakhmostov VS, Theoretical Foundations of Chemical Technology, 1983, 17, 337 (in Russian).

[2] Matros YS, Unsteady Processes in Catalytic Reactors. Amsterdam, Elsevier Science Publishers, 1985.

[3] Benneker AH, Kronberg AE, Westerterp KR, Influence of buoyancy forces on the flow of gases through packed beds at elevated pressures. AIChE Journal, 1998, 44, 263.

[4] Gnielinski V, Berechnung mittlerer Wärme- und Stoffübergangskoeffizienten an laminar und turbulent überströmten Einzelkörpern mit Hilfe einer einheitlichen Gleichung. Forsch Ing Wes, 1975, 41(5), 145–150.

13.7 Three-phase slurry-reactive distillation

13.7.1 Introduction

The purpose of this industrial case is to show that

A: Conversion of solids is feasible in a three-phase slurry-reactive distillation reactor.

B: For a fast consecutive reaction of a product to a by-product, reactive distillation is attractive.

C: A three-phase slurry reactor with product stripping has been designed in the 1950s, without the facility of computer modelling being available, and just by understanding the basics of reactions.

Jan Harmsen recalls:

> I started as a plant technologist at Shell Pernis in 1987. In 1989, I was part of a team to improve an existing process. One of the process units appeared to be a reactive distillation designed and operated as a three-phase slurry reactor. Up till then, I had not been aware of this very special process unit because the operating team called it "The stripper".
>
> Inorganic base particles react away with an organic component to form a salt, which can dissolve in the aqueous phase, and an organic product. The product is sensitive to a consecutive reaction under the liquid reaction conditions. By evaporation, the product it is quickly removed from the liquid, and from the reactor.
>
> The column had many distillation plates. The column has a considerable height and diameter. The column had been in operation since 1951. Since then, the conditions had been optimized to increase the product yield on feed as well to increase the capacity [1].

13.7.2 Takeaway learning points

Learning point 1
Three-phase reactive distillation slurry operation in a distillation column can be done at commercial scale.

Learning point 2
Reactive distillation was chosen in 1950 when no flow sheet models were available. It seemed the obvious choice for the chemical engineer designer to get a high product yield with little by-product formation, as also a deep conversion of the feedstocks due to the many stages involved. Hence, reactor selection and design does not need sophisticated computer modelling.

References: three-phase slurry reactive distillation

[1] Harmsen GJ, Chewter LA, Industrial applications of multi-functional, multiphase reactors. Chemical Engineering Science, 1999 May 1, 54(10), 1541–1545.

13.8 Fluid bed retorting shale oil

13.8.1 Project starting points

In 1984, Harmsen – one of the authors – was asked to work on the development of a process for the conversion of oil shale to crude oil. At that time, he worked at the Shell Research Center, Amsterdam, in the reaction engineering department. Shell top management wanted to have the oil from the shale process superior to the oil from other shale processes and therefore chose the best and most experienced chemical engineer, Heinz Voetter Senior, to define a process concept, which should then be developed into a commercial-scale design. Harmsen had several interesting conversations with Heinz from which he acquired a lot of his reasoning.

Heinz Voetter led a design group at the Shell Central Office, The Hague. Heinz also worked as a designer and developer. He, for instance, developed a generic flow sheeter for Shell (Shell Process Modelling System). Heinz was given the task to generate a promising concept, which would then be developed. Heinz spent two years in reading a large number of patents and publications on oil from shale processes. He noted that the technologies differed enormously from each other. Moving beds and even extrusion were used to heat shale particles to convert kerogen to oil. He also noted that the capacity of a single process train was very limited. The largest capacity found was a process by Lurgi, with a single train shale processing a capacity of 7k ton/day. As the potential oil content of oil shale is in the order of 5–10% by mass, this means a 350–700 ton/day of crude oil production. This is nothing compared to a crude oil refinery, with a typical oil production of 100,000 ton/day. The total cost of crude oil produced by the best known process (the Lurgi process) was estimated by the Shell design office to be far too high. The main item was the investment cost. The crude oil market price at that time was around $40/bbl.

Heinz then concluded from the low oil content in shale that he could only arrive at a low-cost crude oil price if:

- The capacity of a single process train was at least a factor 10 larger than the Lurgi process
- The yield of oil on kerogen was at its maximum
- No external energy would be needed, and all energy would be generated by coke and, perhaps, light hydrocarbons stemming from the oil shale.

Heinz knew that the Fluid Catalytic Cracking (FCC) process in oil refineries could handle solids flow of 100 k t/day and that investment cost per ton of oil treated could be very low. So he decided that the whole process should be based on fluidization.

Heinz also knew that maximum oil field from kerogen was essential to keep the investment cost per barrel of oil low. From the literature, he derived that a rapid heating up of the kerogen and a short gas phase residence time with rapid quenching was needed to obtain this high yield. So he arrived at the cross-flow fluid bed. Near-plug flow of shale particles would further benefit deep kerogen conversion and high oil yield.

Combusting the remaining coke to generate energy for the process in a fluid bed was a no brainer to him as this also was used in the FCC process.

This concept was worked out in sufficient detail to make a first-cost estimate. This revealed that this new concept indeed had a far lower investment cost per barrel of oil. An estimate of the surface mining cost was also included, resulting in an overall acceptable crude oil cost price.

Heinz then obtained a large budget to lead a project group to develop the process and the mining part for the case of Tarfaya oil shale in Morrocco.

The process development parts are summarized in the next sections, with a focus on the reactor selection, design, modelling and experimental data generation, and validation.

13.8.2 Reaction kinetics, reactors, and process concept selections

Some shales in the upper earth crust contain an organic polymer called kerogen inside the shale particles. This kerogen can be cracked to oil in less than a second – around 500 °C. The oil reacts further to light hydrocarbons (methane-butane) and coke. The coke deposits on the shale particles.

Because of the undesired consecutive reaction, the oil vapor should have a low residence time in the reactor and be quenched to low temperature immediately after the reactor. This knowledge led to the horizontal fluid bed retort where the gas phase residence time is short.

The kerogen conversion should be deep so the solid-phase RTD should be narrow. This led to placing baffles in the horizontal fluid bed retort to obtain staged solids flow [1].

The shale particles should heat up quickly to the optimum temperature for kerogen cracking. This led to mixing hot-spent shale particles in the fluid be retort [1].

Coke should be combusted to generate steam and electricity. This led to a fluid bed riser combustor with a narrow RTD for deep conversion of coke [5].

The spent shale was transported by air slides to the fluid be retort and to a preheater fluid bed for the fresh shale.

13.8.3 Shale characteristics

Kerogen (hydrocarbon), containing shale, is found in many parts of the world. The shale types can be very different in hardness. Some, such as the Tarfaya shale of Morocco, are very soft and easy to grind to fine class A powder. Others, such as the Colorado shale of USA, are very hard and grinding is costly. Tarfaya and Colorado shale can be surface mined.

13.8.4 Process concept

The innovation project started with Voetter's conclusions. This was worked out in the following process concept. It consisted of a horizontal fluid bed for preheating and drying with countercurrent flow of spent shale and fresh shale with a heat-exchanger in between [A horizontal fluid bed retort where hot-spent shale was mixed with the dried shale, and kerogen was cracked to oil]. A riser combustor was used for coke combustion and steam generation. The spent shale was sent to the retort and preheater using air-slides for transport.

13.8.5 Process conditions

The technoeconomic study using various process models was used to determine process conditions in each process unit.

The preheater dries and heats particles to around 200 °C. Overheated steam was used as fluidization gas [4].

In the fluid bed retort reactor, shale particles are rapidly mixed with hot-spent shale particles at around 650 °C. The resulting temperature was set at 500 °C. The fluidization gas is mainly the oil vapor resulting from the kerogen cracking. Vertical baffle walls, perpendicular to the solids flow, with small holes through which shale particles flowed ensures staged solids flow. The vapor flows vertically through the bed, and leaves the retort [1, 3].

In the next reactor, coke on the shale particles was combusted in a fluid bed riser reactor. The reaction heat was used to produce steam and hot-spent particles (the steam is used to generate electricity to drive the shale grinders and other process equipment). The spent hot shale was transported by an "air-slide" to heat the retort and to heat the incoming shale in the dryer fluid bed [5].

13.8.6 Process research items

Here are the research items in summary:

- Kinetics of kerogen cracking to hydrocarbons
- Fluidization behavior when gas is produced in the bed
- Heat transfer coefficients in the fluid bed dryer
- Modeling the gas phase RTD and connecting this to kinetics to predict yield and optimizing the fluid bed retort dimensions and flow conditions; also including hot-spent shale subsequent to vapor-phase cracking.
- Shale grinding experiments and cost estimates
- Solids flow behavior for air slides and for flow control devices in a cold flow test unit of 20 ton/h.
- Mixing rates of hot-spent shale in fluid bed retort.

Interesting aspects of these research items are mentioned further.

Fluid bed retorting with the Tarfaya shale was studied in a 2-D fluid bed at 500 °C at University Twente by Wolter Prins in the group of Professor van Swaaij. It revealed normal fluid bed bubble flow behavior by the produced oil vapor.

Techno-economic studies revealed that the heat exchanger in the pre-heater required about \$2 × 10^9 of investment cost and that a high heat transfer coefficient was desired. Therefore, an external study on the heat transfer of class A power was also carried out by Wolter Prins at the Twente University, leading to an improved heat transfer correlation for this class A powder [6]. Additional experiments on heat transfer in class A fluid bed were performed at the Eindhoven University in the group of Professor Rietema.

Rietema also provided background information on the fluidization behavior of the powder at retort conditions. More information on the fluidization behavior of powders is found in Chapter 2.

Krishna developed a reaction engineering retorting model for the horizontal cross-flow baffled fluid bed reactor. Based on his work, patents were obtained [1, 3].

Krishna also formalized the reasoning of Voetter to a general reaction selection method [2].

To convince process designers at the central office that the horizontal fluid bed retort is feasible, the then technologist Jan Harmsen visited a coconut roasting factory of a friend of mine at Zaandam, where chocolate nuts were roasted in a large horizontal fluid bed with perpendicular walls that had small holes at the bottom for the solids to flow through, and in that way obtaining staged flow of particles. The process technologist explained that he had determined the size of the holes experimentally and that he was surprised at how small the holes had to be made to avoid backmixing. The holes were about 10 cm high and 20 cm wide, while the fluid bed was several meters wide, a meter high, and 10 meters long. He also observed that the height of the starting point of the fluid bed was not higher than the ending point of the bed. This removed the fear of the process engineers that the whole bed should be tilted to cause sufficient solids flow.

Grinding experiments were executed by a technology provider. It appeared that the Tarfaya shale was very soft and it was easy to grind to the desired particle size of around 100 microns. In fact, the shale was so soft that a large fraction of fines was formed by which the fluidization behavior changed from class A to Class C (cohesive). Experiments with shale combustion showed that the all the particles fell apart and indeed showed cohesive behavior.

Fluidization experiments with this class C shale power showed that even at high gas velocities (>0.5 m/s), high heat transfer coefficients could be obtained; but at low gas velocities, stagnant sections appeared on the fluid bed.

Research into other shale, such as the Colorado shale, revealed that grinding to class A particles would mean excessive energy for grinding to class A particles. Grinding to larger particles, class B, would mean less rapid mixing in the retort, steeper angles for the air slides, lower heat transfer coefficients in the fluid bed dryer, and much higher air velocities in the riser combustor; requiring larger compressor energy.

After some 5 years, the project was stopped as oil prices dropped to $10/bbl and strategic analysis showed that major oil producers could now and then drop oil prices below the cost price of producing oil from shale. It was, therefore, decided that a long-term oil market price, above the oil-from-shale cost prices, was not likely and the whole project was abandoned.

In an experimental program led by the author, the mixing rate of two class A powders under fluidization in a 0.6 m fluid bed was studied at gas velocities of around 0.5 m/s.

A cold-flow solids processing rig, with a capacity of 20 ton/h of solids flow, consisting of a horizontal fluid bed, a riser, and an air slide (to transport solids over a large distance) was constructed at the Shell research center in Amsterdam to study all the solids processing units.

13.8.7 Takeaway learning points

The learning points concern technical, economical, and organizational aspects.

Learning point 1
A simple reaction scheme and some kinetics information led to the horizontal fluid bed retorting reactor type selection. Chemical reaction engineering concept thinking of residence time and RTDs for the gas and the solids phase were essential for obtaining this selection and configuration.

Learning point 2
Preliminary process concept design by an experienced chemical engineer, in the ideation stage, is of enormous benefit to generate promising options. Knowledge of the FCC process enormously helped to see the advantages of continuous fluid bed processing over moving-bed and fixed bed batch processing. It also helped to convince the management that the concept was feasible and attractive.

Learning point 3
Large-scale solid processing in fluid beds requires far less investment than moving-bed or fixed bed reactors.

Learning point 4
The selection of class A powder for fluidization was economically feasible for one shale type, but not for others, as it strongly depends on the shale grinding cost. Grinding cost of hard shales is prohibitive to grind to 100-micron particle size. Hence, particle size selection based on CRE reasoning alone is not wise. The cost of obtaining the particle size has to be taken into account.

Learning point 5
Carrying out an economic assessment in the early concept stage is very beneficial to decide on stop or go with low research and development cost.

Learning point 6
Fluidization experiments at 500 °C in a 2-D setup real shale helped to convince managers that a process based on fluidization of shale is feasible. Good connections with university research groups enormously helped to quickly get sound knowledge of key critical phenomena.

Learning point 7
Cross-flow horizontal fluid beds are very attractive reactor types if a short gas phase residence time and staged solids flow are desired.

References: fluid bed retorting of shale

[1] Poll I, Krishna R, Voetter H, Van Wechem HMH, The basis of reactor selection for the Shell Shale retorting process. In: Kulkarni BD, Mashelkar RA, Sharma MM, ed. Recent Trends in Chemical Reaction Engineering. Vol. II, New Delhi, Wiley Eastern Limited 1987.

[2] Krishna R, Sie ST, Strategies for multiphase reactor selection. Chemical Engineering Science, 1994 Jan 1, 49(24):4029–4065.

[3] Voetter H, Van Meurs HC, Darton RC, Krishna R, Process for the extraction of hydrocarbons from a hydrocarbon-bearing substrate and an apparatus therefor. United States patent US 4,439,306, 1984.

[4] Voetter H, Van Meurs HC, Darton RC, Krishna R, Method of pre-heating particles of a hydrocarbon-bearing substrate and an apparatus therefore. United States patent US 4,419,215, 1983.

[5] Voetter H, Darton RC, Van Meurs HC, Krishna R, Process for the combustion of coke present on solid particles and for the production of recoverable heat from hydrocarbon-bearing solid particles and apparatus therefor. United States patent US 4,508,041, 1985.

[6] Prins W, Maziarka P, De SJ, Ronsse F, Harmsen J, Van Swaaij WPM, Heat transfer from an immersed fixed silver sphere to a gas fluidised bed of very small particles. Thermal Science, 2019, 23(Suppl. 5), pp.1425–1433.

14 Education case study: polyolefin CRE and scale-up

The purpose of this chapter is to provide a worked-out design case for education purposes. It shows the application of most subjects treated in this book for each innovation stage. The case is a polyolefin reactor. It also shows how a fluid bed polyolefin reactor can be selected, designed, and scaled-up to commercial scale, and how a downscaled pilot plant for product development and troubleshooting for the commercial scale can be designed.

14.1 Introduction

The sections are structured along all the innovation stages – from discovery to commercial-scale implementation. Each stage is treated as if the innovation is executed inside an industrial process company. Publications on polyolefin are used to back up the theoretical approach as well as provide experimental results. So the case study is described as history repeated.

For the discovery stage, a decision between the three-phase slurry and the gas phase fluid bed is made. In the concept stage, a new reaction engineering concept design is made. In the feasibility stage, a new commercial scale design is made. Also, a downscaled pilot plant design is made, taking electrostatic charging into account.

So, the provided general reaction engineering methods are applied to polyolefins and reflected upon using published papers on polyolefins. The case study is focused on polyethylene and polypropylene reactors as these two have many elements in common and many papers are available.

14.2 Discovery-stage reactor family selection

In the discovery stage, reactor options are generated and a preliminary selection of the preferred reaction family is selected as described in Chapter 10.

Manufacturing of polyolefins was discovered by Ziegler and Natta in 1954. Since then, production has grown enormously. First, the catalytic reaction of ethylene to polyethylene, or propylene to polypropylene was carried out in a three-phase mechanically stirred slurry reactor. Some were operated batchwise [1].

Soares provides a complete overview of the reactor configurations [2]. Here are the two main types and their variations:

https://doi.org/10.1515/9783110713770-014

- Gas–solid (G-S) fluidized bed reactors: Single vertical vessel, vessel with draft tube, horizontal multistaged.
- Gas–liquid–solid (slurry) reactors: Mechanically stirred, three-phase bubble column, slurry loop reactor.

Since 1977, gas–solid fluidized bed reactors were introduced and now new processes are mainly based on G-S fluidized beds [3].

A rationale for selecting G-S fluidized beds over three-phase slurry reactors, however, could not be found. It remains hidden in the knowledge cellars of the manufacturing companies. It is likely that the higher investment cost and higher maintenance cost of mechanically stirred tank reactors over G-S fluidized beds was the main reason (For the same reason, mechanically stirred tank reactors are not found in oil refineries and in bulk chemical processes). This argument of low investment and low maintenance cost, probably, became more prominent with the ever increasing scale of polyolefins production.

For a new polyolefins project, for a capacity of 100 kt/a or more, pre-selection of the fluidized beds family is easily made as it is about large-scale solids processing and production. So using the reactor family selection rationale of Chapter 10, results first of all in a fluid-type family and a sub-selection of the G-S two-phase over the three-phase, given the lower overall investment and maintenance cost of the two-phase fluidized bed system over the three-phase.

14.3 Concept stage

14.3.1 Scale-independent basics

In the concept stage, at first, all essential information for making a reactor concept design is generated. The information is structured as follows.

14.3.2 Chemistry and stoichiometry of the reaction

There are two main reactions involved: chain propagation and chain ending. We take polyethylene formation for the description. Other polyolefins have analogue chemistry. A catalyst radical forms an ethylene radical. The radical reacts with an ethylene molecule and forms a polyethylene. This keeps going and this process is called chain propagation.

The reaction scheme and stoichiometry of the chain propagation:

$$C_2H_4 + C_2H_4 = [C_2H_4]_2$$

This chain propagation is repeated until it ends when the catalyst leaves the molecule and starts a different chain. This phenomenon is called chain transfer and affects the molecular weight distribution stochastically. Hydrogen terminates the propagation by the reaction with stoichiometry:

$$[C_2H_4]_n + H_2 = C_{2n}H_{4n+2}$$

14.3.2.1 Kinetics

The average chain length is then determined by the chance of meeting a monomer or a hydrogen molecule. The average chain length is therefore determined by the concentration ratio of monomer/H_2. Meier confirmed this experimentally [4, 5].

The propagation reaction is given as first order in monomer concentration and catalyst concentration (or active radical site concentration). The reaction rate is expressed per unit reactor volume.

The chain propagation rate expression is

$$r_p = k_p M C^*$$

where M is the monomer concentration, and C^* is the sum of the active catalyst and the radical site at the end of the polymer chain. Note that because C^* is in moles catalyst per m^3 reactor, the unit of k_p is $m^3{}_{gas}/s\ mol_{cat}$! And because we have defined the rate r_p per unit reactor volume, the unit of the first order rate constant (k_p C^*) is not 1/s but $m^3{}_{gas}/m^3{}_{reactor}$ s.

The reaction rate expression follows first-order kinetics in hydrogen concentration and in activated chain end concentration [2]:

$$r_e = k_e H_2 C^*$$

The ratio of r_p/r_e determines, among other effects, the chain propagation over the chain termination; hence it affects the molecular weight of the polymers [2]. This ratio shows that the molecular weight can be controlled by the monomer over the hydrogen concentration in the absence of mass transfer limitation of the monomers inside the polymer particle.

The catalyst decays in time. The decay can be described as a first-order decay rate [2], but more complex decay kinetics are also available from [2]. Meier observed experimentally that the presence of hydrogen reduces the decay rate [4]. For our case study, with a focus on the design choice for the solids phase residence time distribution, these additional effects are not important. So we use the first-order decay kinetic equation:

$$r_d = -k_d C^*$$

The effects of temperature on the three reactions are provided by the Arrhenius equation:

$$k_p = k_{p0}\, e^{-E_p/RT}$$

$$k_e = k_{e0}\, e^{-E_e/RT}$$

$$k_d = k_{d0}\, e^{-E_d/RT}$$

The parameter values are determined experimentally. Soares studied experimentally the kinetics of polyethylene formation. The precise analysis of the experimental results to obtain the kinetic parameter values is however not clear; so his deactivation parameter values cannot be used reliably either. In general, analyzing the experimental results to obtain the k_0 and E values is not the reliable way of obtaining parameter values. The value of k_0 will be very uncertain as is the reaction rate constant k at T = infinite. This is far outside the experimental range. It is far better to use a k_{ref} value at a reference temperature within the experimental range and rewrite the expression as

$$\ln(k/k_{ref}) = (E/R)\ (1/T_{ref} - 1/T)$$

So the parameter values of Soares cannot be used reliably. We, therefore, provide in Table 14.1 his experimental parameter values at one temperature to be used for a case illustration and exercise.

Table 14.1: Kinetic parameter values of polyethylene formation [2].

Parameter	Dimension	Value	Description
E_p	J/mol	5.85×10^4	Activation energy propagation reaction
E_d	J/mol	6.3×10^4	Activation energy deactivation reaction
R	J/mol K	8.31434	Universal gas constant
$k_p\ C^*$	$m^3{}_{gas}/m^3{}_{reactor}$ S	1.13×10^5	First-order reaction rate constant propagation at: $T = 393$ K, $P = 20 \times 10^5$ N/m^2
k_p	$m^3{}_{gas}$/S mol_{cat}	160	Reaction rate parameter propagation
k_d	1/s	1.4×10^{-4}	Reaction rate parameter deactivation at $T = 393$ K
M	mol/m^3	700	Monomer gas phase concentration = $n/V = P/RT$
C^* (fresh)	mol/m^3	10^{-7}	Radical catalyst concentration. This is per unit reactor volume; not particle volume
$k_p\ C^*$	1/s	160×10^{-7}	This relates to the rate per unit reactor volume; not particle volume

14.3.3 Heat of reactions

14.3.3.1 Heat of propagation reactions

Roberts [3] provides an overview of all published heat of reaction for many polymers, both for the propagation and also for the chain-end hydrogenation. He states that the latter can be neglected for practical purposes. A summary list is provided in Table 14.2 below, obtained from Ray [6].

Table 14.2: Heat of propagation polymerizations reactions [6].

Monomer	Heat of reaction, kJ/mol $-\Delta H_r$ at 25 °C
Ethylene	88.6
Propylene	81.5
Butadiene	73.5
Styrene	69.8

The actual heat of reaction depends on the reaction temperature via the temperature dependance of the specific heat (C_p) of the reaction components – in this case, the monomer, the old chain, and the new chain. The specific heat dependance of the polymer chains, old and new, will not differ significantly. So, only the temperature dependance of the monomer on the temperature has to be taken into account. As the reaction temperature difference with the standard temperature of 25 °C is less than 100 °C, the temperature effect by the C_p value is in the order of 10% of the reaction enthalpy. For our analysis, we leave this effect out and use the reaction enthalpy at standard conditions.

14.3.3.2 Heat of hydrogenation reactions

The reaction enthalpy at standard conditions for the chain-ending reaction with hydrogen for polyolefins is −30.0 kcal/mol (hydrogen) [7].

The hydrogenation concentration inside the reactor is about a factor 50–100 smaller than the monomer concentration, and on top of that, it is chosen such that the chain-ending reaction is much slower than the propagation reaction. This means that for reactor design the heat generated by the hydrogenation reaction is very small compared to the propagation reaction. An accurate reaction enthalpy for hydrogenation is therefore not needed here.

14.3.4 Physical properties

The most critical physical properties of the reaction engineering models and the fluid bed behavior models are obtained from [2]. The heat conductivity parameter values are conservatively obtained from Eriksson [8]:
- Softening point, polyethylene: 90–110 °C,
- Reactor conditions, gas density: 20 kg/m^3
- Polymer particle density: 900 kg/m^3
- Heat conductivity, porous catalyst particle: 0.1 W/m K
- Heat conductivity, porous polymer particle: 0.01 W/m K

14.3.5 Reaction engineering concept design

14.3.5.1 Polymer and particle formation

The polymer particles are formed as follows [9]. Porous catalyst particles are brought into the reactor. Active radical sites inside the pores form olefin radicals, from which the radical polymerization propagates. Polymer molecules are thus formed, first in the porous catalyst particle. Later, the polymer molecules grow outside the catalyst particles. In the end, each catalyst particle is surrounded by more polymer material (a factor of 1,000) and the original (catalyst) particle diameter grows by a factor of 10 or more. A chain propagation ends temporarily when a radical jumps to another chain and then later the radical may jump backwards. The chain reaction really ends by a reaction with hydrogen.

The reaction rate of propagation over the reaction rate of hydrogenation thus determines the average molecular size (the molecular size distribution around the average is not considered in our analysis. It is often given by the Flory–Schultz expression and is mainly a function of the average molecular weight). So, in the end, the averaged molecular size (and its distribution) is mainly determined by the monomer/hydrogen concentration ratio inside the reactor and the temperature. This means that the averaged molecular weight is the same for each particle (for uniform hydrogen and temperature conditions) and only weakly governed by the particle's residence time distribution.

However, this statement of the averaged molecular weight being nearly independent on the residence time distribution of the solids phase, only holds if the particle temperature is uniform over its diameter. If, for instance, in the early stages, with a fresh and a highly active catalyst, a temperature gradient occurs inside the particle, then smaller (young catalyst containing) particles will have a molecular weight distribution that is different from large (old catalyst containing) particles. Rashedi took samples from a commercial-scale polyethylene fluidized bed reactor and found a correlation between the particle size and the molecular weight distribution [10]. So, this higher temperature with a fresh catalyst can happen in the early part of the reaction and can widen the molecular weight distribution. A study on heat transfer limitation of the reactive particle is found in Section 3.2.5.

14.3.5.2 Gas phase residence time distribution and concept design choices

The residence time distribution of the gas phase in a fluid bed is freakish. Part of the gas flows upward quickly through the reactor, via "bubbles". These so-called bubbles are entities with a high gas hold-up, moving upward through the bed. Part of the gas flows slowly in between the particles in the dense phase. There is exchange of gas between the bubbles and the dense phase. This is, however, very different from gas-mass transfer in the bubble columns. The exchange is by convective flow – from the bubble to the dense phase, and from the dense phase to the bubble. The dense phase has a vertical circulation pattern like the liquid in bubble columns. It flows upward in the central part and downwards in the wall section part.

All these flow phenomena are strongly affected by the particle size distribution and by the reactor dimensions. Numerous researchers have modelled the gas flow behavior on bubble size, bubble velocity, mass transfer between the bubble and the dense phase, dense phase gas flow velocity, and dense phase circulation velocities. Most models have been validated on a small lab-scale. No model has been validated for commercial-scale reactors. So, the reliability of these models is questionable.

For the polymerization reactor design, however, the treatment of these gas residence time distributions for the fluid bed can be treated in a very simple way. The polymer product particles can be easily separated from the feed components; so partial conversion of the feed components with a large recycle can be envisaged – which means that the gas phase, as a whole, can be treated as fully backmixed. Also, mass transfer between the bubbles and the dense phase will not be limited by the small vertical concentration differences in the bubble flow; so the mass transfer happens over the entire height, resulting in negligible concentration gradients. By supplying the monomer and hydrogen, such that the gas phase composition remains uniform, it also means that the product molecular weight distribution is controlled. Also, effects of scale-up on the gas phase residence time distribution are of no consequences as the gas composition is uniform everywhere in the reactor.

14.3.6 Solid-phase residence time distribution

The residence time distribution of the solids phase in a fluid bed is also freakish. Some particles move up at a high velocity in the large bubble core section. The majority of the particles move up and down with the circulation dense phase. For large-scale reactors, the circulation velocity is in the order of 1 m/s. With a reactor height of 10 meters, this means a circulation time in the order of 20 s. The average solids residence time, however, is in the order of 1 h [2]. So the solids phase could be considered completely backmixed for a single fluid bed reactor.

From Alves' description, it is clear that the solids phase is a segregated flow system. Each particle contains a catalyst particle and can be seen as an individual mini-reactor. Small particles mean less polymer is formed per catalyst mass, and so a higher cost of catalyst per product. Very large particles, which have a large residence time, will contain dead catalyst. So, in the last part of their journey in the reactor, the catalyst does not yield product, but still occupies reactor volume.

A fully backmixed reactor for the solids phase, therefore, implies that a large amount of catalyst leaves the reactor early with little polymer production, while other particles, with a very high residence time, contain virtually dead catalyst with almost no production. Hence, a single fully backmixed reactor will use the catalyst far less efficiently than a plug flow reactor. For a plug flow reactor, each catalyst particle forms the same amount of polymer material; so is far more efficient than a backmixed reactor.

This means that to obtain the same overall productivity in the backmixed reactor as in the plug-flow reactor, the backmixed has to be fed with more fresh catalyst than the plug-flow reactor. Or, if the same amount of catalyst is fed to the backmixed reactor as to the plug-flow reactor, then the reactor volume of the backmixed reactor has to be larger than the plug flow reactor.

The reader will have noticed that the solids phase is a segregated flow system, see also Section 4.6. Therefore, here the term backmixed reactor is used and not CISTR. The solids phase is not ideally mixed at all. It only shows backmixed residence time behavior.

The active catalyst profile in the plug flow reactor can simply be calculated for a segregated system, see Chapter 4 and Section 4.6. For the values $k_d = 1.4 \times 10^{-4}$ (1/s) and $\tau = 3{,}600$ s, $k_d\,\tau = 0.50$, and $C/C_{in} = e^{-K_d\tau}$; this yields $C/C_{in} = 0.60$.

So in the plug flow reactor, the active catalyst concentration starts at 100% and then decreases to 60%.

In the segregated solid flow backmixed reactor – for the same reactor volume and thus the same $k_d\,\tau$ – the catalyst concentration (averaged over several particles) is the same as the outlet concentration C everywhere. The outlet concentration is given by a backmixed system: $C/C_{in} = 1/(1 + k_d\,\tau) = 0.66$. So the concentration is 66% of the inlet concentration everywhere. To reach the same productivity, more fresh catalyst (higher C_{in}) is therefore needed for the backmixed reactor than for the PFR.

The number of backmixed stages needed to get to the same active catalyst profile as the plug flow reactor can be easily determined from the provided analytical solutions. For a $k_d\,\tau$ value of 0.1, the PFR and the CISTR differ little in concentration profile. For plug flow, $C/C_{in} = 0.904$. For backmixed reactor, $C/C_{in} = 0.909$. So, there is very small difference between the two. Hence, for the concept design, the number of backmixed stages required to reach a near plug flow profile for the catalyst activity can be obtained by having the value $k_d\,\tau_{stage} = 0.10$ for each backmixed stage. The total number of stages required is then obtained from the value of the plug flow $k_d\tau$. For the case at hand, the PFR value $k_d\tau = 0.5$. The total averaged residence time $\tau = N_{stages}\, \tau_{stage}$. So, 5 stages would be sufficient to require the same amount of catalyst as the plug flow reactor.

As the reaction is carried out under pressure, looking at the investment cost, it is better to go for solid flow staging inside the reactor vessel rather than have several reactors. As the gas phase for such solid flow staging can be common, one external gas recycle can still be used.

This approach however does not take into account the fluidization behavior of the very fine particles present in the first stages. The catalyst particles are typically 50–80 μm in size. The particle density is about 700 kg/m^3, with a porosity of 20%, according to the Geldart classification recently published again by [11]. It is clear that the fluidization, even with the small catalyst particles, still falls in the class A, as the Geldart boundary with cohesive class C is 30 μm. So, the staging, resulting in small particles in the first stage should still be no problem for fluidization behavior. However, the Geldart boundary between classes A and C is less well-defined when electrostatic forces play a role. And these forces play a role in polyolefins fluid bed reactors, due to their non-conductive behavior [12]. This subject will be treated in the development section.

In commercial practice, however, two fluid bed reactors in series are mostly used, each with its own external gas recycle, heat exchanger, and recycle compressor [2]. There is one commercial-scale design that has solids staging inside the reactor. It is called the Horizontally Stirred Bed Reactor, commercialized in 1976 by Innovene. Some details of the reactor are provided in a patent [13] and information about the solids flow staging, as obtained by mechanical stirrers inside the fluid bed, is provided by Soares [2].

We choose here a single pressure vessel with a horizontal orientation. The vessel has internal compartments made by vertical walls, with openings for the fluid bed dense phase to flow through. The number of compartments chosen is 5 to attain the lowest catalyst feed and the lowest reactor volume needed for the capacity.

14.3.7 Mass transfer limitations and concept design choices

The monomer and the hydrogen have to travel by diffusion, first through the porous catalyst particle and later through the porous polymer surrounding the catalyst particle. So this is a case of mass transfer by diffusing through porous particles, governed by the Thiele modulus. The Thiele modulus is the ratio between the reaction rate and the diffusion rate. If the Thiele modulus is smaller than 0.2, no mass transfer limitation occurs. This means that the concentrations of the reaction components inside the porous catalyst are the same as the concentrations outside the catalyst. The reaction rates are then not hampered by diffusion.

For polymerization, this condition of no diffusion limitation is desired for two reasons. The first reason is that the molecular weight distribution is not affected by diffusion limitations, as the concentrations of monomer and hydrogen are uniform inside and outside the particle. The second reason is that the reaction rate has its maximum value; so the reactor volume required is minimal. Hence, this is the preferred condition for reactor design.

The main parameters for determining diffusion limitations are the reaction rates for monomer propagation and the chain-ending reaction by hydrogen. Both have their own kinetic expressions, and for both diffusions, limitation can occur. So, for both, the Thiele modulus expression is needed.

The Thiele modulus Φ (a dimensionless number, see Chapter 5) for the propagation reaction is given by

$$\Phi^2 = L^2k/D_{\text{eff}}$$

where $L = d_p/6$ (for spheres), k is the first-order reaction rate constant for the monomer, D_{eff} is the gas diffusion coefficient for the monomer (or hydrogen), and $D_{\text{eff}} = 10^{-9}$ m^2/s (typical value for gas diffusion in wide pores).

When the pores are filled with polymer, the diffusion coefficient can be much lower. The reaction rate needs to be expressed for the monomer because that is the component diffusing into the particle. Furthermore, the reaction rate needs to be expressed as per unit catalyst particle volume. This is simply obtained from the reaction rate r_p by

$$r_{\text{vcat}} = r_p/\varepsilon_p$$

where ε_p is the volumetric particle holdup fraction of reactor, that is, the particle volume fraction, relative to the total reactor volume.

The r_p expression and parameter values are provided in Section 3.1.2.

We take $\varepsilon_p = 0.001$ as the catalyst volume fraction is 1/1,000 of the reactor volume:

$$r = k\,M$$

where $k = k_p\,C^*/\,\varepsilon_p = 160 \times 10^{-4}\,1/\text{s}$

For the parameter values of Table 14.1, the Thiele modulus is determined as

$$\Phi = 0.016$$

So, for the provided parameter values, it is clear that $\Phi \ll 0.2$; so mass transfer limitation for the monomer diffusing into the porous catalyst particles is absent.

It should however be stressed that when the pores are filled with polymer, the effective diffusion coefficient can become much smaller. No experimental values could be found in literature for that case. Alizada mentions that a fully predictive model for mass transfer inside the particle is still absent [14]. But, from the analysis, it is clear that even if the diffusion coefficient decreases by a factor 10, diffusion limitations would still be absent.

The Thiele modulus analysis has also to be performed for the hydrogenation reaction. The Thiele modulus for the hydrogenation is obtained in the same way because the hydrogenation reaction is also first order in hydrogen. The kinetic rate constant k for the hydrogenation, however, has a different value. Soares does not provide the reaction rate constant for hydrogenation; so the analysis cannot be quantified here.

The particle size, however, is not a constant but increases with time by polymerization. At the end of the polymerization, the polymer particle is a factor 10 larger in diameter and a factor 1,000 more in volume than the catalyst particle. So the Thiele modulus value changes. A simple analysis shows that with increasing particle size, the Thiele modulus value decreases.

The radical concentration per unit particle volume decreases in proportion to d_p^3. So, the reaction rate constant k decreases in proportional to d_p^3. The L value increases proportionally with d_p. Hence, the Thiele modulus decreases with increasing polymer particle size. This means that the analysis for diffusion limitation has to be performed only for the catalyst particle.

It should be stressed that the reaction rates for monomer propagation and for hydrogenation strongly depend on temperature. So, it is recommended to also determine also the temperature for which the Thiele modulus reaches the value of 0.2. This then becomes the maximum temperature for the reactor volume as far as absence of diffusion limitation is concerned.

A check on the Thiele modulus value can be made to see if diffusion limitations play a role. Hydrogen has a much higher self-diffusion coefficient than the monomer; so if diffusion limitations plays a role, then the molecular weight distribution can also be affected due to a different ratio of monomer/hydrogen inside the particle relative to the particle outside.

14.3.8 Heat transfer limitations and concept design

14.3.8.1 Particle heat transfer limitation

There are three locations where heat transfer plays a role. The first location is the fresh catalyst particle in which the reaction occurs. The reaction heat rate is the reaction rate times the reaction enthalpy ($-\Delta H_r\, r$). This reaction heat rate will increase with the particle temperature. By heat conduction, the particle loses this heat. A simple steady state heat balance, assuming the particle itself to be isothermal, results in a particle temperature difference with its surrounding gas flow, and is given by

$$\Delta T = -\Delta H_r\, r/(\alpha\, a)$$

It is important to note that the reaction rate r to be used here is the reaction rate per unit particle volume, and not per unit gas volume. So, it is different from the reaction rates provided by Soares and in the table. It can, however, be easily obtained from the reaction rate r_p using the catalyst volumetric holdup ε_p. The reaction rate for the catalyst particle is then

$$r = r_p/\,\varepsilon_p$$

where ε_p is the volumetric particle holdup; the particle volume fraction relative to the total reactor volume.

As a first conservative approximation, the heat transfer coefficient α for the particle side can be obtained from Nu = 2, see Chapter 7, Section 7.2.4

$$\alpha = 2\,\lambda/d_p$$

The question to answer is: Is the main heat transfer limitation then inside the particle or in the boundary layer outside the particle? The heat conductivity of the particle is 0.1 W/m K [15] and the heat conductivity of the ethylene gas outside the particle is 0.02 W/m K; so it is a factor 5 lower. This means that the main resistance for heat transfer is on the outside of the catalyst particle.

An expert in fluidization will remark that the heat transfer outside the catalyst particle is not just governed by the gas flowing around the catalyst particle, but also by the dense phase, which acts as a liquid medium with a much higher effective heat transfer. Section 7.2.4 provides expressions for determining this heat transfer of the dense phase with the heat exchange objects.

For our first estimate, however, we take the heat conductivity of the gas phase only to obtain a conservative estimate; so we use the heat conductivity of the gas phase.

The value of a is the ratio of (external) particle surface area and the particle volume. For a spherical particle, this is given by

$$a = 6/d_p$$

This results in a temperature difference between the particle and the surrounding gas:

$$\Delta T = -\Delta H_r\, r\, d_p/(6\,\lambda) \tag{14.1}$$

where ΔT is the temperature difference between the particle center and the surrounding fluid, r is the reaction rate per unit particle volume, ΔH_r is the reaction enthalpy, α is the heat transfer coefficient, a is the specific area, particle external surface area per volume particle, λ is the thermal conductivity of the gas phase, d_p is the particle diameter.

This expression (14.1) provides insight into what affects the temperature difference of the reacting particle with is surroundings. Firstly, it shows that the temperature difference is proportional to the reaction rate. The reaction rate is proportional to the radical concentration. Without catalyst decay, this radical number stays constant (neglecting the hydrogenation stopping reaction). So, when the particle grows, the radical concentration drops. The temperature difference is linear with the particle diameter, but the radical concentration drops with the particle diameter cubed: $\sim d_p^3$. So the highest temperature difference occurs inside the fresh catalyst particle, and not later in the growing polymer particles.

The most important reaction rate is the monomer propagation. Therefore, it is best that ΔT be first determined using the kinetics for monomer propagation and its reaction enthalpy, as the chain-ending reaction with hydrogen is only a small fraction of all the reactions taking place.

The first location to determine the value of ΔT is the fresh catalyst particle. The catalyst particle holdup is of the order of $\varepsilon_p = 0.1$. The catalyst particle size is typically, $d_p = 0.0001$ m.

The heat conductivity of the gas phase (ethylene gas) $\lambda = 0.02$ W/m K

The temperature difference between the catalyst particle and the surrounding gas is now calculated from expression 14.1. Table 14.3 contains the parameter values and the calculated temperature difference. The temperature difference appears to be 7 K. This is a significant temperature difference for these parameter values.

Table 14.3: Temperature difference calculation.

Parameter	Dimension	Value	Comment
d_p	m	10^{-4}	Catalyst particle size; typical for polyolefins
ε_p	–	0.1	Catalyst particle volume fraction of gas phase
λ	W/m K	0.02	Conservative value used; gas phase only
ΔH_r	kJ/mol	−84	Reaction enthalpy ethylene polymerization;
r_p	mol/m^3 s	1×10^{-2}	Reaction propagation rate per reactor volume
R	mol/m^3 s	0.1	Reaction propagation rate per unit catalyst volume
ΔT	K	7	Temperature difference catalyst and gas phase

Note: ΔT is catalyst particle and the surrounding gas.

This analysis shows that overheating of the catalyst particle is likely to occur. There are several approaches to tackle this potential problem. One option is to reduce the catalyst activity. This is not attractive as the catalyst is expensive, and by lowering its activity, more catalyst per amount of product will be needed. A second approach is to determine the heat transfer of particles in the fluid bed dense phase, for instance by the simple method of Prins described in section 7.2.4.4 – emerging a silver sphere in the fluid bed and measuring its temperature heat up or cool down in time. With this more accurate heat transfer information for the catalyst particle, overheating should be determined again.

As the propagation reaction rate is strongly affected by temperature, this reactor temperature should then be carefully chosen to avoid overheating of the fresh catalyst particle.

The catalyst particle could also show a so-called temperature runaway when the chosen reaction conditions and temperature are at an unstable point. For such a point, a small increase in temperature causes the reaction to go faster, and the resulting heat production increase is not cooled away by heat transfer. So, steady state is not obtained, but a temperature runaway.

The conditions for such an unstable operation point are given in Chapter 7. Also, accurate kinetics of the polymerization inside the fresh catalyst particles are needed to predict the actual temperature of the catalyst particle. [16] found experimentally that particle overheating occurs in the first 10 s of a polyolefin reaction.

Yiagopoulos [17] indicates by a modelling study that particle heat transfer limitations play a role in polyolefin formation in the early stage – when the catalyst is highly active and when the polymerization rate >15,000 kg/(kg h). His advice is to have a pre-polymerization reactor operating at a lower temperature to overcome this overheating. He, however, does not mention the consequence of such a design decision, namely that polymers are then formed with a different molecular weight distribution than for the other reactor sections. This resulting wider molecular weight distribution is, in general, not preferred. Also, the control of the molecular weight distribution is less reliable as now monomer/hydrogen concentration in two reactor sections need to be controlled .

Our design is to have a reactor system with a uniform temperature and a uniform monomer hydrogen ratio.

14.3.8.2 Heat transfer limitation-aggregated polymer particles

The second heat transfer limitation is when polymer particles aggregate to larger lumps. The expression for temperature difference of a particle where the reaction takes place can also be used for estimating the effect of the polymer particle aggregation. The aggregated lump size should be taken as the value for d_p. For instance, for a lump of 10 cm, for a factor 100 larger than the polymer particle, the temperature difference also increases by a factor 100. This higher temperature can cause polymer melting and, thereby, further particle agglomeration and the so-called sheet formation.

The analysis using the expression has its limitations. When the temperature of the lump increases beyond a certain temperature, steady state is not obtained anymore. The reaction heat production becomes higher that the heat rate removal rate and the reaction then runs away to the adiabatic temperature rise of over 1,000 K. This lump-runaway behavior is not treated here further. In the development stage, a special test program is dedicated to prevent lump formation by using additives.

14.3.8.3 Heat and temperature management reactor

The subject treated here is the reaction heat to be removed from the reactor. First, the adiabatic temperature rise from the monomers feed to the outlet stream of polymers is determined. This is the temperature rise from the reactor inlet to the outlet, without cooling. As the specific heat capacity itself depends on the temperature, no accurate algebraic equation can be defined for the adiabatic temperature rise. For simplicity and order of magnitude calculation, we assume the same C_P value for the inlet and the outlet. We furthermore assume complete conversion of the monomer. The adiabatic temperature rise then follows from the simplified heat balance:

$$F_{\text{total}}\, C_P\, \Delta T_{\text{ad}} = -\Delta H_{\text{r}}\, F_{\text{A}}$$

resulting in

$$\Delta T_{\text{ad}} = \left(-\Delta H_{\text{r}}/C_p\right)\, F_{\text{A}}/\, F_{\text{total}}$$

where F_{total} is the total flow of feed (mol/s) and F_{A} is the flow of monomer feed (mol/s).

Let us take the following values for the parameters:

$$F_{\text{A}}/F_{\text{total}} = 1;\ -\Delta H_{\text{r}} = 80{,}000\,\text{J/mol}\,(20\,\text{kcal/mol});\, C_P = 40\,\text{J/mol K}$$

The adiabatic temperature ΔT_{ad} is then 2,000 K.

So, this means that the adiabatic operation at full single pass conversion is not feasible and cooling is needed. One solution could be to choose a very limited single-pass monomer conversion. To keep the adiabatic temperature rise to say 20 K, which means a single-pass monomer conversion of 0.01 (1%), which means a 100-fold external recycle with cooling of the gas recycle flow, a heat exchanger in the gas recycle loop can be placed.

Even with a low single-pass conversion of 1%, the adiabatic temperature rise is 20 K; hence, a non-uniform temperature could still occur. However, the heat capacity of the solids phase is far higher than that of the gas phase; so the solids heat capacity governs the local heat balances. Thus, the gas fed at a temperature lower than the bed will be quickly heated up by the particles. The cooled down particles are then heated up by contact and mixing.

Let us now explore this particle backmixing in view of potential temperature differences of the reacting polymer particles in the fluid bed. The subject of backmixing in fluid beds is treated in Chapter 4. For gas phase fluid beds of Geldart classification A, the dense phase circulates vertically with a velocity in the order of 1 m/s; so a mixing time is in the order of 10 s. The averaged residence time of the polymer phase is 1 h; so can be considered as back- mixed. This means that a very small fraction of particles in the reactor has a lower temperature than the average temperature. Hence, the effect of small temperature differences in the fluid bed on the molecular size distribution will therefore also be small.

An alternative cooling option is to place a heat exchanger inside the fluid bed. Gas–solid fluid beds have high heat transfer coefficients for heat exchangers, and industrial experience with large-scale fluid beds with heat exchangers is available. However, no technology provider for polyolefins reactors with internal heat exchangers could be found. This means that that option would require a large development effort to determine the feasibility of heat transfer inside the fluid bed in view of fouling by the clogging polymer particles. To avoid the additional development cost and time, it is decided to select cooling by a heat exchanger in the external gas recycle.

14.3.8.4 Choice of particle size related to fluidization class behavior and G/S separation

The choice for a large external gas recycle ratio of a factor 100 over the gas feed to the reactor system means that the monomer and hydrogen concentrations are uniform inside the reactor. This then also means that the actual residence time distribution of a single-pass gas flow has no effect on the reactor performance. This also means that details of the RTD, such as gas components' exchange between the bubbles and the emulsion phase also has no effect on the reactor performance. So the fluidization class choice, which is in fact the choice of the average polymer particle size choice, is also not critical for the concept design.

For the gas–solid separation at the top of the fluid bed, however, the particle size does matter. This topic will be treated in the feasibility stage where a feasible commercial-scale design needs to be made.

14.3.8.5 Overall reaction engineering concept design

The reactor system design criteria and the resulting critical performance phenomena choices are summarized in Table 14.4.

Table 14.4: Reaction engineering concept design criteria and choices.

Criterion	Reaction engineering choice
Molecular weight distribution control	RTD: gas phase backmixed by external recycle, solid-phase free choice
No hydrogen mass transfer limitation, hence Thiele modulus <0.2	Particle size and temperature design
Temperature uniform in fluid bed reactor and at particle scale	Particle heat transfer limitation absent
Uniform monomer concentration	Small single pass monomer conversion
Highest polymer on catalyst yield	Solid phase staged, temperature optimized
Reliable scale-up design and operation	External heat exchanger in gas loop
Low investment cost	Single horizontally staged fluid bed with a high volumetric production rate

14.3.8.6 Molecular weight distribution control

The molecular weight distribution is reliably obtained – first, by controlling the hydrogen-to-monomer ratio everywhere inside the reactor by the large external recycle; measuring the ratio; and feeding hydrogen monomer to keep the ratio, and second, by a uniform temperature over the whole reactor section.

14.3.8.7 Polymer yield on catalyst

The highest polymer yield on a catalyst is obtained by staging the flow of catalyst containing particles.

14.3.8.8 Reliable design for scale-up and operation

A reliable concept design choice is a fluid bed with a low single-pass monomer conversion and a high external gas recycle with external gas cooling.

14.3.8.9 Lowest cost of investment and catalyst per ton of product

First of all, a low investment cost is obtained by have a process with a large capacity. Secondly, the reactor investment cost will be related to the number of vessels required. By having a horizontally internally staged fluid bed, a single reactor vessel is chosen. Furthermore, the reactor volume required is also less than the previous designs due to a higher effectiveness of the catalyst, caused by solids phase staging.

The catalyst per ton of product is also an important cost item. Solids phase staging implies that the yield of a product on a catalyst will be higher than for a solids backmixed system. An optimization study has to be carried out in the feasibility stage to find the optimum combination of catalyst feed rate and reactor volume.

The monomer feed cost per ton of product is fixed as the product yield on the monomer is the same as the theoretical maximum.

14.3.8.10 Final concept design summary

It is a single reactor with internal solids phase staging. Each stage has the same temperature for a uniform molecular weight distribution production, everywhere in the reactor, so that M_W is controlled by the temperature and by the monomer/hydrogen ratio. The external gas recycle ensures uniform gas composition in the reactor and also facilitates gas composition control. The heat exchange control is via the heat exchanger placed in the external gas loop.

14.3.9 Modeling for reactor sizing

The purposes of reactor modelling are to determine the catalyst feed rate, reactor volume, reactor temperature (the same for each stage), heat exchange area, and recycle flow rate. The modelling also provides relations between the product yield on the catalyst and the degree of staging (number of chambers).

The reactor modelling should therefore include reaction kinetics, catalyst decay rate, solids phase mixing, heat transfer inside the catalyst and the polymer particle, heat transfer to the gas phase, and heat transfer to the heat exchanger.

The present status of modelling for polyolefins gas phase reactors is provided by Alves [9]. In his published review of models, Alves comes to the conclusion that these models are still not good enough for segregation, agglomeration, and particle attrition predictions. Dudukovic came to a similar conclusion; so progress in this field is still slow [18].

It is also clear that the present models for polymerization in porous catalyst particles are also not good enough to predict mass transfer and heat transfer limitations. It also means that commercial-scale reactor design cannot completely rely on models. If a new reactor concept is proposed, such as in this case study, then a dedicated kinetic study should be performed to determine what is happening at the particle-size scale.

In the development stage, pilot plant testing will still be needed to validate the new design and models used in the design at the fluid bed scale. The commercial-scale design and the downscaled pilot plant design will be treated in the next section feasibility stage.

14.4 Feasibility stage

14.4.1 Introduction

The purpose of the feasibility stage is explained in Chapter 9.

For that purpose, following items are to be executed:

- a commercial-scale process design
- a downscaled pilot plant design
- economics of commercial scale and pilot plant with a test program
- A risk assessment of commercial scale versus value of information of pilot plant

14.4.2 Commercial-scale design in feasibility stage

14.4.2.1 Introduction: commercial-scale design

In the feasibility stage, a commercial-scale reactor (and process) design is made. This is sometimes called Front End Loading-2 (FEL-2) Design (FEL-1 is made in the concept stage and FEL-3 is made at the end of the development program) [19].

The minimum requirements for commercial-scale design are:

14.4.2.2 Safe, healthy, environmental, economic, technically feasible, and sustainable (SHEETS criteria)

In this section, we will not make a proper commercial-scale design with all details needed for a feasible design; we will, however, highlight the critical aspects of the commercial-scale design. Table 14.5 shows these critical performance phenomena and their relevance.

Table 14.5: Critical performance phenomena and their relevance for the gas phase and the solid phase.

Critical performance phenomenon	Gas phase relevance	Solids and G-S fluid
Feed distribution	High	High
Residence time distribution	Medium	High
Mixing	High	
Shear rate distribution	Low	High
Impulse transfer	High	High
Mass transfer		Medium
Heat transfer		High (particle and reactor)
Electrostatic charging	High	High

14.4.2.3 Feed distribution

Fresh gas feed of monomer and hydrogen to the gas recycle loop: Fresh monomer gas and hydrogen are fed to the recycle loop before re-entering the fluid bed. To ensure complete gas mixing, the pipe length, before entering the fluid bed, is 40 times the pipe diameter.

The gas feed distribution inside the fluid bed has to be very uniform to ensure that all parts of solids are in a fluidized condition. The particle size and the density of the solids, relative to the gas density, ensure that the fluidization is partly in the Geldart classification cohesive C and partly in the classification A. This means that the gas feed distribution has to be very uniform. If nozzles are used, then the distance between the nozzles should be very carefully chosen. This is needed to prevent dead zones of lumps of aggregated particles. These can locally cause overheating, melting, and sheet formation. The total cross-sectional area of the gas nozzles should be such that the linear gas velocity in the nozzle is the same as in the pilot plant so that electrostatic charging by the nozzles is the same as in the pilot plant.

Esmaeili [20] describes the importance of the gas distributor design from a theoretical point of view, using a CFD model. It may give additional design details.

14.4.2.4 Catalyst feed distribution

Catalyst particles entering the bed are critical to the reactor performance [21]. The catalyst particles should be rapidly mixed well with the particles present in the fluid bed so that the heat of the highly active catalyst particles is rapidly transferred to the bulk dense phase particles. The catalyst particles are, in general, fed by pneumatic transport pipes to the bed. The pipe diameter and the gas velocity are critical. Several feed pipes will be necessary for an even catalyst distribution and a rapid mixing with the dense phase to avoid local overheating of the catalyst particles, as shown by the calculations in the concept stage. Also, the linear velocity of the gas plus the catalyst and the lengths of the feed pipes are critical for electrostatic charging. The latter will be avoided by a dedicated pilot plant program using antistatic additives; see the section on pilot plant design and test program. The catalyst is fed to the first chamber of the horizontal fluid bed.

14.4.2.5 Residence time distribution gas phase

The residence time distribution of the gas phase is fully backmixed. It is obtained by a 1% single-pass conversion and an external recycle ratio of 100. The gas phase is then ideally backmixed. The gas composition (olefin/hydrogen ratio) over the reactor is thereby constant in time, and uniform inside the reactor.

14.4.2.6 Residence time distribution solids

The residence time distribution of the solids is very critical for several reasons:
- Maximum product yield on catalyst particles requires staged flow
- Polymer particle distribution should be narrow and controllable for reliable downstream solids handling in transport and extrusion.
- No segregation of small and large particles, causing shortcutting of fine particles and ultra-long residence time large particles, resulting in underused catalyst particles.

The last item needs special attention for the commercial-scale design. Some researchers mention that larger particles are preferentially found near the bottom of the fluid bed and that small particles are found higher up [22]. These researchers also claim that the product outlet of commercial reactors is placed at the bottom part of the reactor. Given that the dense phase shows increasing vertical recirculation with reactor diameter, which will affect particle segregation, predicting particle size segregation for the commercial scale will be unreliable.

Taking all three critical items into account, the multichamber concept design is reconsidered. The multichamber design means that the particle size distribution in each stage is narrower than in a single-stage reactor. Also, the average particle size increases from stage to stage; so, at the outlet of the last chamber, the largest averaged particle size is obtained. Hence, the product flow has the largest average particle size, with a narrower particle size distribution, than when a single- or two-stage reactor system is designed.

So, for the commercial-scale design, a single pressure vessel is chosen with multiple solids phase stages inside. It is a multichamber "horizontal" fluid bed with 5 internal chambers, created by a vertical wall with small openings for the dense phase to flow through. The vessel dimensions are: Diameter = 4 m; Length = 20 m; and Volume = 500 m^3. Such a single reactor would have a lower investment than two reactor vessels in series. Also, the yield of product on catalyst feed is higher, see the concept stage calculations.

For such a horizontal -bed reactor, particle size segregation on size is less of an issue. Due to the staging, the particle size will increase from chamber to chamber and the particle outlet will be at the end of the horizontal fluid bed.

The open area size for the solids dense phase to flow through should be so large that the resistance to flow is minimal and it should also be so low that no backmixing over the area occurs. Cold flow modelling will be applied to experimentally optimize the open area.

14.4.2.7 Shear rate distribution and impulse transfer

At large-scale, the dense phase circulates with high velocities of meters per second. The impulse transfer on internals, such as a header system for the gas feed, heat exchangers or draft tubes, will therefore be considerable. These internals are therefore mechanically designed using information from FCC combustor gas distributor design. These fluid beds have diameters of up to 12 m; so that the dense phase velocity is much higher there. Hence, they also have much higher impulse forces on the gas distributor. In this way, mechanical integrity is ensured. Also, the erosion effects on the gas distributor in the FCC fluid bed are high and are counteracted by thicker and harder construction material. The same construction material may therefore be taken. A check on the material selection will be made, given the oxidative conditions in the combustor and the reducing (hydrogen) conditions in the polyolefins reactor.

The shear rate force for a commercial scale is $\rho_{\text{dense phase}}\, v^2_{\text{circul}}$ The dense phase is typically 1,000 kg/m^3 and the circulation velocity is typically 5 m/s. So the dynamic force on the gas distributor is in the order of 25,000 N/m^2. The static force of the bed should be added to this dynamic force. The static force is equal to $\rho_{\text{dense phase}}\, g\, H_{\text{bed}}$. For a bed height ($H_{\text{bed}}$) of 10 m, the static head is 100,000 N/m^2.

We consider the commercial-scale design with a horizontal-staged fluid bed. The length of the reactor is then 10 m, but the fluid bed height is a little less than the diameter; so the forces on the distributor are reduced. However, the size of the distributor is increased as the length is now 20 m and the width, a little less than 4 m.

Also, the erosion caused by the circulating dense phase shear onto the distributor should be taken into account by selecting the construction material. An EPC contractor with experience in FCC fluid bed design should be involved, although this is no guarantee that he or she will get it right. Often, breakdowns of distributors are not reported by the EPC contractor and also not reported inside the manufacturing company to the design office. Pell [21] provides an interesting paper on fluid bed distributor design.

The shear rate distribution in the fluid bed reactor should exceed the minimum rate needed to avoid lump formation by electrostatic charging, everywhere. But no information is available on the minimum gas velocity to achieve this. Therefore, an alternative approach is followed to combat lump formation by adding an antielectrostatic additive. The additive type and amount is optimized in the pilot plant.

14.4.2.8 Mass transfer

Mass transfer from the gas phase to the catalyst-polymer particle should be sufficient to operate in the kinetic regime so that scale-up effects are absent. The Thiele modulus calculated in the concept stage for the provided kinetics for the averaged particle size, should be checked to see whether mass transfer can be limiting and thereby create a lower conversion rate than anticipated by the kinetics, and whether molecular weight distribution can be different due to a higher mass transfer limitation by the monomer than by hydrogen.

14.4.2.9 Heat transfer

The reaction is highly exothermic; so heat transfer is highly relevant. Heat transfer from the reacting catalytic polymer particle to the gas and dense phases should be so high that the temperature of the reacting particle is less than 0.5 °C, which is higher than the gas phase. Local overheating can easily occur for fresh catalyst particles, as shown theoretically in the concept-stage calculations.

When particles agglomerate to large lumps, overheating can also occur. By filling in the lump size, instead of the particle size, the temperature difference between the lump and its surroundings can be calculated using expression 14.1. Overheating has been practically observed in a polyolefins the test reactor of Meier [5].

This means that, first of all, lump formation should be avoided. A dedicated experimental program will therefore be performed in the pilot plant – testing additives that avoid electrostatic charging and thereby lump formation.

For heat removal from the reactor, two options have been proposed in the concept stage. Option A is a heat exchanger placed inside the reactor and option B is an external gas recycle, chosen with an external heat exchanger. In this feasibility stage, we compare the two options to make a decision for the commercial-scale design and for the downscaled pilot plant design.

Table 14.6 summarizes the key design items and the features of the tow options for reactor heat exchange.

Because reliable operation and reliable product quality are of utmost importance, heat exchanger in the gas recycle is preferred, despite the higher investment cost. Soares reports that the advantages of accurate temperature control, less fouling, and lower erosion have outweighed the lower heat exchange area required, when placing the heat exchanger inside the fluid bed, as all commercial-scale polyolefin process designs contain a heat exchanger in the external gas recycle [2].

Table 14.6: Comparison of heat exchanger in fluid bed versus external gas recycle.

Design item	Heat exchanger in fluid bed	Heat exchanger in gas recycle
Required heat exchange area	Low	High
Accurate temperature control	Low	Medium
Uniform temperature in reactor	Better?	Needs very large recycle
Reliability in view of fouling	Low	Medium
Erosion rate	High	Low
Industrial-scale experience	Absent for polyolefins	Common practice

Having the low -pass monomer conversion with external gas recycle has an additional advantage of no concentration gradients in the reactor, which means that the actual residence time distribution has no effect on the reactions occurring; see the section on gas phase RTD. This increases the reliability of scale-up, as discussed in detail in the section on gas phase residence time distribution. It also shows that design decisions should not hinge on a single critical factor.

As the first chambers have a higher catalytic activity, a higher superficial gas velocity is to be applied. The catalyst activity in the first stage is 40% higher than in the last stage; so, there, a 40% higher gas velocity is then applied in the last stage. For the stages in between 30%, 20%, and 10%, higher gas velocities are applied to ensure equal temperatures in each chamber. A temperature control with gas feed per chamber ensures an accurate uniform temperature over all 5 chambers.

14.4.2.10 Design to combat electrostatic charging

Meier observed electrostatic charging in his pilot plant in 2000 [4]. Since then, many researchers have studied the phenomenon [5, 24–26]. Kahn [26] provides an overview of the experimental studies and the theoretical studies on electrostatic charging of polymer particles inside the fluid bed reactor of polypropylene. This charging occurs by gas–particle shear, particle–particle shear, and particle–wall shear. Also, the catalyst feed with gas can cause this charging [23].

The charged particles (small positive and large negative) cause particle agglomeration. These agglomerations need a higher superficial gas velocity to stay fluidized. The electrostatic charging also causes sticking of particles to the wall, forming a multiparticle layer, which in turn has a lower heat transfer rate, which can cause overheating. This overheating can cause particles to melt and form polymer sheets [26]. The consequence of charging and sheet formation is that pieces of sheets can foul the heat exchanger in the external gas loop [25].

Static charging of the fluid and the particles by the gas inlet velocity is an additional critical performance factor. It can cause sticking of the particles to the wall. This, in turn, can cause lower heat exchange, which can then cause overheating, which can cause local melting and forming sheets of polymer that clog filters, by which the process has to stop for cleaning. Also, the circulating dense phase may be electrostatically charged at the reactor wall (and internals). The circulating velocity increases linearly with the reactor diameter; so it is a larger problem at commercial scale.

Recent review articles describe the charging and its consequences [26–28], but do not provide solutions. However, Taghavivand [23] studied the use of additives to prevent electrostatic charging for gas phase polyethylene manufacturing. So that seems to be the solution.

The effectiveness of antistatic material addition and its possible consequences on catalyst behavior and polymer product quality could be tested at the pilot plant scale. The best way to do this is to design the pilot plant for maximum electrostatic charging by having a draft tube in which part of the gas feed is placed, so that the dense phase circulation velocity is increased. In addition, extra wall surface area inside the reactor will enhance the electrostatic charging.

At zero measurement, the pilot plant is operated without antistatic dosing and electrostatic charging, and its consequences on sheet formation and fouling are measured. An experimental program is to be executed to optimize for the type and amount of antistatic dosing.

14.4.2.11 Gas–solid separation

Unconverted monomer and hydrogen need to be separated from the particles before entering the external recycle loop to avoid fouling of the recycle compressor, and also the heat exchanger. In the present commercial-scale polyolefins fluid bed reactors, this is obtained by having a wider diameter in the upper part of the fluid bed, so that the averaged gas velocity in the reactor is lower, allowing for gravity separation of particles from the gas [2].

By having a horizontal fluid bed, the superficial gas velocity is lower than in the vertical fluid bed, due to the larger horizontal area. So, gas–solid separation in the disengagement zone above the fluid bed can also be obtained. In addition, cyclones can be installed for further separation of the fine particles from the gas outlet.

14.5 Development stage

14.5.1 Pilot plant design

14.5.1.1 Purpose

The chosen purpose of the pilot plant is fourfold:

- Develop novel product grades
- Provide novel product samples for customers
- Test antielectrostatic additives
- Provide validation information for commercial-scale design and models

14.5.1.2 Pilot plant and mock-up model design

Pilot plant downscaled version

The pilot plant is a downscaled version of the commercial-scale design. Having a downscaled pilot plant could be a present practice in the polyolefins industry. Kahn, for instance, mentioned of having a downscaled pilot plant for polypropylene production from a commercial plant in Malaysia. Kahn studied electrostatic charging extensively in this pilot plant and it forms a considerable part of his PhD thesis [28]. Kahn, however, does not provide commercial-scale reactor dimensions or a downscale rationale.

Electrostatic charging study

The effectiveness of antistatic material addition and its possible consequences on catalyst behavior and polymer product quality will be tested in the pilot plant. The best way to do this is to design the pilot plant for maximum electrostatic charging by having a draft tube in which part of the gas feed is placed, so that the dense phase circulation velocity is increased. In addition, extra wall surface area inside the reactor will enhance the electrostatic charging.

At zero measurement, the pilot plant is operated without antistatic dosing and electrostatic charging and its consequences on sheet formation and fouling are measured. An experimental program is to be executed to optimize for the type and the amount of antistatic dosing.

The theory of electrostatic charging is not fully developed, but the hypothesis is that the linear gas velocities in the feed pipes, catalyst feed pipes, and dense phase velocities at the reactor wall induce the charging.

The downscaled pilot plant is then designed with gas feed pipes such that the gas velocity in the pipes is the same as in the commercial scale and also that smaller diameters can be installed to study the effect of the gas inlet velocities on electrostatic charging. The effect of the catalyst feed via a pneumatic transport feed pipe is also investigated. As a reference, it has the same length, diameter, and velocity, as the present commercial-scale reactor. Additionally, a pipe with a smaller diameter can be installed to study its effect on charging.

The superficial gas velocity in the reactor is a variable, and can be chose to be the same as in the commercial scale. A draft tube can be installed in the bed to create additional surface to induce additional charging.

The circulation velocity of the dense phase is most likely proportional to the reactor diameter, with exponent 0.5. This means that the pilot plant circulation velocity will be much lower. To study the effect of the circulation velocity, a vertical inner pipe will be installed in the first chamber, with its own gas feeding, so that the circulation velocity can be influenced by feeding more gas into the central pipe. This will cause a much a higher gas bubble hold-up in the central part, causing a larger density difference between the central upward flowing section and the down-flowing dense phase wall section. Thus, the effect of the dense phase circulation velocity can be studied. The higher gas velocity in the central pipe can also cause enhanced electrostatic charging; thus, mitigation measures such as additives can indeed be tested on electrostatic charging and sheet formation.

Detection measures are installed to determine electrostatic charging and also optical probes are installed to observe sheet formation. Table 14.7 shows the commercial-scale and the downscaled pilot plant reactor parameters to obtain the desired downscaled features, similar to the commercial-scale reactor.

Table 14.7: Commercial-scale and downscaled pilot plant parameter values.

Parameter	Commercial scale	Pilot plant
Diameter, m	4	0.4
Length, m	20	2
Superficial gas velocity	1	1
Production rate, kton/year	100	1
Production rate, kg/s	4	0.04
Dense phase circulation velocity, m/s	4 (estimated)	1.3 (estimated)
Bubble hold-up central core section	0.2	0.2
Bubble hold-up central core section	0.2	0.8 draft tube with gas feed
Dense phase circulation velocity	4	2.3 (induced by internal pipe)

The commercial-scale reactor diameter is obtained by taking a typical commercial-scale production of 100 kt/a, a recycle ratio of 50, and a superficial gas velocity inside the reactor of 1 m/s [2].

14.5.1.3 Mock-up model design

To study the flow pattern and the solids residence time distribution of the horizontal 5 chamber fluid bed, a mock-up model is proposed, with the dimensions same as that of the commercial-scale reactor. The cost of this mock-up model and, in particular, the nitrogen gas compressor is, however, 5 M€.

In the value of information study, it appears that the risk reduction achieved by this mock-up model study is not sufficient to warrant this very large mock-up. Instead a mock-up model, whose size is the size as that of the pilot plant is designed. With this mock-up model, the effect of baffle hole sizes on dense phase backmixing of the stages and on in between stages can still be studied. The mock-up model is planned to be available prior to the hot pilot plant. First results on baffle-holes can then be used to optimize the pilot plant.

The hot pilot plant also has 5 chambers with openings. The opening size is first determined from a theoretical study. Using the mock-up, an optimized hole size may be determined and a fluid flow model for the dense phase flowing through the hole with backmixing is validated.

Results of the mock-up model and the pilot plant solids' residence time distributions will be used to obtain scale-up design rules for the commercial-scale reactor.

14.5.2 Economics commercial scale, pilot plant and mock-up model

The economics of the feasibility stage include the investment cost of the commercial scale process, the investment cost of the pilot plant and mock-up, the variable cost of feedstocks and the fixed cost of personnel. The economics also include the revenues obtained from product sales of the commercial scale. We present here in Table 14.8 the cost and revenue as simple certain values. In the next section, we discus risks involved, assess the value of information obtainable from the development program, and the risk reduction as a consequence of the information obtained.

Table 14.8: Economic parameter values polyolefins case.

Economic Item	Unit	Value
Commercial scale investment	M$	150
Sales – commercial-scale cost	M$/year	50
Net present value – commercial scale only	M$	300
Pilot plant investment and test program cost	M$	8
Full-scale mock-up investment and test program cost	M$	8
Small scale mock-up and test cost	M$	1

14.5.3 Risks and value of information assessment

Here, we present a hypothetical assessment of the risks involved for four cases.

Case A: Directly designing, constructing and starting up the commercial-scale process with a new multichamber fluid bed reactor for new polymer grades, without any development effort.

Case B: Executing a development effort by having a downscaled pilot plant with an experimental program.

Case C: Same as case B, but also includes a mock-up test facility of the commercial-scale reactor size and test program.

Case D: Similar to case C, but with a mock-up test facility of the same size as the pilot plant.

We use the Value of Information analysis method described by Harmsen [29] to determine the effects on the Net Present Value (NPV) of the 4 project options, A, B, C, and D.

Table 14.9 shows the value of information table for the four cases. It is very clear that having a pilot plant reduces the risks considerably. The risk reduction outweighs the increased pilot plant study cost. The net present value increases by $70M. The large mock-up study cost is, however, so large that the further risk reduction is too small to justify this investment. The small mock-up marginally improves the net present value.

It is considered to have the small mock-up study located in a university so that additional information from the university on modelling is quickly obtained. A good relation is built up and also the PhD student can be observed and may be employed later.

Table 14.9: Value of information assessment: polyolefin pilot plant and mock-up studies.

Value of information element	Pilot plant		+ Mock-up	
	No pilot plant	Pilot plant	Full-scale	Small-scale
Design base NPV, M$	300	300	300	300
Consequence failure NPV, M$	−200	−200	−200	−200
Chance of failure, %	50	50	50	50
Expectation failure NPV, M$	−100	−100	−100	−100
Initial net NPV	200	200	200	200
Development NPV, M$	0	−10	−20	−15
New design base NPV, M$	300	290	280	289
New consequence failure	−200	−200	−200	−200
New chance of failure, %	50	10	6	8
New expectation failure NPV	−100	−20	−12	−16
Final net NPV	200	270	268	273

The following conclusions can be drawn from the value of information analysis.

1: Having no pilot plant means an estimated chance of failure of 50%. This is not acceptable to management. It would mean enormous loss of confidence of the customers of the company's capability of developing novel product grades to be produced and delivered in time.

2: Having a pilot plant reduces the risk to an acceptable level.

3: Having a full-scale mock-up model does reduce the risk further but decreases the expected final Net NPV, compared to not having it, although it has a small negative effect on the NPV.

4: Having a small-scale mock-up makes economic sense and has the additional advantage of strengthening ties with the academic world.

14.5.4 Development: front-end engineering design

The development stage for the polyolefins case will mainly be about the pilot plant construction, product-grade development, producing samples for customers, testing antielectrostatic dosing to prevent fouling, and validating the commercial-scale design and models.

At the end of the pilot plant testing, a front-end engineering design (FEED) will be made. All development results will be reported and will be used for the stage-gate decision – to go into the EPC stage or not – and as the main input for the EPC contractor to do detailed engineering, procurement of equipment, and construction [21].

14.6 Commercial-scale implementation (EPC and start-up)

First of all, the EPC contractor will be chosen. Several contractors with experience in polyolefins process EPC execution are available. However, Mahmoud provides a study on a commercial-scale polypropylene Unipol process of 100 kt/a capacity. He shows that the commercial-scale operation has serious problems with several process interruptions per month and a low production per unit catalyst. By systematic parameter studies, cost savings of 6 M$/a could be obtained [27].

This paper shows that even commercial operation of a polyolefin process of on an established Unipol design and with established catalyst and process conditions is still not easy. Start-up of a polyolefins process for a novel catalyst and or novel reactor conditions will therefore also not be easy. All start-up measures reported for new processes [29] should therefore be taken.

14.7 Exercises

14.7.1 Exercise 1: Thiele modulus description and calculation for polyolefin catalyst

Given: This is a case of polyethylene formation, with provided stoichiometry and kinetics of this chapter, a catalyst particle size of 10^{-4} m, and a final polymer particle size of 10^{-3} m. The catalyst concentration on the catalyst particle 10^{-6} mol/m^3

Q1: What is the Thiele modulus description for monomer propagation?
Q2: What is the Thiele modulus value for the fresh catalyst particle (with the parameter values the same as the monomer reaction rate constant of Table 14.1)
Q3: What is the Thiele modulus value for hydrogen for the final polymer particle?

14.7.2 Exercise 2: temperature catalyst particle

Given: A polyethylene reactor is in commercial-scale operation. A catalyst provider proposes to buy his catalyst, which is factor 10 more active than the presently used catalyst. The catalyst salesperson states that therefore a factor 10 less catalyst is needed for the same production. The new catalyst is twice as high in price per kg of catalyst as the presently used catalyst. A young process engineer in the central office 100 km away from the production plant is asked by the plant manager whether he should buy the new catalyst.

Q1: What should be the main concerns of the process engineer?

14.7.3 Exercise 3: polyethylene reactor design

Given: A reaction engineer of a polyethylene manufacturer has to make a reactor design for the concept stage for a capacity of 120 kton/year of polyethylene. His supervisor wants the reactor volume to be the same as the existing reactor, with a capacity of 100 kton/year. The molecular weight distribution of the product should be the same but the average molecular weight distribution should also be tunable. All information for this case is found in this chapter.

Q1: What will be the pressure, temperature, and catalyst concentration for this new design?

Q2: What will be the heat exchange area of the heat exchanger in the gas recycle loop, relative to the existing reactor?

Q3: What is the control parameter to tune the average molecular weight distribution?

14.8 Takeaway learning points

Learning point 1
For polyolefins manufacturing, gas–solid fluid bed reactors are preferred in the industry, over gas–liquid–solid slurry reactors.

Learning point 2
Accurate information about the gas phase residence time distribution of a fluid bed is not needed for low single-pass gas components conversion with a large external recycle. For that design option, the reactor plus recycle can be considered as fully backmixed.

Learning point 3
Solids residence time distribution in a fluid bed matters when the catalyst decays. It affects the amount of catalyst needed per mass of polymer product and on the reactor volume needed for a given polymer production capacity (kton/year).

Learning point 4
The solids residence time distribution in a fluid bed can be reduced in width by having multiple chambers in one so-called horizontal vessel.

Learning point 5
Mass transfer limitations inside the catalyst and the polymer particle using the Thiele modulus can be easily done. They give insight into the maximum temperature allowed for no-mass transfer limitation. However, the analysis only holds for a uniform temperature in the particle. Therefore, the heat transfer limitation has also to be assessed to see whether the particle has a uniform temperature.

Learning point 6
Heat transfer limitation of the reacting particle can be easily estimated using expression 14.1.

Learning point 7
The effect of lump formation on temperature overshoot can also be estimated using expression 14.1.

Learning point 8
Reactor heat and temperature control choice between adiabatic or cooled can be made by simple heat balance calculations.

Learning point 9
Molecular weight distribution control is by reactor design, with uniform temperature and a uniform monomer/hydrogen ratio.

Learning point 10
A new reactor concept for polyolefins has been obtained using basic chemical reaction engineering knowledge.

Learning point 11
Electrostatic charging of particles and its consequences for fouling was observed in pilot plants. It was not observed or predicted in the concept stage.

Learning point 12
A pilot plant can be easily justified using the value of information assessment.

14.9 List of symbols

d_p	Diameter spherical particle	m
D_{eff}	Gas diffusion coefficient for the monomer	m^2/s
E_p	Activation energy propagation reaction	J/mol
E_d	Activation energy Deactivation reaction	J/mol
$C^*_{(fresh)}$	Radical catalyst concentration per unit reactor volume	mol/m^3
H_2	Hydrogen concentration	mol/m^3
k	First-order reaction rate constant for the monomer	1/s
k_p	Reaction rate parameter propagation (based on gas phase volume per mol cat)	$m^3/mol\ s$
k_d	Reaction rate parameter deactivation	1/s
$k_p\ C^*$	First-order propagation reaction rate per unit reactor volume	1/s
L	Length parameter in Thiele modulus	m
M	Monomer concentration in the gas phase	mol/m^3
N_{stages}	Number of mixed-stage solid phase	–
P	Pressure	N/m^2
R	Universal gas constant	J/mol K
r_d	Reaction rate for catalyst deactivation	$mol/m^3\ s$
r_p	Reaction rate for propagation	$mol/m^3\ s$
r_e	Reaction rate for hydrogenation ending the active site for propagation	$mol/m^3\ s$
T	Temperature	K
α	Particle heat transfer coefficient	$W/m^2\ K$
Λ	Heat conductivity porous particle; conservative estimate	J/K m s
ΔH_r	Reaction enthalpy ethylene polymerization	J/mol
ΔT	Temperature difference catalyst particle and gas phase	K
ε_p	Catalyst particle volume fraction	–
Φ	Thiele modulus for the propagation reaction	–
λ	Thermal conductivity	W/m K
τ	Residence time	s
τ_{stage}	Residence time of a stage	s

References

[1] Posch DW, Polyolefins. In: Kutz M, ed. Applied Plastics Engineering Handbook, 3rd ed. Norwich NY USA, William Andrew Publishing, 2020, 27–53.

[2] Soares JB, McKenna TF, Polyolefin Reaction Engineering. Chichester, UK, Wiley-VCH, 2012.

[3] Roberts DE, Heats of polymerization. A summary of published values and their relation to structure. Journal of Research of the National Bureau of Standards, 1950 Mar, 44, 221–232.

[4] Meier GB, Fluidized bed reactor for catalytic olefin polymerization. PhD. thesis, Enschede, University of Twente, 2000.

[5] Meier GB, Weickert G, van Swaaij WPM, FBR for catalytic propylene polymerization: Controlled mixing and reactor modeling. AIChE Journal, 2002, 48(6), 1268–1283.

[6] Ray WH, Dynamic modelling of polymerization reactors. IFAC Proceedings Volumes. 1980 Jan 1, 13 (4), 587–595.

[7] Prosen EJ, Rossini FD, Heats of formation, hydrogenation, and combustion of the mono olefin hydrocarbons through the hexenes, and of the higher 1-alkenes, in the gaseous state at 25 °C. Journal of Research of the National Bureau of Standards, 1946 Jan 1, 36(3), 269–275.
[8] Eriksson EJ, McKenna TF, Heat-transfer phenomena in gas phase olefin polymerization using computational fluid dynamics. Industrial & Engineering Chemistry Research, 2004 Nov 10, 43(23), 7251–7260.
[9] Alves RF, Casalini T, Storti G, McKenna TF, Gas phase polyethylene reactors – A critical review of modeling approaches. Macromolecular Reaction Engineering, 2021 Apr, 1, 2000059.
[10] Rashedi R, Sharif F, Experimental study on the relationship between particles size and properties of polyethylene powder from an industrial fluidized bed reactor. Industrial & Engineering Chemistry Research, 2014 Aug 27, 53(34), 13543–13549.
[11] Cocco R, Karri SR, Knowlton T, Introduction to fluidization. Chemical Engineering Progres, 2014, 110 (11), 21–29.
[12] Khan MJ, Hussain MA, Mansourpour Z, Mostoufi N, Ghasem NM, Abdullah EC, CFD simulation of fluidized bed reactors for polyolefin production – A review. Journal of Industrial and Engineering Chemistry, 2014 Nov 25, 20(6), 3919–3946.
[13] Shepard JW, Jezl JL, Peters EF, Hall RD, Divided horizontal reactor for the vapor phase polymerization of monomers at different hydrogen levels. U.S. Patent 3,957,448, 1976.
[14] Alizadeh A, Sharif F, Ebrahimi M, McKenna TF, Modeling condensed mode operation for ethylene polymerization, part III. Mass and heat transfer. Industrial & Engineering Chemistry Research, 2018, 57(18), 6097–6114.
[15] Soomro M, Hughes R, The thermal conductivity of porous catalyst pellets. The Canadian Journal of Chemical Engineering, 1979 Feb, 57(1), 24–28.
[16] McKenna TF, Tioni E, Ranieri MM, Alizadeh A, Boisson C, Monteil V, Catalytic olefin polymerisation at short times: Studies using specially adapted reactors. The Canadian Journal of Chemical Engineering, 2013 Apr, 91(4), 669–686.
[17] Yiagopoulos A, Yiannoulakis H, Dimos V, Kiparissides C, Heat and mass transfer phenomena during the early growth of a catalyst particle in gas phase olefin polymerization, the effect of prepolymerization temperature and time. Chemical Engineering Science, 2001 Jul 1, 56(13), 3979–3995.
[18] Duduković MP, Mills PL, Scale-up and multiphase reaction engineering. Current Opinion in Chemical Engineering, 2015 Aug 1, 9, 49–58.
[19] Bakker HL, Kleijn JP, eds. Management of Engineering Projects: People are Key. Nijkerk, Netherlands, NAP-The Process Industry Competence Network, 2014.
[20] Esmaeili H, Azizi S, Mousavi SM, Hashemi SA, CFD modeling of polypropylene fluidized bed reactor. Journal of Environmental Treatment Techniques, 2020, 8(1), 272–283.
[21] Pell M, Understanding the design of fluid-bed distributors: Proper design of the distributor will result in better performance. (Feature report). Chemical Engineering, 2002, 109(8), 72–76.
[22] Kim JY, Choi KY, Modeling of particle segregation phenomena in a gas phase fluidized bed olefin polymerization reactor. Chemical Engineering Science, 2001, 56(13), 4069–4083.
[23] Taghavivand M, Role of Catalyst Support and Continuity Additives on the Degree of Electrostatic Charging and Reactor Fouling in Polyethylene Gas-solid Fluidized Bed Reactors. Doctoral dissertation, Ottowa Canada, University of Ottawa, 2021.
[24] Hendrickson G, Electrostatics and gas phase fluidized bed polymerization reactor wall sheeting. Chemical Engineering Science, 2006 Feb 1, 61(4), 1041–1064.
[25] Mahmoud BA, Improved catalytic productivity and performance of polypropylene polymerization plant. Kuwait, Technical Report Kuwait Inst. Sci. Research, 2018.

[26] Kahn MJH, Polypropylene production in a fluidized bed catalytic reactor: Comprehensive modeling, optimization and pilot scale experimental validation. PhD thesis, Malaysia, University of Malaysia, 2016.

[27] Fotovat F, Bi XT, Grace JR, Electrostatics in gas-solid fluidized beds: A review. Chemical Engineering Science, 2017 Dec 14, 173, 303–334.

[28] Lacks DJ, Shinbrot T, Long-standing and unresolved issues in triboelectric charging. Nature Reviews Chemistry, 2019 Aug, 3(8), 465–476.

[29] Harmsen J, Industrial Process Scale-up, 2nd Revised Edition – A Practical Innovation Guide from Idea to Commercial Implementation. Amsterdam, The Netherlands, Elsevier, 2019.

Index

https://doi.org/10.1515/9783110713770-015

Printed in the USA
CPSIA information can be obtained
at www.ICGtesting.com
LVHW081221140624
783111LV00007B/350